METHODS IN MOLECULAR BIOLOGY

For further volumes:
http://www.springer.com/series/7651

For over 35 years, biological scientists have come to rely on the research protocols and methodologies in the critically acclaimed *Methods in Molecular Biology* series. The series was the first to introduce the step-by-step protocols approach that has become the standard in all biomedical protocol publishing. Each protocol is provided in readily-reproducible step-by-step fashion, opening with an introductory overview, a list of the materials and reagents needed to complete the experiment, and followed by a detailed procedure that is supported with a helpful notes section offering tips and tricks of the trade as well as troubleshooting advice. These hallmark features were introduced by series editor Dr. John Walker and constitute the key ingredient in each and every volume of the *Methods in Molecular Biology* series. Tested and trusted, comprehensive and reliable, all protocols from the series are indexed in PubMed.

Group A Streptococcus

Methods and Protocols

Edited by

Thomas Proft and Jacelyn M. S. Loh

School of Medical Sciences, The University of Auckland, Auckland, New Zealand

Humana Press

Editors
Thomas Proft
School of Medical Sciences
The University of Auckland
Auckland, New Zealand

Jacelyn M. S. Loh
School of Medical Sciences
The University of Auckland
Auckland, New Zealand

ISSN 1064-3745 ISSN 1940-6029 (electronic)
Methods in Molecular Biology
ISBN 978-1-0716-0466-3 ISBN 978-1-0716-0467-0 (eBook)
https://doi.org/10.1007/978-1-0716-0467-0

Cover illustration: A scanning electron microscope (SEM) image of Group A Streptococcus provided by Dr. Shaun Lee (Image credit: Kai & Dustin Higashi).

This Humana imprint is published by the registered company Springer Science+Business Media, LLC, part of Springer Nature.
The registered company address is: 1 New York Plaza, New York, NY 10004, U.S.A.

Preface

Group A Streptococcus (GAS; *Streptococcus pyogenes*) is a human pathogen that causes a wide range of diseases that include soft tissue infections, pharyngitis, tonsillitis, cellulitis, necrotizing fasciitis ("flesh-eating disease"), and toxic shock syndrome. The most common GAS disease in high-income countries is pharyngitis ("strep throat") affecting 8–13% of children aged 5–12 years old. Impetigo ("school sores", pyoderma) is more common in tropical climates such as the Northern Territory of Australia and in low resource settings affecting about 19% of 0–9 year olds and 10% of 10–14 year olds. These common diseases if left untreated in susceptible children can result in the autoimmune sequelae acute rheumatic fever (ARF) and eventually rheumatic heart disease (RHD). These diseases carry significant morbidity and mortality globally, with an estimated incidence of 16–20 million cases/year.

GAS produces a wide range of virulence factors, which are involved in the interaction with the host. The presence of genes encoding particular virulence factors can be the predictor of virulence and invasiveness. More recently, several GAS virulence factors were investigated as potential targets for the development of a GAS vaccine. These include M protein, T-antigen (pilus backbone protein), streptolysin O, SpyCEP, and C5a peptidase.

This edition of the Methods in Molecular Biology series provides the reader with an extensive collection of both classic and recent state-of-the-art protocols for GAS research.

The 30 chapters of this book are written by an international panel of experts in GAS research from the United States, Chile, Mexico, United Kingdom, Belgium, Germany, Poland, Russia, Japan, Australia, and New Zealand.

Part I includes methods for the detection of GAS virulence factor genes (including superantigen genes), a recently developed *emm*-typing system (*emm*-cluster typing), methods for bioluminescence labeling of GAS, and novel methods for genetic manipulation.

Part II of the book describes protocols for GAS omics, including genomics (whole genome sequence analysis, WGSA), transcriptomics, proteomics, and glycomics (glycan microarray analysis). The nine chapters in Part III describe methods for the study of GAS virulence factors, including the expression and purification of recombinant collagen-like proteins and fibronectin-binding proteins, and the expression of cell wall-anchored GAS proteins on the surface of the surrogate bacterium *Lactococcus lactis*. This part of the book also includes protocols to investigate GAS biofilm formation, GAS interaction with macrophages, and the detection of the bacteria in GAS-containing autophagosome-like vacuoles (GcAVs), which form after GAS escape into the cytoplasm following internalization through endocytosis. Other chapters in this section provide protocols to analyze the hemolytic effect of streptolysin S (SLS) on erythrocytes and the isolation of GAS-specific monoclonal antibodies from phage libraries.

Four different animal infection models for GAS are described in Part V. The *Galleria mellonella* (wax moth) larvae infection model has recently attracted a lot of interest as an alternative model to study microbial infections and virulence, due to the fact that they can be easily and inexpensively obtained in large numbers and are simple to use. This is followed by three protocols for mouse infection studies: a superficial skin model to mimic GAS skin disease, a nasopharyngeal infection model to analyze GAS colonization, and a humanized plasminogen mouse model to study invasive GAS disease. These infection models are not only used to study GAS infection and virulence, but also present important tools for preclinical in vivo testing of GAS vaccine candidates. This part of the book is completed by two in vitro GAS killing assays for the study of vaccines: a modified and improved protocol of the classic Lancefield whole blood killing assay and a recently developed opsonophagocytic killing assay based on the human neutrophil cell line HL-60. Part V of the book features two chapters on novel potential GAS therapies: a gene antisense therapy with peptide-coupled peptide nucleic acids and a strategy to interfere with the formation of pili using a peptide mimicking β-strand of the backbone pilus protein. The final section of the book (Part VI) contains two chapters describing novel protocols that are derived from GAS research; the cytolysin streptolysin O (SLO) used as a tool to investigate lipid rafts in cellular processes, and a protocol that describes the genetic engineering of an isopeptide bond originally discovered in the backbone pilus protein of a GAS serotype M1 strain into immunoglobulin-like domains for protein stabilization.

We would like to thank the series editor of the book series, Prof. John Walker, for giving us the opportunity to edit this book. We also thank all the contributing authors for sharing their valuable protocols and hope that readers will find those protocols helpful in their research work.

Auckland, New Zealand

Thomas Proft
Jacelyn M. S. Loh

Contents

Part III Assays to Investigate Group A Streptococcus Virulence Factors and Virulence Mechanisms

Part IV Group A Streptococcus Killing Assays and Infection Models

Contributors

A. A. Adzhieva • *Ministry of Health of the Russian Federation, Moscow, Russia*
Edward N. Baker • *School of Biological Sciences and Maurice Wilkins Centre, University of Auckland, Auckland, New Zealand*
Gina Barkowsky • *Institute of Medical Microbiology, Virology and Hygiene, University Medicine Rostock, Rostock, Germany; Institute of Medical Microbiology, Virology and Hygiene, University of Rostock, Rostock, Germany*
Timothy C. Barnett • *Wesfarmers Centre for Vaccines and Infectious Diseases, Telethon Kids Institute, University of Western Australia, Nedlands, WA, Australia; Centre for Child Health Research, University of Western Australia, Nedlands, WA, Australia*
Ashton T. Belew • *Department of Cell Biology and Molecular Genetics and Maryland Pathogen Research Institute, University of Maryland, College Park, MD, USA; Center for Bioinformatics and Computational Biology, University of Maryland, College Park, MD, USA*
Sam Blanchett • *Department of Molecular Medicine and Pathology, School of Medical Sciences and Maurice Wilkins Centre for Biomolecular Discovery, The University of Auckland, Auckland, New Zealand*
Anne Botteaux • *Molecular Bacteriology Laboratory, Université libre de Bruxelles, Brussels, Belgium*
Stephan Brouwer • *Australian Infectious Diseases Research Centre and School of Molecular Biosciences, The University of Queensland, Brisbane, QLD, Australia*
Sok H. By • *Department of Molecular Medicine and Pathology, The University of Auckland, Auckland, New Zealand; School of Medical and Health Sciences and Maurice Wilkins Center, The University of Auckland, Auckland, New Zealand*
Fernanda Contreras • *Centro de Genómica y Bioinformática, Facultad de Ciencias, Universidad Mayor, Santiago, Chile*
G. A. Danilina • *N.F. Gamaleya National Research Center of Epidemiology and Microbiology, Ministry of Health of the Russian Federation, Moscow, Russia*
Tatyana A. Danilova • *N.F. Gamaleya National Research Center of Epidemiology and Microbiology, Ministry of Health of the Russian Federation, Moscow, Russia*
Mark R. Davies • *Department of Microbiology and Immunology, The University of Melbourne at the Peter Doherty Institute for Infection and Immunity, Melbourne, VIC, Australia*
Jessica N. Daw • *Wesfarmers Centre for Vaccines and Infectious Diseases, Telethon Kids Institute, University of Western Australia, Nedlands, WA, Australia*
Christopher J. Day • *Institute for Glycomics, Griffith University, Gold Coast, QLD, Australia*
David M. P. De Oliveira • *Illawarra Health and Medical Research Institute and Molecular Horizons, School of Chemistry and Molecular Bioscience, University of Wollongong, Wollongong, NSW, Australia; Australian Infectious Disease Research Centre, School of Chemistry and Molecular Biosciences, University of Queensland, Brisbane, QLD, Australia*
Anthony R. Flores • *Division of Infectious Diseases, Department of Pediatrics, Center for Antimicrobial Resistance and Microbial Genomics, McGovern Medical School, University of Texas Health Sciences Center at Houston, Houston, TX, USA*

Nour Zahi Gammoh • *Institute of Immunology, Faculty of Medicine, RWTH Aachen University, University Hospital, Aachen, Germany*
David Goldblatt • *Immunobiology, UCL Great Ormond Street Institute of Child Health, London, UK*
Michael F. Good • *Institute for Glycomics, Griffith University, Gold Coast, QLD, Australia*
Lauren E. Hartley-Tassell • *Institute for Glycomics, Griffith University, Gold Coast, QLD, Australia*
Dustin L. Higashi • *Department of Biological Sciences, University of Notre Dame, Notre Dame, IN, USA; Department of Restorative Dentistry, Oregon Health and Science University, Portland, OR, USA*
Anuk Indraratna • *Illawarra Health and Medical Research Institute and Molecular Horizons, School of Chemistry and Molecular Bioscience, University of Wollongong, Wollongong, NSW, Australia*
Taylah B. James • *Department of Microbiology and Immunology, The University of Melbourne at the Peter Doherty Institute for Infection and Immunity, Melbourne, VIC, Australia*
Michael P. Jennings • *Institute for Glycomics, Griffith University, Gold Coast, QLD, Australia*
Scott Jones • *Immunobiology, UCL Great Ormond Street Institute of Child Health, London, UK*
Shigetada Kawabata • *Department of Oral and Molecular Microbiology, Graduate School of Dentistry, Osaka University, Suita, Osaka, Japan*
Adrina H. J. Khemlani • *Department of Molecular Medicine and Pathology, School of Medical Sciences and Maurice Wilkins Centre for Biomolecular Discovery, The University of Auckland, Auckland, New Zealand; School of Medical and Health Sciences and Maurice Wilkins Center, The University of Auckland, Auckland, New Zealand*
Aleksandra Kozińska • *Department of Drug Biotechnology and Bioinformatics, National Medicines Institute, Warszawa, Poland*
Bernd Kreikemeyer • *Institute of Medical Microbiology, Virology and Hygiene, University Medicine Rostock, Rostock, Germany*
Jake A. Lacey • *Doherty Department, The University of Melbourne, at the Peter Doherty Institute for Infection and Immunity, Melbourne, VIC, Australia*
Yoann Le Breton • *Wound Infections Department, Bacterial Diseases Branch, Walter Reed Army Institute for Research, Silver Spring, MD, USA*
Shaun W. Lee • *Department of Biological Sciences, University of Notre Dame, Notre Dame, IN, USA*
Jacelyn M. S. Loh • *Department of Molecular Medicine and Pathology, School of Medical Sciences and Maurice Wilkins Centre for Biomolecular Discovery, The University of Auckland, Auckland, New Zealand; School of Medical and Health Sciences and Maurice Wilkins Center, The University of Auckland, Auckland, New Zealand*
Slawomir Lukomski • *Department of Microbiology, Immunology, and Cell Biology, West Virginia University School of Medicine, Morgantown, WV, USA*
Nicola N. Lynskey • *The Roslin Institute, Royal (Dick) School of Veterinary Studies, University of Edinburgh, Edinburgh, UK*
Reuben McGregor • *Department of Molecular Medicine and Pathology, School of Medical Sciences, The University of Auckland, Auckland, New Zealand; School of Medical and Health Sciences and Maurice Wilkins Center, The University of Auckland, Auckland, New Zealand*

Kevin S. McIver • *Department of Cell Biology and Molecular Genetics and Maryland Pathogen Research Institute, University of Maryland, College Park, MD, USA*
Dudley H. McNitt • *Department of Microbiology, Immunology, and Cell Biology, West Virginia University School of Medicine, Morgantown, WV, USA; Division of Rheumatology and Immunology, Department of Medicine, Vanderbilt University, Nashville, TN, USA*
Heather E. Miller • *Rocky Mountain Laboratories, Laboratory of Bacteriology, National Institute of Allergy and Infectious Diseases, NIH, Hamilton, MT, USA*
Nicole J. Moreland • *Department of Molecular Medicine and Pathology, School of Medical Sciences, The University of Auckland, Auckland, New Zealand; Maurice Wilkins Centre, The University of Auckland, Auckland, New Zealand; School of Medical and Health Sciences and Maurice Wilkins Center, The University of Auckland, Auckland, New Zealand*
Ichiro Nakagawa • *Department of Microbiology, Graduate School of Medicine, Kyoto University, Kyoto, Japan*
Masanobu Nakata • *Department of Oral and Molecular Microbiology, Graduate School of Dentistry, Osaka University, Suita, Osaka, Japan*
Takashi Nozawa • *Department of Microbiology, Graduate School of Medicine, Kyoto University, Kyoto, Japan*
Manisha Pandey • *Institute for Glycomics, Griffith University, Gold Coast, QLD, Australia*
Nadja Patenge • *Institute of Medical Microbiology, Virology and Hygiene, University Medicine Rostock, Rostock, Germany*
Thomas Proft • *Department of Molecular Medicine and Pathology, School of Medical Sciences and Maurice Wilkins Centre for Biomolecular Discovery, The University of Auckland, Auckland, New Zealand; School of Medical and Health Sciences and Maurice Wilkins Center, The University of Auckland, Auckland, New Zealand*
Jeremy M. Raynes • *Department of Molecular Medicine and Pathology, School of Medical Sciences, The University of Auckland, Auckland, New Zealand; Maurice Wilkins Centre, The University of Auckland, Auckland, New Zealand; Monash University, Clayton Campus, VIC, Australia; School of Medical and Health Sciences and Maurice Wilkins Center, The University of Auckland, Auckland, New Zealand*
Mark Reglinski • *Molecular Microbiology, School of Life Sciences, University of Dundee, Dundee, UK*
Angelica Montenegro Riestra • *Division of Host-Microbe Systems and Therapeutics, Department of Pediatrics, School of Medicine, University of California San Diego, La Jolla, CA, USA*
Lothar Rink • *Institute of Immunology, Faculty of Medicine, RWTH Aachen University, University Hospital, Aachen, Germany*
Jaime Andrés Rivas-Pardo • *Centro de Genómica y Bioinformática, Facultad de Ciencias, Universidad Mayor, Santiago, Chile*
Tania Rivera-Hernandez • *Cátedras CONACYT—Unidad de Investigación Médica en Inmunoquímica, Hospital de Especialidades del Centro Médico Nacional Siglo XXI, Instituto Mexicano del Seguro Social, CDMX, Mexico*
Martina L. Sanderson-Smith • *Illawarra Health and Medical Research Institute and Molecular Horizons, School of Chemistry and Molecular Bioscience, University of Wollongong, Wollongong, NSW, Australia*
Misú Sanson • *Division of Infectious Diseases, Department of Pediatrics, Center for Antimicrobial Resistance and Microbial Genomics, McGovern Medical School, University of Texas Health Sciences Center at Houston, Houston, TX, USA*

N. V. Shevlyagina • *N.F. Gamaleya National Research Center of Epidemiology and Microbiology, Ministry of Health of the Russian Federation, Moscow, Russia*
Izabela Sitkiewicz • *Department of Drug Biotechnology and Bioinformatics, National Medicines Institute, Warszawa, Poland*
Pierre R. Smeesters • *Academic Children Hospital Queen Fabiola, Université libre de Bruxelles, Brussels, Belgium; Molecular Bacteriology Laboratory, Université libre de Bruxelles, Brussels, Belgium; Tropical Diseases Research Group, Murdoch Children's Research Institute, Melbourne, Australia; Centre for International Child Health, University of Melbourne, Melbourne, Australia*
T. A. Smirnova • *N.F. Gamaleya National Research Center of Epidemiology and Microbiology, Ministry of Health of the Russian Federation, Moscow, Russia*
Kar Yan Soh • *Department of Molecular Medicine and Pathology, School of Medical Sciences and Maurice Wilkins Centre for Biomolecular Discovery, The University of Auckland, Auckland, New Zealand; School of Medical and Health Sciences and Maurice Wilkins Center, The University of Auckland, Auckland, New Zealand*
John D. Steemson • *School of Biological Sciences, The University of Auckland, Auckland, New Zealand*
Mei Lin Tay • *Department of Molecular Medicine and Pathology, The University of Auckland, Auckland, New Zealand; School of Medical and Health Sciences and Maurice Wilkins Center, The University of Auckland, Auckland, New Zealand*
Charles R. Tessier • *Indiana University School of Medicine, South Bend, IN, USA*
Steven Y. C. Tong • *Doherty Department, The University of Melbourne, at the Peter Doherty Institute for Infection and Immunity, Melbourne, VIC, Australia; Division of Global and Tropical Health, Menzies School of Health Research, Division of Global and Tropical Health, Darwin, NT, Australia; Victorian Infectious Disease Service, The Royal Melbourne Hospital, The University of Melbourne, at the Peter Doherty Institute for Infection and Immunity, Melbourne, VIC, Australia*
Catherine (Jia-Yun) Tsai • *Department of Molecular Medicine and Pathology, School of Medical Sciences and Maurice Wilkins Centre for Biomolecular Discovery, The University of Auckland, Auckland, New Zealand; School of Medical and Health Sciences and Maurice Wilkins Center, The University of Auckland, Auckland, New Zealand*
J. Andrés Valderrama • *Division of Host-Microbe Systems and Therapeutics, Department of Pediatrics, School of Medicine, University of California San Diego, La Jolla, CA, USA*
Mark J. Walker • *Australian Infectious Disease Research Centre, School of Chemistry and Molecular Biosciences, University of Queensland, Brisbane, QLD, Australia*
Laura Wilk • *Division of Infection Medicine, Lund University, Lund, Sweden; Centre for Biological Threats and Special Pathogens, Robert Koch Institut, Berlin, Germany*
Paul G. Young • *School of Biological Sciences and Maurice Wilkins Centre, University of Auckland, Auckland, New Zealand*
V. G. Zhukhovitsky • *N.F. Gamaleya National Research Center of Epidemiology and Microbiology, Ministry of Health of the Russian Federation, Moscow, Russia*

Part I

Genotyping and Genetic Manipulation of Group A Streptococcus

Chapter 1

Detection of *Streptococcus pyogenes* Virulence Factors

Aleksandra Kozińska and Izabela Sitkiewicz

Abstract

Streptococcus pyogenes encodes multiple virulence factors and their presence is often related to the severity of the disease. We designed the system of four low-volume multiplex PCR reactions to detect genes encoding 20 virulence factors: *spd*3, *sdc*, *sda*B, *sda*D, *spe*B, *spy*CEP, *scp*A, *mac*, sic, *spe*L, *spe*K, *spe*M, *spe*C, *spe*I, *spe*A, *spe*H, *spe*G, *spe*J, *sme*Z, and *ssa*. Classification of strains based on the virulence factors absence or presence correlates with PFGE MLST and emm typing results. The typing/detection system is fast and cost-effective, can be used to detect GAS virulence factors and as a rapid tool to effectively differentiate between strains.

Key words *Streptococcus pyogenes*, Group A Streptococcus, Virulence factors, Superantigens

1 Introduction

1.1 *Streptococcus pyogenes Virulence Factors*

Streptococcus pyogenes (Group A Streptococcus, GAS) is an important human pathogen that causes a broad spectrum of infections. GAS diseases range from mild skin and mucosal surface infections, such as pharyngitis or impetigo, to severe, life-threatening toxin-mediated diseases (for a review, *see* [1]). Severe manifestations include necrotizing fasciitis or toxic shock syndrome and post-infectional sequelae [2]. GAS is responsible for over 600 million new infections every year and causes half a million deaths as a result of infections and post-infectional complications [3, 4].

The diseases caused by GAS are multifactorial and include bacterial and host factors. However, the success of GAS during the infection heavily relies on the production of multiple virulence factors involved in various aspects of host–pathogen interactions. Also, the severity of GAS infections is often linked to the production of multiple adhesins, toxins, proteases, DNases, and superantigens encoded by the particular GAS strains [1]. Virulence factors are involved in the interaction with the host on the cellular and organism levels, and it is thought that the presence of genes encoding particular virulence factors can be the predictor of strain

Thomas Proft and Jacelyn M. S. Loh (eds.), *Group A Streptococcus: Methods and Protocols*, Methods in Molecular Biology, vol. 2136, https://doi.org/10.1007/978-1-0716-0467-0_1,

invasiveness. In particular, the presence of the streptococcal pyrogenic exotoxin A (*spe*A) gene is associated with streptococcal toxic like shock syndrome and scarlet fever [5] and streptococcal mitogenic exotoxin Z (*sme*Z) participates in the repression of cognate anti-streptococcal responses [6].

Virulence factors are not equally distributed among *S. pyogenes* strains. Some of the GAS virulence factors, for example, SpeB, streptococcal cysteine protease A (ScpA), the interleukin-8 protease SpyCEP, the IgG-degrading protease Mac, streptodornase B (SdaB), and SpeG, are chromosomally encoded. However, a large fraction of virulence factors such as the majority of DNAses and superantigens, for example, SpeA, SpeC, SpeH, and streptococcal superantigen (SSA) are encoded by mobile genetic elements—phages and conjugative mobile elements integrated into the chromosome (ICEs—integrative and conjugative elements). The profile of the virulence factors encoded by the particular strain can be a part of strain characteristics and their presence/absence can be used as a simple diagnostic method. Distribution of phage-encoded virulence factors could be in the majority of cases attributed to the detected integrated elements known to encode particular virulence factors [7].

Initial contact between bacteria and the host is mediated by multiple adhesins, which bind host proteins and extracellular matrix proteins such as plasminogen, collagen, keratin, and laminin (for a review, *see* [8–10]. The major GAS adhesin and a base for assigning GAS to distinct serotypes is M protein [11]. Those major adhesins, including M protein, are usually variable in size and sequence; therefore, their detection using simple PCR assays is more complex and is not included with the assays presented below. The type of M protein can be detected in the PCR/sequencing assay described in detail by CDCs [12].

After the initial contact, bacteria invade host tissues and disseminate. Multiple classes of GAS virulence factors such as proteases, DNases and pyrogenic toxins (superantigens) are involved in the interaction between bacteria and the host in the post attachment phase [13–15].

1.2 Superantigens

Superantigens are a group of proteins that are able to stimulate T cells to a massive release of cytokines in the absence of any particular antigen. They cross-link the major histocompatibility complex (MHC) class II with the T-cell receptor, resulting in bypassing the normal antigen-presenting process. They are involved in systemic toxicity and it is possibly a mechanism used by the bacteria to escape the immune system [16]. Eleven superantigens, SpeA, SpeC, SpeG, SpeH, SpeI, SpeJ, SpeK, SpeL, SpeM, SmeZ, and SSA, are encoded by *S. pyogenes* and are at least partially characterized. The majority of GAS superantigens are found in other streptococcal species, and three of them, SzeN, SzeP, and SzeF have been detected only in *Streptococcus equi subsp zooepidemicus* [17]. Recently, Reglinski and co-workers described two new putative superantigens found in GAS—SpeQ and SpeR [18]. A detailed review describing the current state of knowledge on the biological properties and the potential role of streptococcal superantigens in the disease was recently published by Proft and Fraser [13].

1.3 DNAses

GAS produces multiple extracellular DNAses that are involved in the dissemination of bacteria and escape from neutrophil extracellular traps (NETs) that are either chromosomally encoded or phage-encoded [19, 20]. Single strains can encode several DNAses. Although all DNAses are distinct, they are often confused, as multiple names are given to identify the same DNAse. Chromosomally encoded SdaB is also named streptodornase B or mitogenic factor 1 (MF1) and is encoded by gene M5005_Spy1738 in strain MGAS5005; Spd3 is encoded by phage 5005.2 and 370.3 in MGAS5005 and SF370 strains, respectively, and is also named Sda3 or mitogenic factor 3 (MF3) (designated as M5005_Spy1169 in MGAS5005); Spd1 is encoded by phage 370.1 in the SF370 strain (Spy_0712) and is also called streptodornase C or mitogenic factor 2; Sdc also named sd alpha, sdn, sda, or streptodornase A is encoded by phage 315.6 (SpyM3_1409 in MGAS315 strain) and SdaD encoded by phage 5005.3 (M5005_Spy1415) is also named Sda1, streptodornase D or SdaD2.

Recently, a novel, chromosomally encoded (locus designation Spy0747 in SF370 strain), cell wall anchored nuclease namely *Streptococcus pyogenes* nuclease A (SpnA) was described and characterized. It destroys NETs and promotes bacterial survival in blood [21]. It is a homolog of *Streptococcus suis* nuclease SsnA which also contributes to the degradation of NETs and evasion of NET-mediated antimicrobial activity [22]. All DNAses described so far contribute to GAS virulence [19].

1.4 Proteases

Proteases in GAS are a distinct group of virulence factors involved in the dissemination of bacteria into the host tissues and the organism. The major protease of GAS is (1) SpeB. It was first described as a pyrogenic exotoxin, hence the name Spe. However, it was later characterized as a cysteine protease. Almost all GAS strains carry the *spe*B gene, the wild-type protein is produced by almost 85% of all strains, and it is a major virulence factor implicated in the severity of the disease [23, 24]. SpeB greatly influences the interaction of the pathogen with the human host. Regardless of host tissue degradation, it cleaves and activates interleukin-1β precursor, activates pro-matrix metalloproteases, degrades plasminogen, fibronectin, vitronectin, and kininogen, cleaves immunoglobins into Fc and Fab fragments, and degrades C3b complement component. SpeB also cleaves bacterial proteins such as M and F1 surface protein, EndoS, SmeZ, fibronectin-binding protein Fba, ScpA, streptokinase, Protein H, Sda1, and activates its zymogen to the active form [25].

Other major GAS-encoded proteases are predominantly involved in the interaction with the host immune system. (2) ScpA is a highly specific, cell wall-anchored peptidase that degrades C5a factor of the complement [26]. It is probably another crucial invasion factor, as all human isolates seem to produce this highly conserved protein. Of note, a close homolog of ScpA is also produced by a related species—*Streptococcus agalactiae* [27]. (3) Mac/IdeS is another cysteine protease that is highly specific for the lower hinge region of human IgG [28, 29]. However, some reports suggest that in the highly virulent M1T1 background, Mac/IdeS is not essential for either phagocyte resistance or virulence. Given the conservation of Mac/IdeS and homologs across GAS strains, it is possible that Mac/IdeS serves another important function in GAS ecology or contributes to virulence in other strain backgrounds [30]. (4) The fourth major protease of GAS, namely SpyCEP/ScpC, is involved in the degradation of chemokines and chemotactic factors such as interleukin-8, granulocyte chemotactic protein-2, growth-related oncogene α, and β and macrophage pro-inflammatory protein 2-α [31, 32].

1.5 Detection of the Virulence Factors as a Typing Tool

The presence of the gene-encoding virulence factors in clinical strains can provide invaluable information for epidemiological investigations. The information can be either used to investigate correlations between the presence of a particular gene and manifestation/severity of the disease or can be also used as a typing tool to establish a relation between strains. Compared with the laborious RFLP-PFGE (restriction fragment length polymorphism—pulse field gel electrophoresis) and expensive MLST (multi-locus sequence typing), PCR-based methods are fast and economical.

We designed primer sequences based on sequences of available *S. pyogenes* genomes. Primer pairs were designed using the modified Primer3 software, available as the Primer-BLAST tool at NCBI (http://www.ncbi.nlm.nih.gov/tools/primer-blast/). Primers were designed to conserve regions of detected genes. Amplicon sizes were designed so that generated PCR products can be easily identified based on their size. Each primer pair listed in Table 1 can specifically detect a virulence factor. However, to streamline the detection, primers were designed to detect 20 virulence factors in four multiplex reactions. We first tested primer pairs individually so they are specific and yield only single PCR product. Primers for each PCR mix were then tested in the multiplex PCR to ensure that they behaved the same way as individual primers. Multiplex mixes 1 and 2 detect superantigens, mix 3 detects gene- encoding DNAses, and mix 4 detects gene-encoding proteases and the sic gene. Amplicons that are intended to be used in a particular mix differ in size and allow easy distinction between detected genes

Table 1
Primer sequences used in the PCR reactions for the detection of GAS virulence factors

Detected gene	Virulence factor	Primer sequence (forward)	Primer sequence (reverse)	Size of the PCR product (bp)	Multiplex Mix number
*mac/ide*S	Ig G-endopeptidase	TCTTGCCCTGTTGAAAGTGT	CGAGGTGGTATTTTTGACGCC	389	4
*scp*A	C5a peptidase	GCTCGGTTACCTCACTTGTCC	CAATAGCAGCAAACAAGTCACC	622	4
*sda*B	Nuclease	TATAGCGCATGCCGCCTTTT	TGATGGCGCAAGCAAGTACC	440	3
*sda*D	Nuclease	TTTACGCTGAATCGGGCACT	GGCTCTGGTTTGCTTTCCCA	295	3
Sdc	Nuclease	AAGCTTAGAAACTCTCTCGCCA	AGTTCCAGTAATAGCGTTTTTCCGT	600	3
Sic	Streptococcal inhibitor of complement	TTACGTTGCTGATGGTGTATATGGT	TTTGATAGAGGGTTTTCAGCTGGC	150	4
*sme*Z	Mitogenic exotoxin Z	TTTCTCGTCCTGTGTTTGGA	TTCCAATCAAATGGGACGGAGAACA	246	1
*spd*3	Nuclease (Streptodornase)	ATCGTCGTACTTGGCAAGGTT	GCCGCTTCTTCAAACTCTTCG	784	3
*spe*A	Superantigen(streptococcal pyrogenic exotoxin A)	AGGTAGACTTCAATTTGGCTTGTGT	GGGTGACCCTGTTACTCACG	576	2
*spe*B	Cysteine protease	AGACGGAAGAAGCCGTCAGA	TCAAAGCAGGTGCACGAAGC	952	4
*spe*C	Superantigen (streptococcal pyrogenic exotoxin C)	GCCAATTTCGATTCTGCCGC	TGCAGGGTAAATTTTTCAACGACA	405	1
*spe*G	Superantigen (streptococcal pyrogenic exotoxin G)	TGGAAGTCAATTAGCTTATGCAG	GCGAACAACCTCAGAGGGCAAA	384	2

(continued)

Table 1 (continued)

Detected gene	Virulence factor	Primer sequence (forward)	Primer sequence (reverse)	Size of the PCR product (bp)	Multiplex Mix number
*spe*H	Superantigen (streptococcal pyrogenic exotoxin H)	TGAGATATAATTGTCGC TACTCACAT	CCTGAGCGGTTAC TTTCGGT	480	2
*spe*I	Superantigen (streptococcal pyrogenic exotoxin I)	TTCATAGACGGCG TTCAACAA	TGAAATC TAGAGGAGCGGCCA	176	1
*spe*J	Superantigen (streptococcal pyrogenic exotoxin J)	TCCTTGTACTAGA TGAGGTTGCAT	GGTGGGGTTACACCA TCAGT	286	2
*spe*K	Superantigen (streptococcal pyrogenic exotoxin K)	CCTTGTGTGTGTATCGC TTGC	TTGCTGTCCCCCA TCAAACT	568	1
*spe*L	Superantigen (streptococcal pyrogenic exotoxin L)	CCTGAGCCGTGAAA TTCCCA	ACACCAGAATTGTCG TTTGGT	657	1
*spe*M	Superantigen (streptococcal pyrogenic exotoxin M)	ATCGCTCATCAAAC TTTTCCT	CCTTGTGTGTGTATCGC TTGC	496	1
*spy*CEP/*scp*C/*prt*S	Interleukin 8 protease	GATCCGGCCCA TCAAAGCAT	AGCTGCCACTGATG TTGGTG	786	4
Ssa	Streptococcal superantigen A	AAGAATACTCGTTG TAGCATGTGT	AATATTGCTCCAGG TGCGGG	678	2

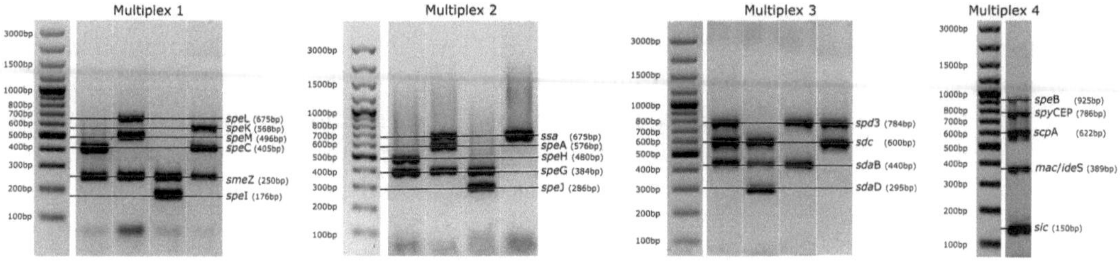

Fig. 1 Examples of products detected in four multiplex PCR reaction

(Fig. 1). The composition of all primer mixes and the size of the amplicons generated in multiplex reactions are listed in Table 1 and the examples of detected products are shown in Fig. 1.

Initially, the method described below was tested on over 650 strains, and we were able to detect 95 distinct profiles of virulence factors among 40 different M types. The method has high resolution with Simpson's index of diversity 0.943 (0.936–0.951) and can be used for both typing and detection of virulence factors [33]. We used it successfully for the detection of the virulence factors in almost 3000 GAS strains (Kozinska and Sitkiewicz, unpublished).

2 Materials

2.1 Reagents and Kits Needed for the DNA Extraction

1. Selected *S. pyogenes* strain grown on Columbia 5% sheep blood agar plate.
2. 20 mg/ml Lysozyme stock.
3. 10 U/μl Mutanolysin stock.
4. 10 mg/ml RNase stock.
5. Tris-EDTA buffer (TE; 2×): add 1 ml of 1 M Tris, pH 8 to 0.2 ml of 0.5 M EDTA and top up solution to 50 ml with water.
6. Cell lysis buffer (per single sample): 100 μl of the 2×TE buffer, 80 μl of water, 5 μl of RNase, 10 μl lysozyme, 2.5 μl of mutanolysin stocks and 2.5 μl of Triton X-100 (can be omitted from the reaction).
7. Chromosomal DNA isolation kit (*see* **Note 1**).

2.2 Reagents Needed for PCR Reaction (See Note 2)

1. 1 mM stock of dNTP.
2. Taq polymerase.
3. Primers (*see* Table 1).

2.3 Reagents Needed for Product Separation and Visualization

1. Tris-borate-EDTA (TBE; 1×) electrophoresis buffer: 90 mM Tris-borate, 2 mM EDTA. The buffer stock solution can be either purchased from sellers such as Thermo Scientific or Sigma or prepared from individually purchased reagents. To prepare 1 L of 5× stock solution of the buffer, 54 g of Tris base and 27.5 g of boric acid should be dissolved in ~750 ml of water, then add 20 ml of 0.5 M EDTA pH 8.0 and adjust the volume to 1 L.
2. 1.5% SeaKemLE agarose (Lonza) in 1× TBE buffer (*see* **Note 3**).
3. Gel loading buffer can be purchased from various providers as 6× concentrated solution or prepared "in house." We routinely use 2× concentrated gel loading buffer that contains 10% glycerol with traces of bromophenol blue and orange G (up to 0.2% w/v).

2.4 Equipment

The protocol requires centrifuges (for 1.5 ml tubes and 96-well plates), a thermal cycler, and a gel electrophoresis tank (*see* **Note 4**). We also recommend the use of multichannel electronic pipettes that allow precise dispensing of the small volumes of liquids.

3 Methods

3.1 Template Preparation

Any DNA template can be used for detection. However, we recommend to keep the DNA concentration below 10 ng/μl, preferably in the range 1–3 ng/μl.

1. Collect *S. pyogenes* cells grown on half of the Columbia 5% sheep blood agar plate and wash the bacteria with 500 μl of TE buffer. To maximize DNA yield, pellets should be frozen and thawed prior to the isolation.
2. Isolation of the DNA from *S. pyogenes* requires initial digestion of the cell wall prior to the manufacturer's recommended protocol. We use a mix of TE with RNAse, lysozyme, and mutanolysin (cell lysis buffer). However, if the DNeasy Blood & Tissue Kit by Qiagen is used, the manufacturer recommends the use of Triton X-100 in the lysing buffer.

The thawed bacterial pellet should be resuspended in 180 μl of such lysis solution and incubated at 37 °C for about 45–60 min. Bacteria with predigested cell wall can be directly used for DNA isolation with the manufacturer's recommended protocol. For the PCR setup, we usually dilute the isolated DNA ten-fold and keep it at 4 °C to avoid multiple freezing-thawing cycles.

3.2 Primer Mix Preparation Reaction Setup (See Note 5)

1. For the ease of use, equal volumes of 100 μM primer stocks should be mixed into appropriate mixes, according to Table 1. To avoid degradation, primer premixes should be aliquoted into 50–75 μl portions, so the single aliquot is sufficient to run the whole 96-well PCR plate without multiple freezing-thawing cycles.

   All mixes, templates, and reactions during setup should be kept on ice.
2. Mix together all reagents for PCR master mixes. The final composition of each PCR reaction is presented in Table 2.
3. Dispense 2.2 μl (Mix 1), 2.1 μl (Mix 2 and 4), or 2.0 μl (Mix 3) of the prepared master mix to each well of the reaction plate or PCR tubes. Using multichannel pipette, add 2.8 μl of DNA template to wells/plates with master Mix 1, 2.9 μl to wells/plates with master Mixes 2 and 4, 3.0 μl to wells/plates with master Mix 3. All PCR reactions are carried out in a total volume of 5 μl.
4. Close tubes or seal the plate.
5. Mix well and spin for 15 s (tubes) or 1 min at 2500 rpm (plates).
6. Place plate/tubes in a thermocycler and run program 1 or 2. (Table 3).

Table 2
Composition of PCR multiplex reactions used for the detection of GAS virulence factors

Reagent	Mix 1per reaction	Mix 1master mix per 96 well plate (×110)	Mix 2 and 4per reaction	Mix 2 and 4master mix per 96 well plate (×110)	Mix 3per reaction	Mix 3master mix per 96 well plate (×110)
100 μM primers mix	0.6 μl	66 μl	0.5 μl	55 μl	0.4 μl	44 μl
10× Taq polymerase buffer with $(NH_4)_2SO_4$	0.5 μl	55 μl	0.5 μl	55 μl	0.5 μl	55 μl
25 mM $MgCl_2$	0.5 μl	55 μl	0.5 μl	55 μl	0.5 μl	55 μl
1 mM dNTP	0.5 μl	55 μl	0.5 μl	55 μl	0.5 μl	55 μl
10× diluted chromosomal DNA template	2.8 μl	–	2.9 μl	–	3.0 μl	–
Taq polymerase	0.1 μl (0.5 U)	11 μl	0.1 μl (0.5 U)	11 μl	0.1 μl (0.5 U)	11 μl

Table 3
Conditions used for amplification

	Program 1 (Mixes 1, 2, and 3)		Program 2 (Mix 4)	
	Temperature	Time	Temperature	Time
Initial denaturation	95 °C	3 min	95 °C	3 min
Denaturation	95 °C	15 s	95 °C	15 s
Annealing	60 °C	20 s	52.5 °C	45 s
Elongation	72 °C	2 min	72 °C	3 min
Final elongation	72 °C	7 min	72 °C	7 min

7. Amplification reactions are performed for 40 cycles using two different programs. Reactions with primer Mix 1–3 are amplified using program 1 and reactions with Mix 4 are amplified using program 2.

3.3 Gel Electrophoresis

1. After amplification, 5 μl of the 2× concentrated gel loading buffer is added to the sample and the whole 10 μl is run on the gel. We routinely analyze products of multiplex PCR reactions by electrophoresis in 1.5% (wt/vol) SeaKemLE agarose gel in TBE buffer, with the addition of the ethidium bromide. The conditions that allow efficient separation and visualization of amplicons are the following: current 160 mV, time ~90–100 min.
2. Gels are photographed and analyzed (*see* **Note 6**).

4 Notes

1. For the isolation of chromosomal DNA, we tested commercially available kits such as DNeasy Blood and Tissue Kits by Qiagen or Genomic Mini (A&A Biotechnology). However, any method of chromosomal DNA isolation that yields PCR products after amplification can be used.
2. We recommend the test of the reagents used for multiplex reaction prior to proceeding with a large number of samples. Although all tested reagents gave positive results, we noted that use of some reagents increased the yield of the PCR amplification. For example, we observed the best amplification results for the recombinant polymerase (5 U/μl) with Taq buffer with $(NH_4)_2SO_4$ (Thermo Scientific). We usually prepare 1 mM nucleotide solutions from 100 mM stock solutions of individual dNTPs that can be purchased from multiple suppliers.

3. We routinely use 1.5% SeaKem agarose, but any other high-performance agarose that allows efficient separation of fragments below 1000 bp can be used. The adjustment of agarose (1.2–1.8%) and buffer (1–0.5×) concentration may be needed to achieve distinct separation of amplified products.
4. For DNA isolation, we use various centrifuges and it does not influence downstream applications. For the amplification, we tested Veriti or 9700 PCR by Life Technologies (formerly Applied Biosystems) or T100™ Thermal Cycler (BioRad). All tested instruments give universal and efficient amplification in small volumes of the multiplex reactions with the use of 96-well plates. We usually use the gel electrophoresis Sub-Cell tank Model 192 (BioRad,) with 26-well combs, as it allows simultaneous analysis of 96 samples from a single 96-well plate and combs are compatible with multichannel pipettes.
5. Using plate format for DNA templates noticeably speeds up the reaction setup. In the majority of cases, we run 96-well plate for each of four mixes using the same DNA template plate. Therefore, we recommend either DNA isolation using 96-well format kits and protocols, or preparation of the diluted DNA in 96-well format.

   We strongly recommend aliquoting all reagents such as lysozyme, mutanolysin, dNTPs, and primer mixes in volumes that are required to run single 96-well or half of the reaction plate. We observed the loss of PCR efficiency after multiple freezing-thawing cycles.

   Reaction volume is small (5 μl). Preparation of single reaction that requires pipetting of volumes below 0.5 μl is very difficult and rather inaccurate because of manual errors. We strongly recommend using master mixes and running multiple reactions at the same time. We also recommend using electronic pipettes to avoid pipetting errors caused by small reaction volume. To minimize errors, dispense master mixes on the side of the tube/well, as small droplets are more visible on the side than on the bottom of the tube.

   The whole procedure can be performed during a single day, including DNA isolation.

   (a) DNA isolation: ~90 min.

   (b) Reaction setup: ~15 min for 96 well plate, but only using multichannel pipettes and pre-dispensed template master plate.

   (c) PCR amplification: Program 1, ~2 h 15 min; Program 2, ~4 h.

   (d) Gel electrophoresis: ~100 min.

6. As a result of each multiplex reaction, a set of multiple bands of various sizes can be observed (Fig. 1). Each PCR product corresponds to the detected virulence factor. We usually note the absence/presence of a band as a 0 or 1 in a string of digits. As a result of four multiplex reactions, we generate a string of 20 digits. Digitizing the results makes them easy to compare between strains. For example, during the analysis of products generated with mix 1, code 000011 denotes that we did not detect genes *spe*L, *spe*K, *spe*M, and *spe*C but detected products *smeZ* and *spe*I. The generated strings of 20 digits can be used as a marker of each strain and be further used for strain clustering and the construction of phylogenetic trees. Categorization of strains based on the virulence factors correlates with *emm* typing, and superantigens profile is a good predictor of *emm* type. Primers from mix 4 and *sdaB* primers from mix 3 detect chromosomal genes that are present in virtually all GAS strain and they can serve as positive controls for PCR amplification. In our laboratory, we typed thousands of GAS strains and only in two cases, we were unable to detect *spe*B. Products that correspond with the sic gene are detected in few serotype strains such as M1, M12, M55, and M57. Usually, with M77 serotype strains, we do not detect any genes from Mix2 and in some cases, M81 strains do not yield any products after amplification with primers from mix1. It is caused by a lack of particular virulence factors in M77 and M81 strains, not by fault PCR amplification.

   The method can be used when rapid strain comparison is required; it is fast and simple when compared with standard PFGE. The method is also inexpensive because it uses small reaction volumes (5 μl) and the cost of reagents is markedly low. The use of virulence factor detection can be used as an extension of any *S. pyogenes* typing and detection scheme. However, if the researchers want to mix our primers with other primers they use routinely, we recommend testing each individual primer pair they want to add to their typing scheme, whether it performs well in a mix. For laboratories that use a method that detects 11 superantigens [34], we recommend the use of mix 3 and mix 4 separately, as described in our study, to extend sensitivity and specificity of their detection. Chromosomally encoded proteases SpeB, SpyCEP, ScpA, and Mac are detected in virtually all strains, and the detection of virulence factors, if used for typing, can be simplified. Abbreviated method (without Mix 4) has an almost identical resolution as a not abbreviated method. As an alternative approach, primers detecting sic gene can be added to mixes 1, 2, or 3.

Acknowledgments

The work was supported by grants 2017/27/B/NZ7/00040 and 2018/29/B/NZ6/00624 to I.S.

References

1. Sitkiewicz I (2018) How to become a killer, or is it all accidental? Virulence strategies in oral streptococci. Mol Oral Microbiol 33:1–12
2. Sitkiewicz I, Hryniewicz W (2010) Pyogenic streptococci—danger of re-emerging pathogens. Pol J Microbiol 59:219–226
3. Carapetis JR, Steer AC, Mulholland EK et al (2005) The global burden of group a streptococcal diseases. Lancet Infect Dis 5:685–694
4. Sims Sanyahumbi A, Colquhoun S, Wyber R et al (2016) Global disease burden of group a streptococcus. In: Ferretti JJ, Stevens DL, Fischetti VA (eds) Streptococcus pyogenes: basic biology to clinical manifestations. University of Oklahoma Health Sciences Center, Oklahoma City, OK
5. Hauser AR, Stevens DL, Kaplan EL et al (1991) Molecular analysis of pyrogenic exotoxins from Streptococcus pyogenes isolates associated with toxic shock-like syndrome. J Clin Microbiol 29:1562–1567
6. Unnikrishnan M, Altmann DM, Proft T et al (2002) The bacterial superantigen streptococcal mitogenic exotoxin Z is the major immunoactive agent of Streptococcus pyogenes. J Immunol 169:2561–2569
7. Beres SB, Musser JM (2007) Contribution of exogenous genetic elements to the group a streptococcus metagenome. PLoS One 2:e800
8. Brouwer S, Barnett TC, Rivera-Hernandez T et al (2016) Streptococcus pyogenes adhesion and colonization. FEBS Lett 590:3739–3757
9. Rohde M, Cleary PP (2016) Adhesion and invasion of streptococcus pyogenes into host cells and clinical relevance of intracellular streptococci. In: Ferretti JJ, Stevens DL, Fischetti VA (eds) Streptococcus pyogenes: basic biology to clinical manifestations [Internet]. University of Oklahoma Health Sciences Center, Oklahoma City, OK. PMID: 26866223, NBK333420
10. Ryan PA, Juncosa B (2016) Group a streptococcal adherence. In: Ferretti JJ, Stevens DL, Fischetti VA (eds) Streptococcus pyogenes: basic biology to clinical manifestations. University of Oklahoma Health Sciences Center, Oklahoma City, OK
11. Fischetti VA (2016) M protein and other surface proteins on streptococci. In: Ferretti JJ, Stevens DL, Fischetti VA (eds) Streptococcus pyogenes: basic biology to clinical manifestations. University of Oklahoma Health Sciences Center, Oklahoma City, OK
12. Beall B.W. Protocol for emm typing. https://www.cdc.gov/streplab/groupa-strep/emm-typing-protocol.html. Accessed 08 Jul 2019
13. Proft T, Fraser JD (2016) Streptococcal Superantigens: biological properties and potential role in disease. In: Ferretti JJ, Stevens DL, Fischetti VA (eds) Streptococcus pyogenes: basic biology to clinical manifestations. University of Oklahoma Health Sciences Center, Oklahoma City, OK
14. Hynes W, Sloan M (2016) Secreted extracellular virulence factors. In: Ferretti JJ, Stevens DL, Fischetti VA (eds) Streptococcus pyogenes: basic biology to clinical manifestations. University of Oklahoma Health Sciences Center, Oklahoma City, OK
15. Walker MJ, Barnett TC, McArthur JD et al (2014) Disease manifestations and pathogenic mechanisms of group a streptococcus. Clin Microbiol Rev 27:264–301
16. Llewelyn M, Cohen J (2002) Superantigens: microbial agents that corrupt immunity. Lancet Infect Dis 2:156–162
17. Commons RJ, Smeesters PR, Proft T et al (2013) Streptococcal superantigens: Categorization and clinical associations. Trends Mol Med 20:1–15
18. Reglinski M, Sriskandan S, Turner CE (2019) Identification of two new core chromosome-encoded superantigens in Streptococcus pyogenes; speQ and speR. J Infect 78(5):358–363
19. Sumby P, Porcella SF, Barbian KD et al (2005) Evolutionary origin and emergence of a highly successful clone of serotype M1 group a streptococcus involved multiple horizontal gene transfer events. J Infect Dis 192:771–782
20. Walker MJ, Hollands A, Sanderson-Smith ML et al (2007) DNase Sda1 provides selection pressure for a switch to invasive group a streptococcal infection. Nat Med 13:981–985
21. Chang A, Khemlan A, Kang H et al (2011) Functional analysis of Streptococcus pyogenes

nuclease a (SpnA), a novel group a streptococcal virulence factor. Mol Microbiol 79:1629–1642

22. de Buhr N, Neumann A, Jerjomiceva N et al (2014) Streptococcus suis DNase SsnA contributes to degradation of neutrophil extracellular traps (NETs) and evasion of NET-mediated antimicrobial activity. Microbiology 160:385–395
23. Ly AT, Noto JP, Walwyn OL et al (2017) Differences in SpeB protease activity among group a streptococci associated with superficial, invasive, and autoimmune disease. PLoS One 12: e0177784
24. Olsen RJ, Raghuram A, Cantu C et al (2015) The majority of 9,729 group a streptococcus strains causing disease secrete SpeB cysteine protease: pathogenesis implications. Infect Immun 83:4750–4758
25. Chiang-Ni C, Wu J-JJ (2008) Effects of streptococcal pyrogenic exotoxin B on pathogenesis of Streptococcus pyogenes. J Formos Med Assoc 107:677–685
26. Chen CC, Cleary PP (1990) Complete nucleotide sequence of the streptococcal C5a peptidase gene of Streptococcus pyogenes. J Biol Chem 265:3161–3167
27. Brown CK, Gu Z, Matsuka Y et al (2005) Structure of the streptococcal cell wall C5a peptidase. Proc Natl Acad Sci U S A 102:18391–18396
28. Pawel-Rammingen U, Björck L (2003) IdeS and SpeB: immunoglobulin-degrading cysteine proteinases of Streptococcus pyogenes. Curr Opin Microbiol 6:50–55
29. Lei B, DeLeo FR, Hoe NP et al (2001) Evasion of human innate and acquired immunity by a bacterial homolog of CD11b that inhibits opsonophagocytosis. Nat Med 7:1298–1305
30. Okumura CYM, Anderson EL, Döhrmann S et al (2013) IgG protease mac/IdeS is not essential for phagocyte resistance or mouse virulence of M1T1 group a streptococcus. MBio 4(4):e00499–e00413
31. Kurupati P, Turner CE, Tziona I et al (2010) Chemokine-cleaving Streptococcus pyogenes protease SpyCEP is necessary and sufficient for bacterial dissemination within soft tissues and the respiratory tract. Mol Microbiol 76:1387–1397
32. Hidalgo-Grass C, Mishalian I, Dan-Goor M et al (2006) A streptococcal protease that degrades CXC chemokines and impairs bacterial clearance from infected tissues. EMBO J 25:4628–4637
33. Borek AL, Wilemska J, Izdebski R et al (2011) A new rapid and cost-effective method for detection of Phages , ICEs and virulence factors encoded by Streptococcus pyogenes. Pol J Microbiol 60:187–201
34. Lintges M, Arlt S, Uciechowski P et al (2007) A new closed-tube multiplex real-time PCR to detect eleven superantigens of Streptococcus pyogenes identifies a strain without superantigen activity. Int J Med Microbiol 297:471–478

Chapter 2

Closed-Tube Multiplex Real-Time PCR for the Detection of Group A Streptococcal Superantigens

Nour Zahi Gammoh and Lothar Rink

Abstract

Conventional PCR techniques are laborious and usually not suited for fast screening of large sample numbers in a clinical or research setting. Using this closed-tube multiplex real-time PCR, the presence of all 11 *Streptococcus pyogenes* superantigen (SAg) genes can be rapidly and accurately characterized. Identifying whether a strain contains a SAg can be done within 4 h compared to conventional methods which would take 11 times as long. This method provides an excellent diagnostic tool as well as a screening tool to help researchers clarify the role of SAgs in *S. pyogenes* infections.

Key words Superantigens, Multiplex PCR, *Streptococcus*, T-cell proliferation, T-cell receptor, Toxin genes, MHC class II

1 Introduction

Streptococcus pyogenes bacteria produce a range of exotoxins which function as superantigens (SAgs) and are considered to be important for the pathogenesis of invasive streptococcal infections [1, 2]. SAgs are a family of proteins produced by certain viruses and bacteria that are able to stimulate large portions of T lymphocytes via the interaction with the variable domain on the β-chain of the T-cell receptor (TCR-Vβ). It bypasses conventional antigen processing and directly bridges the TCR with the major histocompatibility complex (MHC) class II molecule on antigen presenting cells, instigating effector cell proliferation and differentiation and release of cytokines (IL-1β, IL-2, and TNF-α). A high number of T cells share a Vβ domain, which is a more conserved region on the TCR. Therefore, many cells may be activated by SAgs regardless of antigen specificity [3].

Currently, 11 SAgs from *S. pyogenes* have been characterized: streptococcal pyrogenic exotoxin A (SPEA), SPEC, SPEG, SPEH, SPEI, SPEJ, SPEK, SPEL, SPEM, streptococcal superantigen (SSA), and streptococcal mitogenic exotoxin Z (SMEZ) [4–

Thomas Proft and Jacelyn M. S. Loh (eds.), *Group A Streptococcus: Methods and Protocols*, Methods in Molecular Biology, vol. 2136, https://doi.org/10.1007/978-1-0716-0467-0_2,

13]. Some of the SAg genes have several subtypes, such as *spea* which consists of five alleles; alleles 1–3 and 5 are very homologous with only one base difference and allele 4 has 26 base pair exchanges [5, 14, 15]. Additionally, more than 20 different alleles were reported for *smez* [16]. The *speb* and *spef* genes will not be analyzed here, because they do not encode for superantigens. The gene *speb* codes for a protease with mitogenic activity, whereas *spef* codes for a deoxyribonuclease [17]. Identification of SAg genes by conventional PCR techniques using agarose gel electrophoresis to detect amplification products is strenuous and time consuming. However, using multiplex real-time PCR assay, which measures more than one gene in a single reaction, provides rapid results and reduces the amount of sample required per test. Testing 96 samples requires around 4 h and information on whether the bacterial strain contains a SAg gene can be immediately identified. Multiplex PCR assay followed with agarose gel electrophoresis requires 8 h. Without the multiplex method, conventional PCR assay takes 11 times longer to complete, which is not suited for rapid screening of large numbers of samples. Furthermore, this proposed closed system is free of any post-PCR contamination and is able to identify strains that contain SAg genes irrespective of their expression status [18]. Previous attempts to repurpose a multiplex PCR method initially designed for *Staphylococcus aureus* failed to detect the SAg genes *spei*, *spel*, and *spem* [19], but here, we demonstrate how a closed-tube multiplex PCR can detect all 11 streptococcal SAg genes, hence providing clinicians with a robust tool to aid them in their diagnosis [18].

2 Materials

2.1 Bacterial DNA Isolation

1. Sheep blood agar plates prepared with 5% defibrinated sheep blood and agar according to manufacturer's instructions.
2. Todd-Hewitt broth (THB) prepared according to manufacturer's instructions.
3. CO_2-incubator set at 37 °C.
4. Spectrophotometer to measure OD_{600}.
5. DNeasy Kit.
6. Enzymatic lysing buffer: 20-mM Tris–Cl, pH 8.0; 2-mM sodium EDTA; 1.2% Triton® X-100. Immediately before use, add mutanolysin to 1 μg/100 ml.

2.2 Multiplex PCR

1. Multiplex PCR kit.
2. Primers mixes designed according to Table 1.
3. PCR thermal cycler.
4. 2.5% agarose gel.

Table 1
Composition of primer mixes for SAg PCR

Primer mix	Target gene	Primer sequence (5′–3′)
Mix 1	16S rRNA	f: GTGAGTAACGCGTAGGTAACCTACCTCATAG r: CCCAGGCGGAGTGCTTAATG
	Spec	f: GGTAAATTTTTCAACGACACACACATTAAA r: TGTTGAGATTCTCCCGAAATAAATAGAT
	Speg	f: GCTATGGAAGTCAATTAGCTTATGCAGAT r: TTATGCGAACAGCCTCAGAGG
	Spej	f: CAATTAAATTACGCATACGAAATCATACCAGTA r: ACGAGTAAATATGTACGGAAGACCAAAAATA
	Spek	f: TATCGCTTGCTCTATACACTACTGAGAGT r: CCAAACTGTAGTATTTTCATCCGTATTAAA
	Spel	f: GGACGCAAGTTATTATGGATGCTCA r: TTAAATAAGTCAGCACCTTCCTCTTTCTC
	speal-3, 5 (*see* **Note 4**)	f: GGTATTTGCTCAACAAGACCCCGAT r: TGTGTTTGAGTCAAGCGTTTCATTATCT
Mix 2	16S rRNA	f: GGTGAGTAACGCGTAGGTAACCTACC r: GCCCAACTTAATGATGGCAACTAACA
	Speh	f: TCTATCTGCACAAGAGGTTTGTGAATGTCCA r: GCATGCTATTAAAGTCTCCATTGCCAAAA
	Spei	f: AAGGAAAAATAAATGAAGGTCCGCCAT r: TCGCTTAAAGTAATACCTCCATATGAATTCTTT
	Spem	f: GCTTTAAGGAGGAGGAGGTTGATATTTATGCTCTA r: CAAAGTGACTTACTTTACTCATATCAATCGTTTC
	Smez	f: CAATAATTTCTCGTCCTGTGTTTGGAT r: GATAAGGCGTCATTCCACCATAG
	Ssa	f: AATTATTATCGATTAGTGTTTTTGCAAGTA r: AGCCTGTCTCGTACGGAGAATTATTGAACTC
	speal-4 (*see* **Note 4**)	f: CAAGAAGTATTTGCTCAACAAGACCCCA r: TTAGATGGTCCATTAGTATATAGTTGCTTGTTATC

f: forward primer; r: reverse primer

2.3 SYBR-Green PCR

1. Real-time thermal cycler.
2. 10× SYBR-Green (Cambrex).
3. 100× 6-carboxy-X-rhodamine (ROX, TIB Molbiol).

3 Methods

3.1 Bacterial DNA Isolation

1. Culture bacterial isolates on sheep blood agar overnight in a CO_2 incubator at 37 °C.
2. Transfer three colonies into 6-ml THB and incubate overnight with 6% CO_2 at 37 °C.
3. Transfer 0.5 ml of this culture into 11.5-ml THB and incubate again for 3 h. Determine OD_{600} and calculate number of bacteria (*see* **Note 1**).

Table 2
Composition of reaction mixture for PCR

Component	Volume/reaction	Final concentration
Master mix		
10× PCR buffer	5 µl	1×
25 mM $MgCl_2$	15 µl	7.5 mM[a]
dNTP mix (10 mM each)	1 µl	
SAg primers F + R (mix 1 or 2)	Variable[b]	200 nM from each Primer pair
16S rRNA primers F + R	Variable[b]	20 nM
Q-solution	5 µl	1×
Taq DNA polymerase	Variable[b]	3.75 U/µl[a]
RNase-free water	Variable[c]	–
Template DNA	Variable[b]	100 ng
Total volume	**50**	–

[a]Optimal concentration should be determined empirically
[b]Depending on stock concentration
[c]Depending on volume of other components

4. Isolate DNA using DNeasy Kit or similar. Use enzymatic lysis buffer containing mutanolysin (*see* **Note 2**).

3.2 Multiplex PCR

1. Use a multiplex PCR kit along with the two sets of different primer mixes (*see* **Note 3**). As an internal positive control, use an additional primer pair targeting the 16S rRNA gene (Table 1).
2. Prepare PCR mixture (50 µl) according to Table 2.
3. Carry out PCR amplification in a standard PCR thermocycler with initial denaturation step at 95 °C for 15 min, followed by 35 cycles of 30 s at 94 °C, 90 s at 57 °C, 90 s at 72 °C, and a final extension step of 72 °C for 10 min.
4. Analyze samples on 2.5% agarose gel. Figure 1 provides an example of SAg gene profiles of different Group A Streptococcus isolates.

3.3 Real-Time PCR Using SYBR-Green

1. For PCR in a real-time cycler, add 1.3-µl SYBR-Green (10×) and 0.5-µl ROX (100×) to the reaction mix (see above) (*see* **Note 5**).
2. Use same PCR conditions as described above, but with 40 cycles.
3. Carry out melting (dissociation) curve analysis after every amplification (*see* **Note 6**).

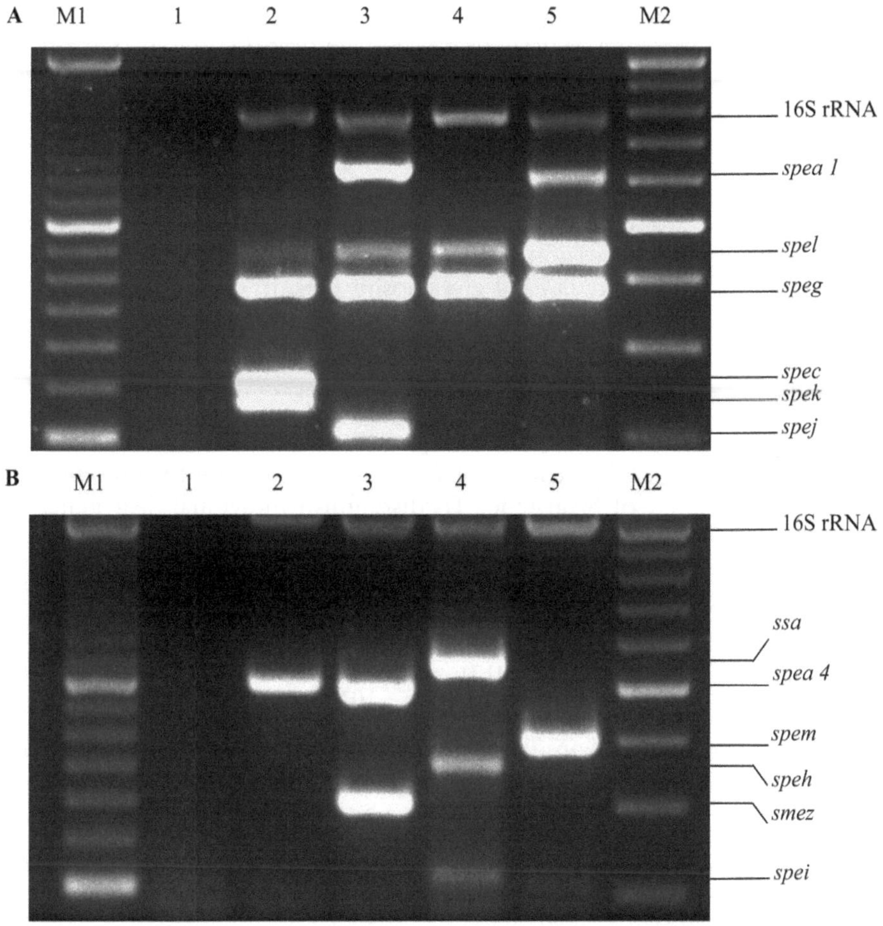

Fig. 1 SAg gene profiles obtained by agarose gel electrophoresis of multiplex PCRs. (**a**) Multiplex PCR reaction 1. (**b**) Multiplex PCR reaction 2. **M1** 50-bp DNA ladder marker; **M2** 100-bp DNA ladder marker; **1** Negative control, without DNA; **2–5** representative GAS isolates with different SAg profiles

4 Notes

1. There are several methods to determine the number of bacteria. If using a reference bacteria strain, plot a standard curve with OD_{600} versus colony-forming units per milliliter (CFU/ml) and use it to calculate the number of bacteria in the sample. If reference strain is not available, count bacteria using a microscopic counting chamber (dilute sample if necessary) or use the DNA extraction kit manufacturer's recommendations.
2. DNA concentration after isolation should range from 10 to 200 μg/ml.

3. Multiplex PCR kits from various suppliers or single reagents can be used; however, denaturation time and temperature must be optimized for the specific *Taq* DNA polymerase used.
4. The primer *spea*1-3,5 detects the *spea* alleles 1, 2, 3, and 5, whereas the primer *spea*1-4 detects the *spea* alleles 1, 2, 3, and 4. So if *spea*1-3+5 and *spea*1-4 both yield positive results, then the isolate contains *spea* allele 1, 2, or 3. If only *spea*1-3+5 gives positive results, then the isolate contains the allele *spea*5; if only *spea*1-4 gives positive results, then the isolate contains the allele *spea*4.
5. During SYBR-Green PCR, add DNA template to the mix as the last component.
6. Analysis of melting curves enables the differentiation between SAg genes and 16S rRNA, confirming the presence or absence of SAg gene. To distinguish individual SAg genes, the PCR products must be further analyzed on agarose gel.

References

1. Imöhl M, Fitzner C, Perniciaro S et al (2017) Epidemiology and distribution of 10 superantigens among invasive Streptococcus pyogenes disease in Germany from 2009 to 2014. PLoS One 12(7):e0180757. https://doi.org/10.1371/journal.pone.0180757
2. Lintges M, van der Linden M, Hilgers R-D et al (2010) Superantigen genes are more important than the emm type for the invasiveness of group a streptococcus infection. J Infect Dis 202(1):20–28. https://doi.org/10.1086/653082
3. Actor JK (2012) Elsevier's integrated review. In: Immunology and microbiology: 5—role of major histocompatibility complex in the immune response, Elsevier's integrated series, 2nd edn. Elsevier/Saunders, Philadelphia, PA
4. Beres SB, Sylva GL, Barbian KD et al (2002) Genome sequence of a serotype M3 strain of group A Streptococcus: phage-encoded toxins, the high-virulence phenotype, and clone emergence. Proc Natl Acad Sci U S A 99 (15):10078–10083. https://doi.org/10.1073/pnas.152298499
5. Bessen DE, Izzo MW, Fiorentino TR et al (1999) Genetic linkage of exotoxin alleles and emm gene markers for tissue tropism in group a streptococci. J Infect Dis 179(3):627–636. https://doi.org/10.1086/314631
6. Ferretti JJ, McShan WM, Ajdic D et al (2001) Complete genome sequence of an M1 strain of Streptococcus pyogenes. Proc Natl Acad Sci U S A 98(8):4658–4663. https://doi.org/10.1073/pnas.071559398
7. Igwe EI, Shewmaker PL, Facklam RR et al (2003) Identification of superantigen genes speM, ssa, and smeZ in invasive strains of beta-hemolytic group C and G streptococci recovered from humans. FEMS Microbiol Lett 229(2):259–264. https://doi.org/10.1016/S0378-1097(03)00842-5
8. Kamezawa Y, Nakahara T, Nakano S et al (1997) Streptococcal mitogenic exotoxin Z, a novel acidic superantigenic toxin produced by a T1 strain of Streptococcus pyogenes. Infect Immun 65(9):3828–3833
9. McCormick JK, Pragman AA, Stolpa JC et al (2001) Functional characterization of streptococcal pyrogenic exotoxin J, a novel Superantigen. Infect Immun 69(3):1381–1388. https://doi.org/10.1128/IAI.69.3.1381-1388.2001
10. Proft T, Moffatt SL, Berkahn CJ et al (1999) Identification and characterization of novel Superantigens from Streptococcus pyogenes. J Exp Med 189(1):89–102. https://doi.org/10.1084/jem.189.1.89
11. Proft T, Webb PD, Handley V et al (2003) Two novel Superantigens found in both group a and group C streptococcus. Infect Immun 71 (3):1361–1369. https://doi.org/10.1128/IAI.71.3.1361-1369.2003
12. Proft T, Arcus VL, Handley V et al (2001) Immunological and biochemical characterization of streptococcal pyrogenic exotoxins I and

J (SPE-I and SPE-J) from Streptococcus pyogenes. J Immunol 166(11):6711–6719. https://doi.org/10.4049/jimmunol.166.11.6711

13. Smoot LM, McCormick JK, Smoot JC et al (2002) Characterization of two novel pyrogenic toxin Superantigens made by an acute rheumatic fever clone of Streptococcus pyogenes associated with multiple disease outbreaks. Infect Immun 70(12):7095–7104. https://doi.org/10.1128/IAI.70.12.7095-7104.2002
14. Baker M, Gutman DM, Papageorgiou AC et al (2001) Structural features of a zinc binding site in the superantigen streptococcal pyrogenic exotoxin a (SpeA1): implications for MHC class II recognition. Protein Sci 10 (6):1268–1273. https://doi.org/10.1110/ps.330101
15. Nelson K, Schlievert PM, Selander RK et al (1991) Characterization and clonal distribution of four alleles of the speA gene encoding pyrogenic exotoxin a (scarlet fever toxin) in Streptococcus pyogenes. J Exp Med 174 (5):1271–1274
16. Proft T, Moffatt SL, Weller KD et al (2000) The streptococcal Superantigen Smez exhibits wide allelic variation, mosaic structure, and significant antigenic variation. J Exp Med 191 (10):1765–1776. https://doi.org/10.1084/jem.191.10.1765
17. Fagin U, Hahn U, Grötzinger J et al (1997) Exclusion of bioactive contaminations in Streptococcus pyogenes erythrogenic toxin a preparations by recombinant expression in Escherichia coli. Infect Immun 65 (11):4725–4733
18. Lintges M, Arlt S, Uciechowski P et al (2007) A new closed-tube multiplex real-time PCR to detect eleven superantigens of Streptococcus pyogenes identifies a strain without superantigen activity. Int J Med Microbiol 297 (6):471–478. https://doi.org/10.1016/j.ijmm.2007.03.015
19. Schmitz F-J, Beyer A, Charpentier E et al (2003) Toxin-gene profile heterogeneity among endemic invasive European group a streptococcal isolates. J Infect Dis 188 (10):1578–1586. https://doi.org/10.1086/379230

Chapter 3

The *emm*-Cluster Typing System

Pierre R. Smeesters and Anne Botteaux

Abstract

emm-cluster typing system allows to classify most *Streptococcus pyogenes* variants into 48 different *emm* clusters. The system correlates nicely with the host serum binding capacities of the M proteins and has been used in epidemiological surveys, strain selection, and vaccine development. Here we describe the allocation of the *emm* cluster based on the *emm*-typing defining region.

Key words *emm*-typing, *emm*-cluster typing, M protein, *Streptococcus pyogenes*, PCR

1 Introduction

Typing of *Streptococcus pyogenes* can be done using different typing methods. The surface-exposed streptococcal M protein, encoded by the *emm* gene, has been the typing antigen for decades [1–4]. More than 230 *emm* types have been described using a PCR-based method developed by the American reference laboratory at the CDC. The *emm* gene size ranges from 690 to 1569 bp. A specific *emm* type being allocated based on the sequence of 180 nucleotides encoding for the 10 residues of the signal peptide and the 50 first amino acids of the mature M protein. A new *emm* type is assigned for changes in the first 30 codons encoding the mature M protein and leading to less than 92% of identity to a known *emm* type. A new subtype is assigned for any change in the 180 bp and noted, for example, as *emm4.1* (*emm*-type 4, subtype 1).

The *emm* gene is encoded in a genetic locus which spans from the ubiquitous genes for Mga (*mga*), a transcriptional regulator, and the C5a peptidase (*scpa*) [5]. In some strains, mainly isolated from low-income countries, other genes can be present in the Mga regulon, including genes encoding for the M-like proteins (*emm*-like genes) Mrp (*mrp*) and Enn (*enn*) [6].

emm-cluster typing has been described in 2014 based on the complete *emm* gene sequences, selective pressures, and binding

Thomas Proft and Jacelyn M. S. Loh (eds.), *Group A Streptococcus: Methods and Protocols*, Methods in Molecular Biology, vol. 2136, https://doi.org/10.1007/978-1-0716-0467-0_3, © Springer Science+Business Media, LLC, part of Springer Nature 2020

Table 1
***emm*-type, *emm*-cluster, and predicted binding capacities**

	CLUSTER	EMM-TYPES	BINDING CAPACITIES* PLASM	IGG	IGA	FIBRI	ALB	C4BP
CLADE X	E1	4-60-78-165 -176		■	■		■	■
CLADE X	E2	13-27-50 (50/62)-66-68-76-90-92-96-104-106-110-117-166-168		■			■	
CLADE X	E3	9-15-25-44 (44/61)-49-58-79-82-87-103-107-113-118-144-180-183-209-219-231		■			■	■
CLADE X	E4	2-8-22-28-73-77-84-88-89-102-109-112-114-124-169 -175-232		■				■
CLADE X	E5	34-51-134-137-170-174 -205		■	■			
CLADE X	E6	11-42-48-59-63-65 (65/69)- 67-75-81-85-94-99-139-158-172-177-182-191		■	■		■	■
CLADE X	Single protein cluster clade X	164-185- 211-236						
CLADE Y	D1	36- 54- 207				■	■	
CLADE Y	D2	32- 71-100-115-213						
CLADE Y	D3	123-217						
CLADE Y	D4	33-41-43-52-53-56-56.2-64-70-72-80-83-86-91-93-98-101-108-116-119-120-121-178-186 -192 -194 -208 -223 -224 -225 -230 -242	■			■		
CLADE Y	D5	97-157 -184				■	■	
CLADE Y	A-C1	46-142				■		
CLADE Y	A-C2	30-197						
CLADE Y	A-C3	1-163- 227 - 238 (1-2)- 239 (1-4)		■		■		
CLADE Y	A-C4	12-39-193-228-229				■	■	
CLADE Y	A-C5	3-31-133				■	■	
CLADE Y	Single protein cluster clade Y	5-6-1417-18-19-23-24-26-29-37-38 (38/40)-47-55-57-74-95-105-111-122-140 -179 - 215 - 218 - 221-222 - 233 - 234						

PLASM plasminogen, *IGG* immunoglobulins G, *IGA* immunoglobulins A, *FIBRI* fibrinogen, *ALB* albumin, *C4BP* C4-binding protein. *Based on experimental data or motifs prediction. (Detailed information can be found in [7])

capacity analyses of the M proteins to 6 host serum proteins (IgG, IgA, C4BP, fibrinogen, plasminogen, and albumin) [7–10]. Forty-eight *emm* clusters have been defined (*emm*-cluster E1-E6, D1-D5, AC1-AC-5, and 32 single M-protein *emm* cluster) [7]. *emm* clusters have been used to select representative strains in genomic studies [11], assays development [12], epidemiological studies [13–19], and vaccine development [20–22]. *Emm*-cluster allocation can be deduced by the *emm*-typing result (*see* Table 1).

2 Materials

2.1 DNA Extraction

1. GAS strains.
2. Sterile pipette tips/wooden stick.

3. Blood agar plates.
4. Sterile/nucleases free water.
5. Vortex.
6. Thermocycler.
7. 0.85% NaCl.
8. Heating block.
9. Microfuge.
10. Tris–EDTA (TE): 20-mM Tris, 3-mM EDTA pH 8.
11. THY: Todd Hewitt broth with 0.5% yeast extract.
12. Screw cap conical polypropylene tube, 30 mL.
13. CO_2 incubator.
14. Lysis buffer: 50-μL TE, 10-μL mutanolysin (3000 U/mL), 2-μL hyaluronidase (300–750 U/mg).
15. Genomic DNA extraction kit (NEB).
16. Microvolume spectrophotometer.

2.2 PCR

1. PCR Master Mix: 10 μL of Taq buffer (with $MgCl_2$), 2 μL of each primer (at 20 mM) (*see* Table 2), 2 μL of dNTPs (at 10 mM), 0.2 μL of Taq polymerase, nuclease free water to 45 μL.
2. Vortex.
3. Centrifuge.
4. Electrophoresis system for agarose gels.
5. UV transilluminator.
6. Tris–acetate–EDTA buffer (TAE; 50×): 242 g Tris–HCl, 100 mL of 0.5 M EDTA, 57.1-mL glacial acetic acid. Make up to 1 L with deionized water.
7. Agarose gels (1%): add 1 g of agarose into 100 mL of 1× TAE and melt it for 5 min in a microwave at 600 W. Add 5 μL of a fluorescent DNA staining.

Table 2
Primer sequences

Name	Sequence
CDC1	5′-TATT(C/G)GCTTAGAAAATTAA-3′
CDC2	5′-GCAAGTTCTTCAGCTTGTTT-3′
MR1	5′-GGATCCATAAGGAGCATAAAAATGGCTA-3′
MF2	5′-GGATCCATAAGGAGCATAAAAATGGCTA-3′
Emmseq2	5′-TATTCGCTTAGAAAATTAAAAACAGG-3′

8. Molecular weight marker for DNA.
9. Gel extraction kit.
10. PCR clean up kit.
11. Sequencing plates or tubes.

3 Methods

3.1 DNA Template Preparation

To assign an *emm* type and an *emm* cluster from an isolate, three different methods of DNA template preparation are provided as required (*see* **Note 1**).

3.1.1 Using Heat to Lysate GAS Colonies from a Plate

1. Using a sterile pipette tips or wooden stick, pick one beta-hemolytic colony from the blood agar plate into 10 μL of sterile/nucleases free water and mix by vortexing to suspend the colony.
2. Heat for 10 min at 95 °C in a thermocycler to lyse bacteria.
3. Take 5 μL of the lysate to work within the PCR reaction (*see* Subheading 3.2).

3.1.2 Using Enzymatic Lysis of GAS Colonies from Plate

1. Pick a fair amount, that is, 1/5 of fresh bacterial colony from the plate, with a pipette tip and re-suspend in 300 μL of 0.85% NaCl.
2. Incubate at 70 °C in a heating block for 15 min.
3. Centrifuge the samples at 13,000 × *g* for 2 min in microfuge and pipet out supernatant.
4. Resuspend pellet in 50 μL TE (*see* **Note 2**).
5. Vortex thoroughly and incubate at 37 °C for 30 min.
6. Incubate on a heating block at 100 °C for 10 min to inactivate the enzymes.
7. Proceed with PCR or store the lysates at −20 °C until use.

3.1.3 Using DNA Extraction of GAS from Liquid Culture

1. Make an overnight culture of GAS by inoculation of 1 colony in 5 mL of THY at 37 °C and 5% CO_2.
2. Harvest 2×10^9 GAS by centrifugation at >8000 × *g*. Remove the supernatant.
3. Use the genomic DNA extraction kit (NEB) according to manufacturer's instructions for Gram-positive bacteria.
4. Measure the DNA concentration by spectrophotometry (*see* **Note 3**).
5. Take 100 ng of the eluted DNA to work within the PCR reaction (*see* Subheading 3.2).

3.2 Emm Amplification by PCR

1. Make a Master Mix for the PCR (45 μL/tube).
2. Add 5 μL of bacterial lysate or 100 ng of DNA extraction in 5 μL of nuclease free water to 45 μL of master mix to each tube.
3. Mix and spin down tubes.
4. Use standard GoTaq PCR reaction parameters (Hot Start): 94 °C for 5 min, 25 cycles of: 94 °C for 15 s, 47 °C for 30 s (for CDC1/CDC2 primers), 70 °C for 1 min 15 s, and a final step at 70 °C for 10 min.
5. Check the purity and size of your amplicon. Load 5 μL of the PCR reaction on a 0.8–1% agarose gel and run the gel for 20 min at 200 V. A single band should be visible (*see* **Note 4**).
6. Purify your PCR product using a PCR clean up kit according to the manufacturer's instructions and measure the concentration by spectrophotometry.

3.3 Sequencing

Send the tubes containing purified PCR products or the plate containing unpurified PCR product to a sequencing company, or facility, using primer Emm seq 2.

3.4 emm-Type and emm-Cluster Assignation

1. Go to the CDC website and paste your sequence in FASTA format: https://www2a.cdc.gov/ncidod/biotech/strepblast.asp.
2. Choose "blastn," subtype-specific sequence, 10 matches of alignments.
3. You will receive an email with the BLAST results from your query against the *emm* database (10/50 alignments per sequence).
4. If the matching result of the first 180 nucleotides is 180/180, you can be confident in the *emm*-type and *emm*-cluster result.
5. If the matching result of the first 180 nucleotides is less than 180/180, check the chromatogram before concluding of a new *emm* type or subtype (*see* Subheading 1).
6. Add a new *emm* subtype or type to the CDC database by sending the untrimmed sequence by email to Srinivasan@cdc.gov.

4 Notes

1. You can often simply boil (95 °C, 10 min) very freshly grown GAS with good results for DNA extracts. Sometimes, we encounter isolates that are not easy to lyse. In these cases, make lysate using enzymatic lysis from plate or liquid culture.
2. Adding hyaluronidase helps to break down the hyaluronic capsule of GAS. Mutanolysin is a *N*-acetylmuramidase that

cleaves the *N*-acetymuramyl-(1–4)-*N*-acetylglucosamine linkage of the GAS cell wall.

3. Yield of DNA extraction can range from 100 to 3000 ng/μL depending on the strains.
4. A double band may be visible in some isolates, mainly isolated from low-income settings. In these isolates, CDC1 and CDC2 primers also amplify the *emm*-like genes due to their related sequences. This problem is generally overcome by using another pair of primers (MF2 and MR1) with an annealing temperature of 52 °C. In rare isolates, the presence of double band persists whatever the primers used. Extraction of both bands from the agarose gel using a DNA extraction kit before sequencing can help address this problem.

References

1. Smeesters PR, McMillan DJ, Sriprakash KS (2010) The streptococcal M protein: a highly versatile molecule. Trends Microbiol 18:275–282. https://doi.org/10.1016/j.tim.2010.02.007
2. Walker MJ, Barnett TC, McArthur JD et al (2014) Disease manifestations and pathogenic mechanisms of group a streptococcus. Clin Microbiol Rev 27(2):264–301. https://doi.org/10.1128/CMR.00101-13
3. Beall B, Facklam R, Thompson T (1996) Sequencing emm-specific PCR products for routine and accurate typing of group A streptococci. J Clin Microbiol 34(4):953–958
4. Li Z, Sakota V, Jackson D et al (2003) Array of M protein gene subtypes in 1064 recent invasive group A streptococcus isolates recovered from the active bacterial core surveillance. J Infect Dis 188(10):1587–1592. https://doi.org/10.1086/379050
5. Hollingshead SK, Readdy TL, Yung DL, Bessen DE (1993) Structural heterogeneity of the emm gene cluster in group A streptococci. Mol Microbiol 8(4):707–717
6. Frost HR, Sanderson-Smith M, Walker M et al (2018) Group A streptococcal M-like proteins: from pathogenesis to vaccine potential. FEMS Microbiol Rev 42(2):193–204. https://doi.org/10.1093/femsre/fux057
7. Sanderson-Smith M, De Oliveira DM, Guglielmini J et al (2014) A systematic and functional classification of Streptococcus pyogenes that serves as a new tool for molecular typing and vaccine development. J Infect Dis 210 (8):1325–1338. https://doi.org/10.1093/infdis/jiu260
8. Smeesters PR, Dramaix M, Van Melderen L (2010) The emm-type diversity does not always reflect the M protein genetic diversity—is there a case for designer vaccine against GAS. Vaccine 28(4):883–885. https://doi.org/10.1016/j.vaccine.2009.10.137
9. Smeesters PR, Mardulyn P, Vergison A et al (2008) Genetic diversity of group A streptococcus M protein: implications for typing and vaccine development. Vaccine 26 (46):5835–5842. https://doi.org/10.1016/j.vaccine.2008.08.037
10. McMillan DJ, Dreze PA, Vu T et al (2013) Updated model of group A streptococcus M proteins based on a comprehensive worldwide study. Clin Microbiol Infect 19(5): E222–E229. https://doi.org/10.1111/1469-0691.12134
11. Davies MR, McIntyre L, Mutreja A et al (2019) Atlas of group A streptococcal vaccine candidates compiled using large-scale comparative genomics. Nat Genet 51(6):1035–1043. https://doi.org/10.1038/s41588-019-0417-8
12. Jones S, Moreland NJ, Zancolli M et al (2018) Development of an opsonophagocytic killing assay for group a streptococcus. Vaccine 36 (26):3756–3763. https://doi.org/10.1016/j.vaccine.2018.05.056
13. Giffard PM, Tong SYC, Holt DC et al (2019) Concerns for efficacy of a 30-valent M-protein-based Streptococcus pyogenes vaccine in regions with high rates of rheumatic heart disease. PLoS Negl Trop Dis 13(7):e0007511. https://doi.org/10.1371/journal.pntd.0007511

14. Gergova R, Muhtarova A, Mitov I et al (2019) Relation between emm types and virulence gene profiles among Bulgarian Streptococcus pyogenes clinical isolates. Infect Dis (Lond) 51(9):668–675. https://doi.org/10.1080/23744235.2019.1638964
15. Smeesters PR, Laho D, Beall B et al (2017) Seasonal, geographic, and temporal trends of emm clusters associated with invasive group A streptococcal infections in US multistate surveillance. Clin Infect Dis 64(5):694–695. https://doi.org/10.1093/cid/ciw807
16. Williamson DA, Smeesters PR, Steer AC et al (2015) M-protein analysis of Streptococcus pyogenes isolates associated with acute rheumatic fever in New Zealand. J Clin Microbiol 53(11):3618–3620. https://doi.org/10.1128/JCM.02129-15
17. Shulman ST, Tanz RR, Dale JB et al (2014) Added value of the emm-cluster typing system to analyze group A streptococcus epidemiology in high-income settings. Clin Infect Dis 59 (11):1651–1652. https://doi.org/10.1093/cid/ciu649
18. Baroux N, D'Ortenzio E, Amédéo N et al (2014) The emm-cluster typing system for group A streptococcus identifies epidemiologic similarities across the Pacific region. Clin Infect Dis 59(7):e84–e92. https://doi.org/10.1093/cid/ciu490
19. Seale AC, Davies MR, Anampiu K et al (2016) Invasive group A streptococcus infection among children, Rural Kenya. Emerg Infect Dis 22(2):224–232. https://doi.org/10.3201/eid2202.151358
20. Frost HR, Laho D, Sanderson-Smith ML et al (2017) Immune cross-opsonization within emm clusters following group A streptococcus skin infection: broadening the scope of type-specific immunity. Clin Infect Dis 65 (9):1523–1531. https://doi.org/10.1093/cid/cix599
21. Dale JB, Smeesters PR, Courtney HS et al (2017) Structure-based design of broadly protective group a streptococcal M protein-based vaccines. Vaccine 35(1):19–26. https://doi.org/10.1016/j.vaccine.2016.11.065
22. Steer AC, Carapetis JR, Dale JB et al (2016) Status of research and development of vaccines for Streptococcus pyogenes. Vaccine 34 (26):2953–2958. https://doi.org/10.1016/j.vaccine.2016.03.073

Chapter 4

Protocols for Tn-seq Analyses in the Group A Streptococcus

Yoann Le Breton, Ashton T. Belew, and Kevin S. McIver

Abstract

Transposon-sequencing (Tn-seq) has revolutionized forward-genetic analyses to study genotype–phenotype associations and interrogate bacterial cell physiology. The Tn-seq approach allows the *en masse* monitoring of highly complex mutant libraries, leveraging massive parallel DNA sequencing as a means to characterize the composition of these mutant pools on a genome-scale with unprecedented nucleotide-level high resolution. In this chapter, we present step-by-step protocols for Tn-seq analyses in the human pathogen *Streptococcus pyogenes* (Group A Streptococcus or GAS) using the *mariner*-based *Krmit* transposon. We detail how to generate highly complex *Krmit* mutant libraries in GAS and the *en masse* production of *Krmit* insertion tags for Illumina sequencing of the transposon–genome junctions for Tn-seq analyses. Most of the protocols presented here were developed and implemented using the *S. pyogenes* M1T1 serotype clinical isolate 5448, but they have been successfully applied to multiple GAS serotypes as well as other pathogenic Streptococci.

Key words Tn-seq, *Mariner*, *Krmit* transposon, Isolate 5448

1 Introduction

Since its development for *Streptococcus pneumoniae* (*S. pneumoniae*) by the Camilli group in 2009 [1], transposon-sequencing (Tn-seq) and similar approaches (e.g., HITS [2, 3], INSeq [4], or TraDIS [5]) have revolutionized forward-genetic screens in bacteria to employ functional genomics through interrogating the contribution of genes in phenotype screens of different aspects of bacterial cell physiology and pathogenesis without a priori knowledge (for reviews, *see* [6–8]). The Tn-seq method represents a powerful approach to monitor *en masse* the composition of comprehensive mutant pools by tracking the transposon flanking junctions (or insertion tags) using massive parallel DNA sequencing to produce whole-genome datasets with nucleotide-level high resolution.

In this chapter, we present detailed protocols to implement Tn-seq analyses in the human pathogen *Streptococcus pyogenes* (also known as the Group A Streptococcus, or GAS) using the

Thomas Proft and Jacelyn M. S. Loh (eds.), *Group A Streptococcus: Methods and Protocols*, Methods in Molecular Biology, vol. 2136, https://doi.org/10.1007/978-1-0716-0467-0_4,

Krmit system (for *K*anamycin-*r*esistant cassette for *m*assive *i*dentification of *t*ransposants) (Fig. 1). GAS is a Biosafety Level 2 (BSL-2) pathogen and it is critical that the procedures presented here be conducted following all appropriate guidelines and regulations for the use and handling of pathogenic microorganisms, as described in the fifth Biosafety in Microbiological and Biomedical Laboratories manual from the Centers for Disease Control and Prevention (www.cdc.gov/labs/BMBL.html).

The *Krmit* system and methods presented here have been successfully used to perform multiple Tn-seq screens in GAS (Fig. 1), to characterize GAS essential genes (set of nonmutable genes, also known as minimal genome) [9], to identify genes required for survival during soft-tissue infection [10], as well as for determining the susceptibility and/or resistance to disease-relevant physicochemical stresses (e.g., zinc, antibacterial phospholipase A2) [11–13]. The protocols presented here cover in detail the procedures to (1) generate comprehensive *Krmit* mutant libraries suitable for Tn-seq screens, (2) produce *Krmit* flanking junctions for Illumina sequencing, and (3) the first steps for Tn-seq data recovery and analysis. Tn-seq screens can be conducted in multiple conditions; and those to interrogate GAS pathogenesis would likely require using in vivo models of GAS infections, thus requiring the investigator to follow animal BSL-2 (ABSL-2) guidelines and regulations not discussed in this chapter.

A key component for the success of Tn-seq and similar approaches is the ability to generate complex mutant libraries in the bacteria being tested. *S. pyogenes* can be quite challenging for genetic manipulations and, unlike naturally competent organisms such as *S. pneumoniae*, requires an in vivo transposon delivery system. *Krmit* is a Tn-seq-compatible transposon based on the first *mariner* system developed for GAS, *Oskar*, initially designed for in vivo transposon mutagenesis in GAS and for transposon-site hybridization (TraSH) screens [14, 15]. This method tracked the composition of mutant libraries in GAS using DNA microarrays; however, several technical bottlenecks inherent to TraSH complicated mutant identification and prompted us to modify *Oskar* to suit the requirements described by van Opijnen et al. [1] to conduct Tn-seq screens in GAS, creating *Krmit* [9].

The *Krmit* Tn-seq system in GAS relies on the transposon delivery plasmid pKRMIT (Fig. 1a and b) to produce complex mutant libraries. Briefly, pKRMIT harbors a composite *mariner* mini-transposon, *Krmit*, that consists of a Kanamycin-resistance (Km^r) cassette flanked with *mariner* inverted terminal repeats (ITRs) and modified to include an MmeI restriction site for Tn-seq purposes (Fig. 1a). MmeI modifications within the ITRs were verified not to significantly affect the in vivo delivery of the transposon in GAS when compared to transposition efficiency by its parental *Oskar* [9]. The *Krmit* mini-transposon does not encode

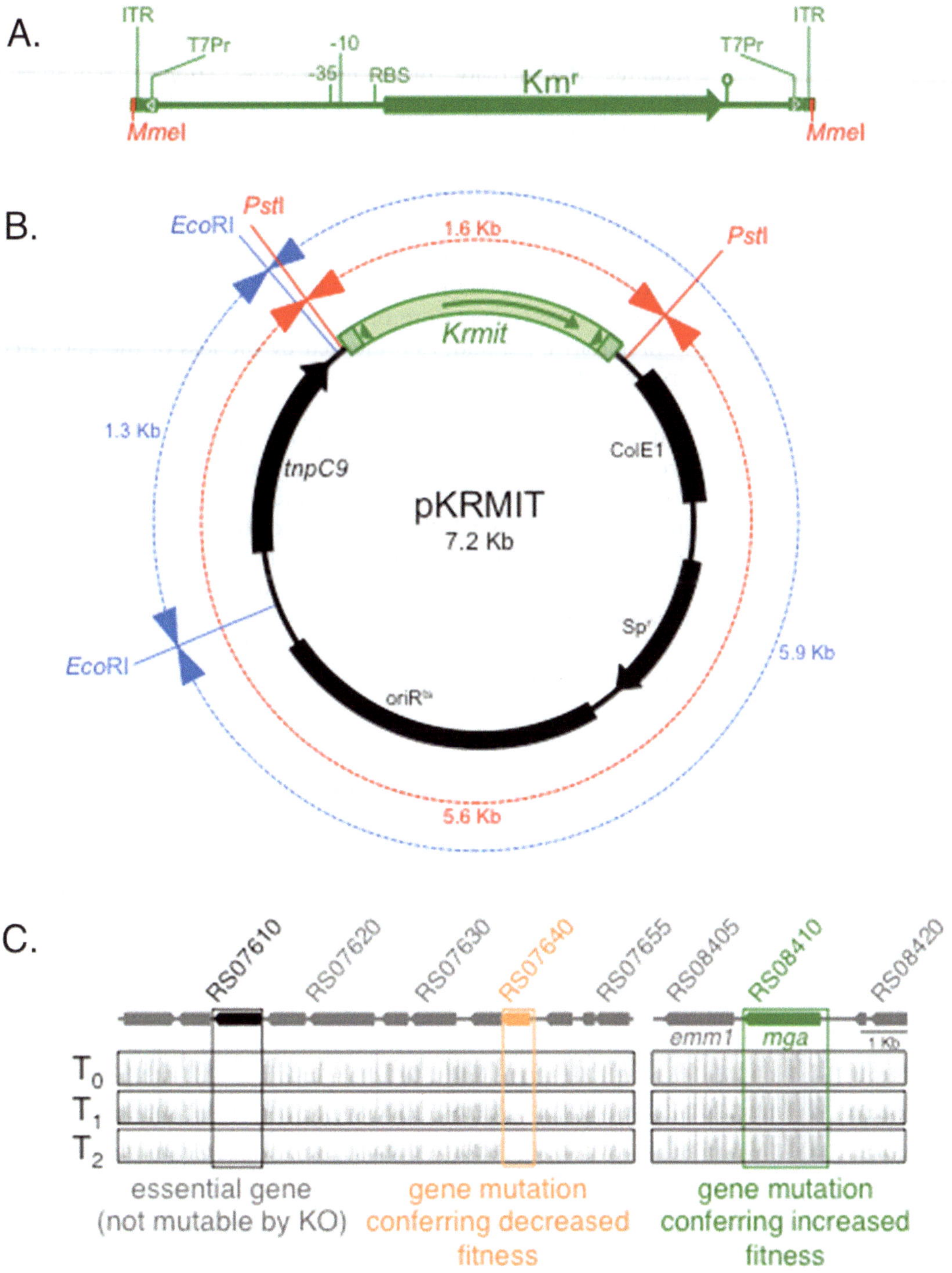

Fig. 1 The *Krmit* system for Tn-seq analyses in *S. pyogenes*. (**a**) Representation of main features on the *Krmit mariner* transposon, with MmeI sites incorporated within the *Krmit* inverted terminal repeats (ITRs) for Tn-seq compatibility. Km^r, Kanamycin-resistance gene; T7Pr: T7 promoter; −35/−10/RBS: promoter region for Km^r expression in GAS. (**b**) Map of the pKRMIT plasmid with key features (see text for details), and EcoRI and PstI restriction sites used to ensure proper preparation of the pKRMIT plasmid. *tnpC9*: hyperactive allele of the

the transposase activity to ensure that *Krmit* is unable to transpose on its own, making *Krmit* mutations stable. *Krmit* recruitment requires the activity of a hyperactive *Himar1* transposase (C9) expressed in GAS cells in trans from the pKRMIT plasmid. pKRMIT also contains a ColE1 replicon for propagation in *Escherichia coli* (*E. coli*), a spectinomycin-resistance (Sp^r) cassette and a thermosensitive pWV01 origin of replication that is functional in GAS at 30 °C, but not at 37 °C (Fig. 1b). *Krmit* transposition is implemented at the permissive temperature of 30 °C; and the selection of *Krmit* transposants (Km^r clones) is achieved at the nonpermissive temperature of 37 °C, along with the loss of the nonreplicative plasmid, as evidenced by the loss of the associated Sp^r phenotype.

This chapter presents detailed step-by-step protocols to prepare and use the *Krmit* system for Tn-seq analyses in GAS. These procedures have been specifically optimized to produce highly complex mutant libraries and implement Tn-seq analyses mainly in the GAS strain 5448, a clinical isolate representative of the invasive M1T1 serotype, and a strain that can be challenging for genetic manipulations. However, with minor modifications, these methods have proven efficient in multiple GAS strains (e.g., M49 NZ131, M1T1 MGAS5005, M3 MGAS315) and even other pathogenic Streptococci (Le Breton and McIver, unpublished).

In addition to the *Krmit* Tn-seq system in GAS presented here, two recent reports describe an alternative mutagenesis system that utilizes the IS*S1* transposon to produce complex mutant libraries in the M1T1 GAS strain MGAS2221 and allows for analysis by the TraDIS approach [16]. This system was used to identify GAS genes required for survival in human saliva and in the soft tissue of nonhuman primates [16, 17]. As using IS*S1* and TraDIS in GAS is not described here, we recommend these references 16, 17 for additional details.

Fig. 1 (continued) Himar1 transposase (C9); ColE1: origin of replication for *E. coli* propagation; Sp^r: Spectinomycin-resistance gene; $oriR^{ts}$: thermosensitive pWV01 origin of replication for GAS propagation. (**c**) Schematic representation of chromosomal regions of M1T1 GAS 5448 with the contribution (below) of GAS genes (arrows) during growth in THY determined by Tn-seq illustrated with location (horizontal axis) and depth (vertical axis) of all *Krmit* transposon insertion sites identified in the GAS 5448 *Krmit* library initially (T0) and after 24-h and 48-h passages, respectively. Highlighted (boxed) are genes found to be nonmutable (black), or associated with decreased (orange), or increased (green) fitness in the screen. (Portions of this figure are modified versions of original materials presented by Le Breton et al., 2005 [9] and Le Breton et al., 2017 [10], licensed under the Creative Commons Attribution 4.0 International License)

2 Materials

2.1 Strains, Media, and Solutions

1. *Escherichia coli* C43[DE3] [18].
2. Luria-Bertani (LB) agar: 1% tryptone, 0.5% yeast extract, 0.8% NaCl, 1.5% agar.
3. LB liquid media: 1% tryptone, 0.5% yeast extract, 0.8% NaCl.
4. Ice-cold 10% glycerol solution.
5. Sterile distilled water.
6. SOC medium: 0.5% yeast extract, 25 tryptone, 10 mM NaCl, 2.5 mM KCl, 10 mM $MgCl_2$, 10 mM $MgSO_4$, 20 mM glucose.
7. *Streptococcus pyogenes* strain of choice (e.g., 5448 [[19]]).
8. Blood agar plates: 1.7% pancreatic digest of casein, 0.5% NaCl, 0.3% papain digest of soybean meal, 0.25% dextrose, 0.25% Na_2HPO_4, 5% sheep blood, 1.5% agar.
9. THY agar: 1% pancreatic digest of casein, 1.2% yeast extract, 0.31% beef heart infusion, 0.2% dextrose, 0.25% Na_2CO_3, 0.2% NaCl, 0.04% Na_2HPO_4, 1.5% agar.
10. THY liquid medium: 1% pancreatic digest of casein, 1.2% yeast extract, 0.31% beef heart infusion, 0.2% dextrose, 0.25% Na_2CO_3, 0.2% NaCl, 0.04% Na_2HPO_4.
11. 2 M glycine solution.
12. 25% glycerol solution in phosphate-buffered saline (PBS).
13. Saline solution: 0.9% NaCl.
14. 80% glycerol solution.
15. 10 mM Tris, pH 8.0.
16. 100 mg/ml lysozyme in 10 mM Tris, pH 8.0.
17. 50 mg/ml Proteinase K in water.
18. 100% isopropanol.
19. Phenol solution, saturated with 0.1 M citrate buffer, pH 4.3.
20. Phenol:chloroform:isoamyl alcohol mixture, pH 6.7/8.0 (25:24:1).
21. Chloroform.
22. 100% ethanol.
23. 3 M sodium acetate, pH 5.2.
24. RNase/DNase-free water.
25. 5 mg/ml RNase A.

2.2 Reagents, Kits, and Consumables

1. Plasmid pKRMIT (*see* **Note 1**).
2. 0.025 μm dialysis membranes (Millipore).
3. 2 mm electroporation cuvettes (Bio-Rad).

4. Antibiotics: Kanamycin, Spectinomycin.
5. Plasmid DNA miniprep kit.
6. Restriction enzymes: EcoRI, PstI, MmeI.
7. Plasmid DNA midiprep kit.
8. MasterPure Complete DNA and RNA Purification Kit (Epicentre).
9. Taq DNA polymerase.
10. Wizard SV Gel and PCR Clean-Up System (Promega).
11. Modification enzymes: alkaline phosphatase (CIP), T4 DNA ligase.
12. Accuprime Pfx kit (Promega).
13. Agarose gels, using 1% or 2% agarose.

2.3 Equipment

1. Temperature-controlled microbiology incubators: one at 37 °C, one at 30 °C.
2. Temperature-controlled CO_2 microbiology incubators: one at 37 °C, one at 30 °C.
3. Temperature-controlled shaking incubators: one at 37 °C, one at 30 °C.
4. Spectrophotometer.
5. Nanodrop spectrophotometer.
6. Centrifuge and tubes for centrifugation (15 ml and 50 ml conical tubes, 250 ml centrifugation bottles).
7. Gene Pulser Xcell Electroporation System (Bio-Rad) and 2 mm electroporation cuvettes.
8. Temperature-controlled water baths: one at 37 °C, one at 30 °C.
9. Speedvac DNA concentrator (Savant).
10. BSL-2 Type IIA biosafety cabinet.
11. Temperature-controlled heat block.
12. Temperature-controlled thermomixer.
13. Thermocycler.
14. Electrophoresis apparatus and consumables.
15. 2100 Bioanalyzer instrument (Agilent).

3 Methods

3.1 Generating Complex Transposon Mutant Pools

3.1.1 Producing the pKRMIT Plasmid

For pKRMIT propagation (*see* **Note 1**), prepare electrocompetent cells of the *E. coli* strain C43[DE3] as follows, using sterile techniques throughout the entire procedure.

1. On *Day 1*, streak a LB agar plate with the *E. coli* C43[DE3] strain and incubate overnight at 37 °C to obtain isolated colonies.
2. On *Day 2*, using a freshly grown colony from the LB agar plate, inoculate 10 ml of fresh LB broth in a 50 ml flask and grow the culture overnight at 37 °C in a shaking incubator (250 rpm).
3. On *Day 3*, using the overnight *E. coli* C43[DE3] culture, inoculate 100 ml of fresh LB broth in a 500 ml flask to an initial OD_{600} of ~0.05. Grow the *E. coli* C43[DE3] culture at 37 °C in a shaking incubator (250 rpm) to midexponential phase corresponding to an OD_{600} of 0.5. Place the culture flask on ice for 30 min. The *E. coli* C43[DE3] cells need to stay at 4 °C (or on ice) for the rest of the procedure. Transfer the cells into four ice-cold 50 ml conical tubes (25 ml per tube).
4. To wash the *E. coli* C43[DE3] cells, pellet the culture by centrifugation at 10,000 × *g* for 20 min, promptly remove the supernatant (*see* **Note 2**), and wash the cell pellets by adding 40 ml of ice-cold 10% glycerol to each tube and vigorously vortex to completely resuspend the pellets. Repeat the cell wash step two more times, then centrifuge at 10,000 × *g* for 20 min, discard the supernatant, and resuspend the *E. coli* C43[DE3] cells in 500 μl of 10% glycerol. Finally, prepare 40 μl aliquots of the electrocompetent *E. coli* C43[DE3] cells into prechilled 1.5 ml microtubes and promptly freeze at −80 °C. For best results, electrocompetent *E. coli* cells should be stored and used for transformation within 6 months to 1 year.
5. Dialyze the pKRMIT DNA by carefully placing a large drop (50 μl maximum) in the center of a dialysis membrane (0.025 μm) (*see* **Note 3**) floating on sterile distilled water in a small 2″ petri dish and incubate at room temperature (RT) for 30 min. Carefully remove the dialyzed pKRMIT sample from the membrane and transfer the DNA into a sterile 1.5 ml microtube.
6. Transform the pKRMIT plasmid into *E. coli* C43[DE3], using sterile techniques throughout the entire procedure. Thaw an aliquot of electrocompetent *E. coli* C43[DE3] cells on ice (*see* Subheading 3.1.1, **step 4**). Place a 1.5 ml microtube and a 2 mm electroporation cuvette on ice; prepare a 1.5 ml microtube with 1 ml of SOC media at RT. Add 1–2 μl of a dialyzed

pKRMIT DNA miniprep into the prechilled 1.5 ml microtube and, using a micropipette, transfer the thawed electrocompetent cells, mix well. Transfer the DNA–cells mix into the ice-cold 2 mm electroporation cuvette and tap the suspension to the bottom. Wipe the cuvette dry, quickly place it into the electroporation pod, and pulse the suspension (parameters: $C = 25$ μF; PC $= 200$ Ω; $V = 3.0$ kV). Immediately, add 1 ml of RT SOC to the cells, transfer them into a 1.5 ml microtube, and outgrow the transformants for 1 h in a 30 °C water bath. Make tenfold serial dilutions (10^{-1}–10^{-3}) using saline and plate 100 μl aliquots onto LB agar with 100 μg/ml spectinomycin and 50 μg/ml kanamycin (LB + Sp^{100} + Km^{50}). Incubate at 30 °C overnight to produce isolated colonies (*see* **Note 4**).

7. Produce large amounts of pKRMIT (*see* **Note 1**) for GAS transformation. From the LB + Sp^{100} + Km^{50} agar plates (*see* Subheading 3.1.1, **step 6**), use individual colonies to start 8–10 separate cultures by individually inoculating 200 ml of LB broth +Sp^{100} + Km^{50} in a 1 L flask and incubate in a shaking incubator (250 rpm) overnight at 30 °C (*see* **Note 1**).
8. Test each individual culture for the presence of unmodified pKRMIT: use sterile technique to collect 10 ml of culture from each flask, pellet cells by centrifugation, and perform a plasmid DNA preparation using a plasmid miniprep kit, following the manufacturer's recommendations. In preparation to **step 9** (see below), harvest the cells from the overnight cultures by centrifugation and freeze down the cell pellets. Using the purified plasmid DNA, set up two separate restriction digests: one with EcoRI and the other with PstI. The expected pKRMIT digest profile should reveal two bands with EcoRI (~5.9 kb and ~1.3 kb) and two bands with PstI (~5.6 kb and ~1.6 kb) (Fig. 1b; *see* **Note 1**).
9. When the presence of pKRMIT is confirmed, purify the plasmid from the stored pellet using a plasmid midiprep kit, following the manufacturer's recommendations. Assay the quantity and quality of the purified pKRMIT plasmid using a spectrophotometer or a Nanodrop.
10. Repeat the steps for the preparation of pKRMIT (*see* Subheading 3.1.1, **steps 7–9**) until obtaining the needed DNA amount required for the transformation of *S. pyogenes.* Typically, 300 μg of plasmid is required for the transformation of GAS strain 5448 with pKRMIT: pool together the different pKRMIT preparations into a 1.5 ml microtube and reduce the total sample volume using a speedvac concentrator.
11. Dialyze the pKRMIT DNA by carefully placing a large drop (50 μl maximum) in the center of a dialysis membrane

(0.025 μm) (*see* **Note 3**) floating on distilled water in a small 2″ petri dish and incubate at RT for 30 min. Carefully remove the dialyzed pKRMIT sample from the membrane and transfer the DNA into a sterile 1.5 ml microtube.

3.1.2 Transformation of pKRMIT into Streptococcus pyogenes

Prepare electrocompetent cells of *S. pyogenes* as follows, using sterile technique throughout the entire procedure. It is critical that these steps be performed in BSL-2 microbiology facilities with proper laboratory equipment and required BSL-2 training for personnel.

1. On *Day 1*, using a freezer stock, streak the GAS strain on a blood agar plate or a THY plate with antibiotic(s), if required. Incubate the plate at 37 °C (*see* **Note 5**) until isolated colonies are observed (typically 24 h). On *Day 2*, use an isolated GAS colony to inoculate 10 ml of THY broth containing 20 mM glycine (and optional Hyaluronidase, *see* **Note 6**) with antibiotic(s), if required, in 15 ml conical tube; and incubate overnight at the proper temperature (30 °C or 37 °C).
2. On *Day 3*, inoculate 150 ml of THY broth containing 20 mM glycine (and optional Hyaluronidase, *see* **Note 6**) and antibiotics if needed, in a 200 ml screw-cap media bottle using 7.5 ml of the GAS overnight culture. Grow GAS cells at 37 °C until OD_{600} reaches 0.2–0.4. Check culture under the microscope to confirm pure culture and rule out contamination (*see* **Note 7**). Store the GAS cultures on ice for 30 min. From now on, the GAS cells need to stay at 4 °C (or on ice) for the rest of the procedure.
3. To wash GAS cells, transfer the GAS cultures into sterile ice-cold centrifugation bottles and pellet GAS cells at 10,000 × *g* for 20 min, 4 °C. Carefully pour off and discard the supernatant (*see* **Note 2**) and resuspend the GAS cells in 20 ml of ice-cold 10% glycerol solution. Repeat the wash step three times before resuspending the washed GAS cells into 1 ml of ice-cold 10% glycerol solution. Transfer 200 μl aliquots (single reactions) into prechilled and labeled microfuge tubes and store at −80 °C (*see* **Note 7**). For best results, electrocompetent GAS cells should be stored and used for transformation within 6 months to 1 year.
4. Transform electrocompetent GAS cells with pKRMIT as follows, using BSL-2 sterile technique throughout the entire procedure. Thaw a 200 μl aliquot of electrocompetent GAS cells on ice (*see* Subheading 3.1.2, **step 1**). Place a 1.5 ml microtube and a 2 mm electroporation cuvette on ice; prepare a 15 ml conical tube with 10 ml of THY medium at RT. Add the dialyzed pKRMIT DNA into the prechilled 1.5 ml microtube and, using a micropipette, transfer the thawed electrocompetent GAS cells, mix well. Transfer the DNA–GAS cells

mix into the ice-cold 2 mm electroporation cuvette and tap the suspension to the bottom. After wipe-drying cuvette, quickly place it into the electroporation pod and pulse the suspension (parameters: $C = 25$ μF; PC = 400 Ω; $V = 1.75$ kV). Immediately, pipette 1 ml of THY medium from the 15 ml conical tube, transfer it into the electroporation cuvette, mix well, and transfer the cell suspension back into the 15 ml conical tube.

5. Grow the 10 ml cell suspension static at 30 °C for 2 h. Pellet the cell suspension at 6000 × *g* for 10 min, 4 °C. Resuspend the transformants into 500 μl of THY and make tenfold serial dilutions (10^{-1} to 10^{-5}) using saline. Plate 100 μl aliquots onto THY agar plates containing 100 μg/ml spectinomycin (Sp^{100}) and 300 μg/ml kanamycin (Km^{300}). Incubate the plates at 30 °C in 5% CO_2 incubator until colonies appear (24–48 h).

3.1.3 Selection of S. pyogenes Krmit Mutant Libraries for Tn-seq

Conduct phenotype analyses to confirm that GAS transformants display pKRMIT-associated growth properties, using BSL-2 sterile technique throughout the entire procedure. We recommend screening a minimum of 60–80 individual GAS transformants by groups of 20 for resistance traits to verify the presence of pKRMIT at permissive temperature and *Krmit* transposition events at non-permissive temperatures.

1. Streak GAS colonies onto the following: three plates to incubate at 30 °C: plate (A) (THY agar +Km^{300} +Sp^{100}), plate (B) (THY agar +Sp^{100}), and plate (C) (THY agar +Km^{300}); and three different plates to incubate at 37 °C: plate (D) (THY agar +Sp^{100}), plate (E) (THY agar +Km^{300}), and plate (F) (blood agar) (*see* **Note 8**). Incubate plates at the proper temperature for 48 h in 5% CO_2 incubator. GAS clones containing the pKRMIT plasmid should grow on plates (A), (B), (C), (E), and (F), but not (D). Indeed, at the permissive temperature of 30 °C, the pKRMIT plasmid replicates in GAS conferring resistance to Sp and Km, as reflected by growth on plates (A), (B), and (C). However, at the restrictive temperature of 37 °C, the thermosensitive replicase $RepA^{ts}$ is no longer functional, leading to the loss of the plasmid pKRMIT; if *mariner* transposition occurred, then the resulting *Krmit* transposants can grow on Km (E), but ideally should not grow on Sp (D). Plate (F) is a control to confirm that the colonies correspond to GAS cells, as evidenced by β-hemolytic activity surrounding the colonies.
2. Once the phenotypes are confirmed for a given colony, add 5 ml of 25% glycerol solution in PBS onto the corresponding plate A (THY agar +Km^{300} +Sp^{100} incubated at 30 °C), and using a sterile glass cell spreader, scrape and resuspend the colonies that grew on the agar. Transfer 1 ml aliquots of the

cell suspension into four 2 ml screw-cap vials and freeze down at −80 °C for GAS strain collection: these freezer stocks represent the pKRMIT source stocks.

3. Perform additional phenotype screens of the selected pKRMIT source stocks, with the remaining volume of cell suspension from plate (A) (*see* Subheading 3.1.3, **step 2**) using BSL-2 sterile technique throughout the entire procedure. Prepare a tenfold serial dilution of the GAS cell suspension (10^{-1} to 10^{-6}) using saline. Plate a 100 μl aliquot of each dilution onto THY agar + Km^{300} and THY agar + Sp^{100} plates and incubate for 24–48 h at 37 °C and 30 °C, respectively.
4. Determine viable cell counts (CFU) from the THY +Km^{300} plates incubated at 37 °C (*Krmit* transposants) and the THY +Sp^{100} plates incubated at 30 °C (clones containing pKRMIT). The frequency of *Krmit* transposition is calculated by dividing the number of CFU from the plate THY Km^{300} at 37 °C by the number of CFU from the plate THY $Spec^{100}$ at 30 °C. *Krmit* transposition frequency usually varies from 10^{-2} to 10^{-3} depending on the GAS strain used.
5. Next, from the THY agar +Km^{300} plates incubated at 37 °C obtained above (*see* Subheading 3.1.3, **step 4**), patch 100–200 GAS colonies onto plates containing either THY agar +Km^{300} or THY agar +Sp^{100} and incubate both plates at 37 °C overnight in 5% CO_2 incubator. Determine the number of GAS transposants (thus, Km^r) that are also Sp^r: the percentage of Sp^r-Km^r CFU reflects the amount of nonproductive integration of pKRMIT that occurs by single cross-over into a *Krmit* transposon already inserted into the GAS chromosome. Ideally, *Krmit* mutant libraries should only contain Sp^s-Km^r clones; and our experience shows that *Krmit* mutant libraries harbor Sp^r-Km^r clones. Typically, we discard *Krmit* mutant libraries with Sp^r-Km^r percentage greater than 5%.
6. Analyze the genotype of *Krmit* transposants by arbitrary-primed polymerase chain reaction (AP-PCR) (Fig. 2) [14] from the pKRMIT source stocks selected through the different screens mentioned above (**step 5**). For cultures, use BSL-2 sterile techniques. Grow individually a minimum of 20–40 Sp^s-Km^r clones (*Krmit* transposants) into separate sterile 2 ml screw-cap microtubes containing 1 ml of THY Km^{300} broth, incubating at 37 °C overnight. Using BSL-2 sterile technique, transfer 0.3 ml of each overnight culture into a sterile 1.5 ml microtube. Centrifuge to pellet cells, discard the supernatant, and proceed to a small-scale genomic DNA extraction (*see* Subheading 3.1.3, **step 7** below). To the remaining 0.7 ml of culture in the 2 ml screw-cap microtube, add 0.3 ml of 80% glycerol solution and freeze down at −80 °C for GAS *Krmit* mutant collection.

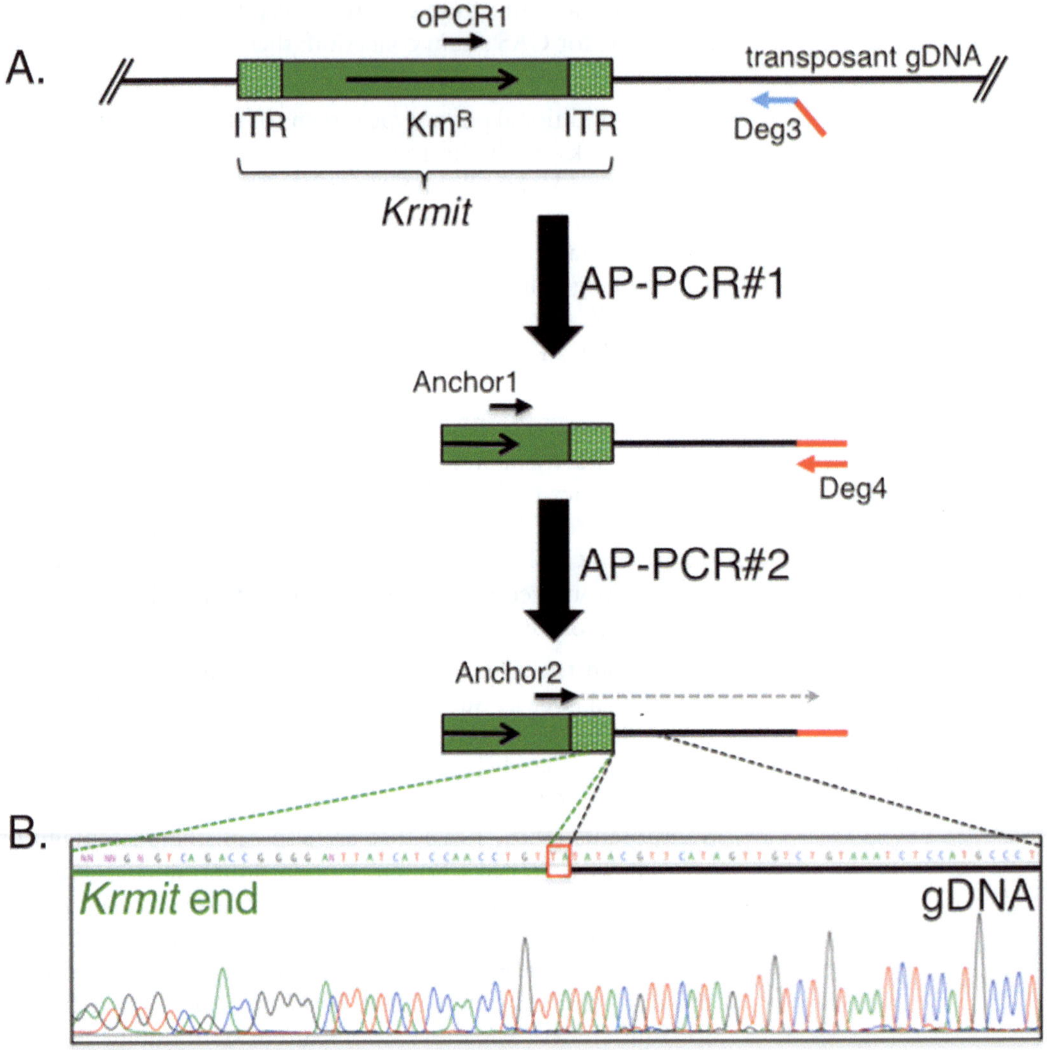

Fig. 2 The AP-PCR procedure to identify *Krmit* insertion site within the transposant genome. (**a**) A PCR product is produced through a two-step (AP-PCR#1 and AP-PCR#2) reaction for Sanger DNA sequencing. (**b**) Electrophoregram depicting the expected result of the AP-PCR procedure implemented on a *Krmit* mutant: the beginning of the sequence (green line, left) corresponds to the *Krmit* ITR region, followed by the *Krmit* insertion site TA (red box) and the adjacent chromosomal region (black line, right)

7. Prepare genomic DNA (gDNA), using the following small-scale genomic DNA extraction, adapted from the MasterPure Complete DNA and RNA Purification Kit (Epicentre) (*see* **Note 9**). Prepare a fresh solution of 100 mg/ml lysozyme solution in 10 mM Tris, pH 8.0 (not provided in the Epicentre kit), as needed: 50 μl of lysozyme solution is required for each cell pellet. Resuspend the cell pellet in 50 μl of lysozyme solution and incubate the suspension at 37 °C for 2 h. During

the incubation, gently shake the cell suspension (do not use vortex). During the incubation, prepare 2× T&C solution as needed: for each 150 μl (this represents three extractions) of 2× T&C solution, add 1 μl of Proteinase K (50 mg/ml) (from the kit or from the lab stock). Add 50 μl of 2× T&C to the cell suspension and mix well (do not vortex). Incubate at 70 °C for 30 min. Add 60 μl of MPC solution and mix very well (do not vortex). Centrifuge at 4 °C at 6000 × *g* for 1 h. Transfer 80 μl of supernatant into a new microfuge tube. Add 80 μl of 100% isopropanol and mix by inverting. Centrifuge at 4 °C for at 6000 × *g* for 1 h. Discard isopropanol without disturbing the cell pellet. Resuspend DNA in 50 μl of double-distilled water (ddH$_2$O) and check 5–10 μl of the genomic DNA preparation on a 1% agarose gel. DNA is stored at 4 °C.

8. Using the gDNA (*see* Subheading 3.1.3, **step** 7), set up arbitrary-primed PCR (AP-PCR) to determine the *Krmit* flanking junction (Fig. 2a), that is, the transposon insertion point within the transposant genome, as follows. The first PCR round (AP-PCR#1) (*see* **Note 10**) is carried in a 50 μl reaction combining 1 μl of transposant genomic DNA (see the protocol above) with 1 μl of primers oPCRl and Deg3 (10 μM each) (Table 1) using *Taq* as DNA polymerase, as per manufacturer's recommendations. Thermocycle is as follows: an initial denaturation for 5 min at 95 °C, followed by 6 cycles of 30 s at 95 °C, 30 s at 30 °C, and 2 min at 72 °C, followed by 30 cycles of 30 s at 95 °C, 30 s at 45 °C, and 2 min at 72 °C, and a final extension of 5 min at 72 °C. Purify the resulting AP-PCR#1 products using the Wizard SV Gel and PCR Clean-Up System (Promega) as per manufacturer's recommendations and resuspend DNA in 50 μl of RNase/DNase-free water.
9. Set up the second PCR (AP-PCR #2) (*see* **Note 10**) in a 50 μl reaction mixture using 1 μl of the AP-PCR#1, 1 μl of primers Anchorl and Deg4 (10 μM each) (Table 1) with *Taq*. Thermocycle is as follows: an initial denaturation for 5 min at 95 °C, followed by 30 cycles of 30 s at 95 °C, 30 s at 50 °C, and 2 min at 72 °C, and a final extension of 5 min at 72 °C. Purify the resulting AP-PCR#2 products using the Wizard SV Gel and PCR Clean-Up System (Promega) as per manufacturer's recommendations and resuspend in 50 μl of RNase/DNase-free water.
10. Analyze the purified AP-PCR#2 product using Sanger DNA sequencing service, mixing the appropriate amount of AP-PCR product #2 and primer Anchor2, as recommended by the DNA sequencing provider.
11. Open the resulting read in a sequence analysis application (e.g., MacVector, SnapGene): Fig. 2b provides a typical example of the DNA sequencing result, with the sequence starting with

Table 1
Oligonucleotide primer sequences

Name	Sequence (5′ to 3′)
AP-PCR	
oPCR1	TACTGGATGAATTGTTTTAGTACC
Deg3	TAGAGTTATTAATGGAATTGCTGATNNNNNNNNNN
Anchor1	CGCAACTGTCCATACTCTG
Deg4	TAGAGTTATTAATGGAATTGCTGAT
Anchor2	GCCTACGAGGAATTTGTATCG
Tn-seq adapters	
Adapter501A	TTCCCTACACGACGCTCTTCCGATCTTATAGCCTNN
Adapter501B	/5P/AGGCTATAAGATCGGAAGAGCGTCGTGTAGGGAAAGAG/3P/
Adapter502A	TTCCCTACACGACGCTCTTCCGATCTATAGAGGCNN
Adapter502B	/5P/GCCTCTATAGATCGGAAGAGCGTCGTGTAGGGAAAGAG/3P/
Adapter503A	TTCCCTACACGACGCTCTTCCGATCTCCTATCCTNN
Adapter503B	/5P/AGGATAGGAGATCGGAAGAGCGTCGTGTAGGGAAAGAG/3P/
Adapter504A	TTCCCTACACGACGCTCTTCCGATCTGGCTCTGANN
Adapter504B	/5P/TCAGAGCCAGATCGGAAGAGCGTCGTGTAGGGAAAGAG/3P/
Adapter505A	TTCCCTACACGACGCTCTTCCGATCTAGGCGAAGNN
Adapter505B	/5P/CTTCGCCTAGATCGGAAGAGCGTCGTGTAGGGAAAGAG/3P/
Adapter506A	TTCCCTACACGACGCTCTTCCGATCTTAATCTTANN
Adapter506B	/5P/TAAGATTAAGATCGGAAGAGCGTCGTGTAGGGAAAGAG/3P/
Adapter507A	TTCCCTACACGACGCTCTTCCGATCTCAGGACGTNN
Adapter507B	/5P/ACGTCCTGAGATCGGAAGAGCGTCGTGTAGGGAAAGAG/3P/
Adapter508A	TTCCCTACACGACGCTCTTCCGATCTGTACTGACNN
Adapter508B	/5P/GTCAGTACAGATCGGAAGAGCGTCGTGTAGGGAAAGAG/3P/
Tn-seq PCR	
KrmitTnseq2	CAAGCAGAAGACGGCATACGAAGCGCCTACGAGGAATTTGTATCG
AdapterPCR	AATGATACGGCGACCACCGAGATCACACTCTTTCCCTACACGACGCTCTTCC

The primers to produce the Tn-seq adapters contain barcode sequences (underlined) for multiplexing up to eight experiments (*see* **Note 11**). N: any nucleotide; /5P/: primer is 5′ phosphorylated; /3P/: primer is 3′ phosphorylated the end of the *Krmit* transposon ITR, followed by the *Krmit*-targeted dinucleotide TA and the precise *Krmit* insertion point within the GAS genome.

12. Select the DNA sequence corresponding to the TA and the *Krmit* insertion point and identify its position on the corresponding genome of the mutated GAS strain using the downloaded genome of interest or a genome resource website (e.g., BLAST, MicrobesOnline).

13. For a given *Krmit* mutant library, compare the results for the *Krmit* insertion sites in the different mutants analyzed by AP-PCR to assess the randomness of *Krmit* insertions. Libraries selected for Tn-seq should contain less than 5% redundancy: if sequencing a minimum set of 20 clones, then all *Krmit* insertion sites should be unique.

3.2 Synthesis and Sequencing of Transposon Junctions from Mutant Pools

Figure 3 depicts the different steps to produce the *Krmit* insertion sites from the GAS mutant pools. The following procedure, specifically adapted for Tn-seq using *Krmit*, is a modification from the protocol described by van Opijnen et al. [1, 20]. The procedure

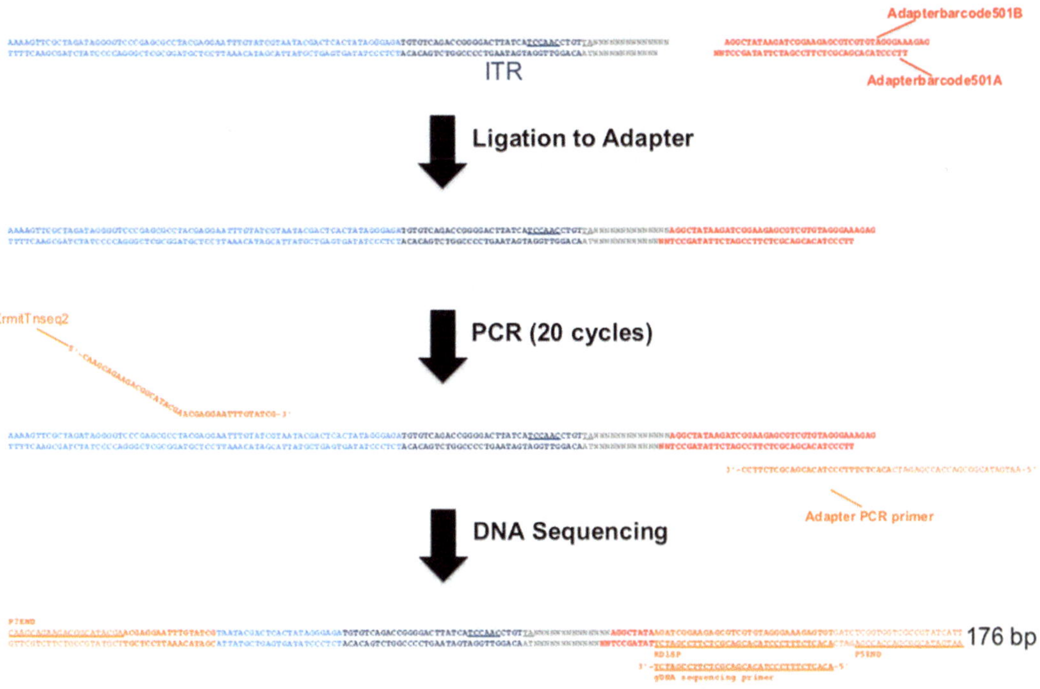

Fig. 3 Production of the *Krmit* insertion tags by MmeI-mediated ligation PCR for Illumina DNA sequencing. Nucleotide resolution representation of the different steps leading to the production of *Krmit* insertion tags compatible with the Illumina DNA sequencing platform. ITR: inverted terminal repeat. P7end and P5end correspond to nucleotide sequences complementary to the primers on the Illumina flowcell lane required for the clustering step. RD1SP: site where the primer for DNA sequencing is located

presented here allows for multiplexing up to eight different Tn-seq experiments on one lane for the Illumina sequencing (*see* **Note 11**).

3.2.1 Preparation of the Adapters

1. Produce Tn-seq adapters as follows. Order synthesis of the primers (Table 1) is required to produce the different adapters used in this protocol (Adapter 501–Adapter 508). All primers are ordered at a quantity of 100 nmole, and some will require 3′ and 5′ phosphorylation (*see* Table 1 for details). Polyacrylamide gel electrophoresis (PAGE) purification is required for all primers. Upon reception, resuspend each primer to a 200 μM solution using 1 mM Tris, pH 8.0, and store at −20 °C.
2. Prepare the different adapters by mixing equivalent volume of two complementary primers (primers A and B, Table 1) in microtubes (in total, a set of eight tubes). Place the microtubes on heat block at 99 °C for 10 min, then take the microtubes out, and let them cool down at RT for 1 h.

3.2.2 Total Genomic DNA Extraction

1. Prepare gDNA from the mutant pool sample as follows. The protocol is adapted from the MasterPure Complete DNA and RNA Purification Kit (Epicentre), with the following modifications. Before proceeding with the kit reagents, an initial step to treat the GAS cells with lysozyme is required. Prepare a fresh solution of lysozyme by dissolving 100 mg/ml of lysozyme in 10 mM Tris, pH 8.0. For each GAS cell pellet, 150 μl of lysozyme solution is required. Carefully resuspend the GAS cell pellet in the lysozyme solution, transfer into a 1.5 ml microtube, and incubate the suspension on the ThermoMixer at 37 °C/2,000 × *g* for 2 h. During this incubation, prepare 2× T&C solution (provided with the kit), by mixing 150 μl of 2× T&C solution with 1 μl of 50 mg/ml Proteinase K per sample. Switch the ThermoMixer temperature to 70 °C.
2. Add 150 μl of 2× T&C solution to the cell suspension, mix well (do not vortex), and incubate on the ThermoMixer at 70 °C/2,000 × *g* for 30 min. Cool the sample to 37 °C in a water bath.
3. Add 1 μl of 5 mg/ml RNaseA, mix well (do not vortex), and incubate in the 37 °C water bath for 30 min. Place samples on ice for 5 min, add 175 μl of MPC solution, and mix very well (do not vortex). Centrifuge at 4 °C for 10 min at 15,000 × *g*.
4. Transfer the supernatant into a new 1.5 ml microfuge tube and add 500 μl of 100% isopropanol. Mix gently by inverting the tube until the gDNA aggregates. Centrifuge at 4 °C for 10 min at 15,000 × *g* to pellet the genomic DNA.
5. Discard the isopropanol, wash the DNA pellet with 1 ml of 70% ethanol, and centrifuge at 4 °C for 10 min at 15,000 × *g*. Discard the ethanol and dry the DNA pellet by pipetting as

much as ethanol as possible (use centrifuge pulses to collect samples at the bottom of tube if needed).

6. Gently dry the DNA pellet in a biosafety cabinet (BSC), add 100 μl of RNase/DNase-free water for DNA resuspension (*see* **Note 12**).

3.2.3 Digestion of gDNA by MmeI and Ligation to MmeI Adapter

1. Set up the digestion reaction with the MmeI enzyme using 3 μg of total gDNA as per supplier's recommendations. Use a final reaction volume of 200 μl. Incubate at 37 °C for 2.5 h. Check 10% volume of the reaction by electrophoresis using a 1.5% agarose gel. Add 2 μl of calf alkaline phosphatase (CIP) and incubate at 37 °C for 1 h.
2. To purify the restricted dephosphorylated DNA fragments, perform a phenol extraction, then a phenol/chloroform extraction, and finally a chloroform extraction. Perform a sodium acetate/ethanol precipitation. Vortex to mix.
3. Centrifuge at 15,000 × *g* at 4 °C for 30 min. Remove the supernatant. At this stage, a small DNA pellet can be seen: wash the DNA with 70% ethanol. Centrifuge at 15,000 × *g* at 4 °C for 30 min. Remove the supernatant, then use the speedvac concentrator to dry DNA pellet. Resuspend in 25 μl of RNase/DNase-free water.
4. Set up a ligation reaction using the T4 DNA ligase as recommended by manufacturer using the entire 25 μl volume of CIP-treated MmeI DNA fragments and 1 μl of the appropriate MmeI adapter. Incubate at 16 °C overnight.

3.2.4 Production of the Krmit Insertion Tag Library

1. Prepare a PCR mix using the Accuprime Pfx kit. The reaction volume is 100 μl: use 8 μl of the ligation reaction mixture and 2 μl of primer KrmitTnseq2 and 2 μl of primer AdapterPCR. Set up the PCR as per manufacturer's recommendations using a primer melting temperature (T_m) of 55 °C and an extension time of 20 s and a total of 22 cycles.
2. Use a speedvac to concentrate the PCR product to 20 μl and load the entire sample on a 2% agarose gel. Use a gel extraction kit to purify the band around 180 bp (We recommend the Wizard SV Gel and PCR Clean-Up System, Promega).
3. After purification, concentrate DNA by speedvac concentrator and resuspend in 20 μl of RNase/DNase-free water. Check the DNA concentration using a NanoDrop (We usually have a concentration of around 10 ng/μl).
4. Verify the proper PCR amplification of the *Krmit* insertion tags by sending 8 μl of the *Krmit* insertion tag library for Sanger sequencing using the Anchor2 primer (*see* Table 1). In this condition, the resulting sequence should show the end of the *Krmit* transposon and then a messy sequence that is the result

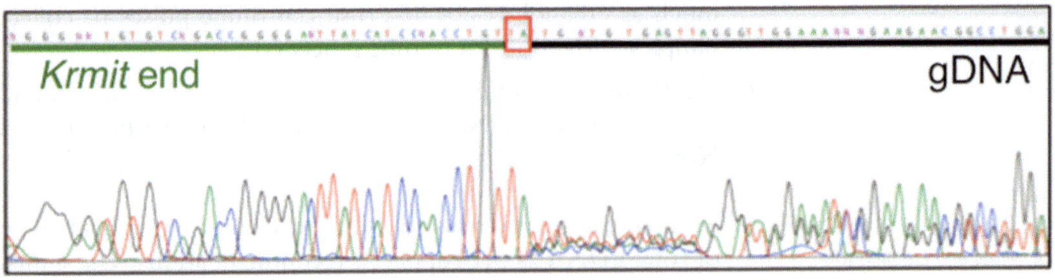

Fig. 4 Quality control using Sanger DNA sequencing of the *Krmit* insertion tags prior to the Illumina run. The beginning of the sequence (green line, left) corresponds to the *Krmit* ITR region, followed by the *Krmit* insertion site TA (red box) and a smear corresponding to the *en masse* sequencing of the adjacent chromosomal region in the *Krmit* mutant library (black line, right)

of the sum of sequences from the different *Krmit* clones (Fig. 4). This is a cost-effective way to visualize the result of the procedure.

3.2.5 Illumina Sequencing and Data Analysis

1. Prior to the Illumina sequencing per se, perform a last quality control analysis of each library of *Krmit* insertion tags using a Bioanalyzer. A unique pick with a size of approximately 180 bp is expected (Fig. 5a), corresponding to the *Krmit* insertion tags. Determine the DNA concentration for each library to sequence. If analyzing multiple libraries in an Illumina run (multiplexing), calculate the amount of each library to include in the sample to be loaded on the lane as per Illumina recommendation.
2. Sequence DNA of the libraries of *Krmit* insertion tags by massive parallel sequencing using a setting for 50-nt single end reads on an Illumina HiSeq 1500 platform (*see* **Note 13**).
3. Bioinformatically process the data. Different bioinformatics tools for carrying out the data analyses described here can be found at the following link: https://github.com/abelew/CYOA/blob/master/lib/Bio/Adventure/TNSeq.pm.
Retrieve the output files from the Illumina sequencer. The format of the reads generated with our procedure to produce the *Krmit* insertion tags is not standard as per Illumina protocols and causes the default *bcl2fastq* demultiplexer used for Illumina sequencing to fail. As a result, when acquiring the raw data from the sequencing core, the reads will usually be stored in the data that failed the automatic demultiplexing step. Importing the Tn-seq reads is performed by copying the data into a staging directory and writing a delimited text file "*indexes.txt*," with the first column containing the index sequence (*see* **Note 14**) and the second containing the sample ID. The "cyoa" script is then invoked with options to demultiplex: *cyoa --task tnseq --method sortindex --input <raw_filename>*. This invokes the *Sort_Indexes()* function

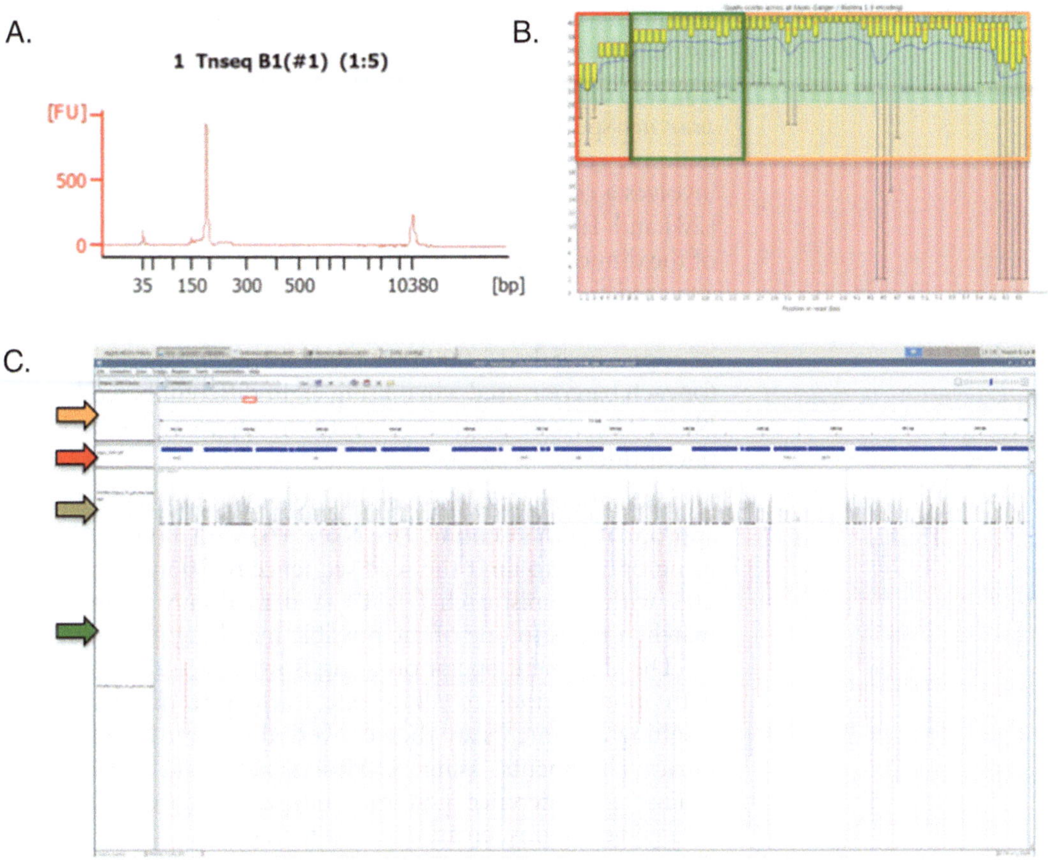

Fig. 5 Tn-seq analyses of GAS *Krmit* libraries. (**a**) Electrophoregram obtained by Nanodrop revealing the size of the *Krmit* insertion tags. (**b**) FastQC analyses showing the quality of the *Krmit* library reads representing quality scores across reads for each nucleotide position. A score between 28 and 40 represents high-quality results. From the left to the right, the read displays the experiment index (red box), followed by the GAS genome fragment flanking the *Krmit* transposon (green box), which corresponds to the sequence of interest in the Tn-seq experiment, and on the right the sequence of the *Krmit* transposon ITR (orange box). When analyzing the reads for the Tn-seq experiments, sequence sections from the red and orange box are trimmed prior to the alignment of the remaining sequence onto the reference GAS genome. (**c**) Screenshot of a session in the IGV data visualization tool. Individual reads are aligned to their matching nucleotide sequences onto the reference GAS genome (orange arrow), with the option to zoom-in to specific genomic regions (red arrow). Orientation of the reads on the Watson or Crick strand coded by color (red or blue, respectively) is displayed (green arrow). A $\log_2$ transformation (brown arrow) provides a summary display to quickly visualize the density (amount of unique insertions represented by a bar) and depth (amount of clones per unique insertion, height of the bar)

within cyoa. The resulting raw reads (fastq format) are first checked for quality using FastQC [[21]], with a subset of the quality metrics extracted and written into a csv file. The html fastqc outputs are manually viewed and/or combined with multiqc.

4. Demultiplex the different experiments against the indexing adapters with Bio::SeqIO [22] (*see* **Note 14**) and again query

for quality using FastQC [21] (Fig. 5b), with quality metrics extracted and written into csv files.

5. For each experiment, trim the reads to remove the adapters and *mariner*-specific ITR sequences in order to extract only the genomic sequence next to the transposon (Fig. 3). The step involves a combination of Biopieces (biopieces.org), Cutadapt [23] and Trimmomatic [24], and also checks for the presence of the dinucleotide TA. In the process, the sets of reads are also analyzed to visualize the distribution of the read lengths and quality. This process typically produces individual read libraries of approximately 15,000,000 reads for each experiment (i.e., for each barcode used when using eight different barcodes) (*see* **Note 15**), with sequence lengths of approximately 12–15 nucleotides.
6. For each library, align the trimmed reads (*see* **step 5**) against the appropriate GAS genome (*see* **Note 16**) with Bowtie [25–27] using options to limit the seed sequence to 10 nucleotides (−l 10), with no mismatch (−v 0), and randomly assign multi-matched reads to one of the possible matching positions (−M 1). The alignment operation using Bowtie [25–27] is repeated with modifications to allow one nucleotide mismatch. The resulting alignments are then converted to sorted/compressed binary alignments using SAMtools [28] and HTseq [29] against the reference genome coding sequence (CDS) and intergenic regions, creating count tables.
7. The reads are mapped to the GAS reference genome (*see* **Note 16**) using Bowtie [25–27] and data visualized using the Integrative Genomics Viewer (IGV) browser (broadinstitute.org/igv/home) (Fig. 5c). Typically, about 90% of the reads align to the reference genome, the remaining 10% matching to the pKRMIT plasmid. On average, the *Krmit* libraries produced in GAS 5448 contained from ca. 9000 to 90,000 unique transposon insertion sites (*see* **Note 17**).
8. **Steps 1–7** provide the initial steps to analyze the composition of *Krmit* mutant libraries by Tn-seq and verify how comprehensive the produced libraries are. For further Tn-seq analyses, multiple bioinformatics tools are now available to study gene essentiality and gene fitness, and/or to perform more sophisticated data analyses (normalization, statistical analyses). We recommend consulting the appropriate literature [30–35] to identify the proper bioinformatics tools that fit best the type of Tn-seq analyses to be conducted.

4 Notes

1. The pKRMIT plasmid was initially developed for *mariner* transposition in *S. pyogenes* and was successfully tested in different GAS strains. The pKRMIT plasmid was found unstable in *E. coli*, making its propagation and maintenance challenging. We found that the use of *E. coli* strain C43[DE3] for pKRMIT propagation made it easier to produce the plasmid. Growth of the pKRMIT-containing *E. coli* C43[DE3] at 30 °C was found critical to obtain colonies and/or cultures harboring the proper pKRMIT plasmid. However, when preparing pKRMIT, it is essential to perform quality control steps (digestion by restriction enzymes) to verify the presence of the proper pKRMIT. We also do not recommend producing freezer stocks of pKRMIT in *E. coli* C43[DE3] for its storage, but rather keep a validated mini-preparation of the plasmid in the −20 °C freezer for future transformations. The pKRMIT plasmid is available upon request to Dr. Y. Le Breton (yoann.s.lebreton.ctr@mail.mil) and Pr Kevin McIver (kmciver@umd.edu).
2. When preparing electrocompetent cells of *E. coli* or *S. pyogenes* (GAS), it is better to sacrifice yield by pouring off a few cells than to leave any supernatant behind.
3. Dialysis across a 0.025 μM membrane Millipore filter removes excess salts found in samples to prevent arcing during electroporation.
4. Plates with *E. coli* C43[DE3] transformants containing pKRMIT can be stored at 4 °C up to a week. Better results are obtained when using fresh transformants.
5. GAS strains used to develop these protocols (e.g., 5448, MGAS5005) grow best at 37 °C. For other strains, the researcher may need to optimize growth conditions.
6. The addition of free glycine acts to reduce cell wall cross-bridges and makes the GAS envelope more permeable for DNA uptake during electroporation. Hyaluronidase solution (85 mg/ml in water) should be added to the GAS culture, if the strain expresses significant amount of capsule.
7. Under the microscope, GAS appears as cocci in clusters, short chains, diplococci, and single cocci. During growth conditions required to produce electrocompetent GAS cells, these form long chains. GAS grows best on complex "rich" medium such as trypticase soy agar (TSA) supplemented with 5% sheep blood, where it typically produces large zones of β-hemolysis (the complete disruption of erythrocytes and the release of hemoglobin).

8. The phenotype screens to identify the GAS clones containing the proper pKRMIT plasmid require testing large numbers of colonies. We recommend reducing the amount of plates used by streaking up to four different clones per plate, by dividing the surface into four equal areas.
9. The "small-scale" genomic DNA extraction described was optimized using the MasterPure Complete DNA and RNA Purification Kit (Epicentre), but other protocols for genomic DNA extraction can be used. The researcher will need to optimize conditions to ensure gDNA is not sheared during the preparation step.
10. The arbitrary-primed (AP)-PCR protocol presented here was optimized using the Taq DNA polymerase for PCRs. Different DNA polymerase and PCR methodology could be used, requiring adjustments of the protocol.
11. In the Tn-seq procedure presented here, we used up to eight different barcodes for experiment multiplexing. It is possible to include additional barcodes, so that more experiments can be analyzed simultaneously by Illumina sequencing. However, the more experiments multiplexed on an Illumina lane, the less depth is associated with the resulting datasets.
12. The gDNA extraction for the Tn-seq procedure was optimized with the MasterPure Complete DNA and RNA Purification Kit (Epicentre), as the described procedure produces high-quality nonsheared DNA. Additional protocols can be applied to produce the gDNA from the mutant libraries: these protocols must generate high-quality nonsheared DNA.
13. We had the Illumina sequencing serviced and carried out at the Institute for Bioscience and Biotechnology Research (IBBR) Sequencing Core Facility located at the University of Maryland, College Park, using the Illumina HiSeq 1500 platform with a setting for 50-nt single end reads. Other platforms can be used to analyze the *Krmit* insertion tags; however, the researcher will need to ensure the standard Illumina protocols associated with the instrumentation used.
14. The sort indexes function uses a Levenstein distance approximate match (with an option for 1 substitution, provided by String::Approx) in order to address the somewhat increased likelihood of errors in the first bases of a read. We used eight different indexes: D501 (TATAGCCT), D502 (ATAGAGGC), D503 (CCTATCCT), D504 (GGCTCTGA), D505 (AGGCGAAG), D506 (TAATCTTA), D507 (CAGGACGT), and D508 (GTACTGAC).
15. The procedure described here was designed to multiplex a total of eight individual *Krmit* read libraries, leading to approximately 15,000,000 reads for each experiment (i.e., for each

barcode used). The amount of reads generated will depend on the amount of libraries multiplexed in the Illumina run. If multiplexing less than eight libraries, the read count per experiment will be higher; if multiplexing more than eight libraries, the read count per experiment will be lower. The researcher needs to keep this in mind as multiplexing can potentially impact the Illumina sequencing results and the type of Tn-seq analyses one can conduct.

16. Genome of GAS 5448: GenBank Accession CP008776.
17. The amount of unique transposon insertion sites found in the *Krmit* mutant libraries can vary between libraries; and this had a direct impact on the type of Tn-seq analyses that can be performed later. For example, high mutant counts are critical when conducting analyses looking at essential genes. Analyses looking at gene fitness will not necessary require highly saturated mutant libraries. A library of over 10,000 transposon insertions is typically the minimum requirement for Tn-seq analyses (T. van Opijnen and A. Camilli, personal communication).

Acknowledgments

Material has been reviewed by the Walter Reed Army Institute of Research. There is no objection to its presentation and/or publication. The opinions and assertions contained herein are the private views of the author and are not to be construed as official or as reflecting true views of the Department of the Army or the Department of Defense. This work was supported by grants (AI047928 and AI094773) from the National Institute of Allergy and Infectious Diseases (NIAID) at the National Institutes of Health (NIH).

References

1. van Opijnen T, Bodi KL, Camilli A (2009) Tn-seq: high-throughput parallel sequencing for fitness and genetic interaction studies in microorganisms. Nat Methods 6 (10):767–772. https://doi.org/10.1038/nmeth.1377
2. Gawronski JD, Wong SM, Giannoukos G et al (2009) Tracking insertion mutants within libraries by deep sequencing and a genome-wide screen for *Haemophilus* genes required in the lung. Proc Natl Acad Sci U S A 106 (38):16422–16427. https://doi.org/10.1073/pnas.0906627106
3. Akerley BJ, Rubin EJ, Camilli A et al (1998) Systematic identification of essential genes by *in vitro mariner* mutagenesis. Proc Natl Acad Sci U S A 95(15):8927–8932. https://doi.org/10.1073/pnas.95.15.8927
4. Goodman AL, McNulty NP, Zhao Y et al (2009) Identifying genetic determinants needed to establish a human gut symbiont in its habitat. Cell Host Microbe 6(3):279–289. https://doi.org/10.1016/j.chom.2009.08.003
5. Langridge GC, Phan MD, Turner DJ et al (2009) Simultaneous assay of every *Salmonella* Typhi gene using one million transposon mutants. Genome Res 19(12):2308–2316. https://doi.org/10.1101/gr.097097.109
6. van Opijnen T, Camilli A (2013) Transposon insertion sequencing: a new tool for systems-

level analysis of microorganisms. Nat Rev Microbiol 11(7):435–442. https://doi.org/10.1038/nrmicro3033

7. Barquist L, Boinett CJ, Cain AK (2013) Approaches to querying bacterial genomes with transposon-insertion sequencing. RNA Biol 10(7):1161–1169. https://doi.org/10.4161/rna.24765
8. Kwon YM, Ricke SC, Mandal RK (2016) Transposon sequencing: methods and expanding applications. Appl Microbiol Biotechnol 100(1):31–43. https://doi.org/10.1007/s00253-015-7037-8
9. Le Breton Y, Belew AT, Valdes KM et al (2015) Essential genes in the Core genome of the human pathogen Streptococcus pyogenes. Sci Rep 5:9838. https://doi.org/10.1038/srep09838
10. Le Breton Y, Belew AT, Freiberg JA et al (2017) Genome-wide discovery of novel M1T1 group a streptococcal determinants important for fitness and virulence during soft-tissue infection. PLoS Pathog 13(8):e1006584. https://doi.org/10.1371/journal.ppat.1006584
11. van der Beek SL, Le Breton Y, Ferenbach AT et al (2015) GacA is essential for group a streptococcus and defines a new class of monomeric dTDP-4-dehydrorhamnose reductases (RmlD). Mol Microbiol 98(5):946–962. https://doi.org/10.1111/mmi.13169
12. van Hensbergen VP, Movert E, de Maat V et al (2018) Streptococcal Lancefield polysaccharides are critical cell wall determinants for human group IIA secreted phospholipase A2 to exert its bactericidal effects. PLoS Pathog 14(10):e1007348. https://doi.org/10.1371/journal.ppat.1007348
13. Edgar RJ, van Hensbergen VP, Ruda A et al (2019) Discovery of glycerol phosphate modification on streptococcal rhamnose polysaccharides. Nat Chem Biol 15(5):463–471. https://doi.org/10.1038/s41589-019-0251-4
14. Le Breton Y, McIver KS (2013) Genetic manipulation of Streptococcus pyogenes (the group a streptococcus, GAS). Curr Protoc Microbiol 30:9D.3.1–9D.3.29. https://doi.org/10.1002/9780471729259.mc09d03s30
15. Le Breton Y, Mistry P, Valdes KM et al (2013) Genome-wide identification of genes required for fitness of group a streptococcus in human blood. Infect Immun 81(3):862–875. https://doi.org/10.1128/IAI.00837-12
16. Zhu L, Charbonneau ARL, Waller AS et al (2017) Novel genes required for the fitness of Streptococcus pyogenes in human saliva. mSphere 2(6):e00460–e00417. https://doi.org/10.1128/mSphereDirect.00460-17
17. Zhu L, Olsen RJ, Beres SB et al (2019) Gene fitness landscape of group a streptococcus during necrotizing myositis. J Clin Invest 129(2):887–901. https://doi.org/10.1172/JCI124994
18. Miroux B, Walker JE (1996) Over-production of proteins in Escherichia coli: mutant hosts that allow synthesis of some membrane proteins and globular proteins at high levels. J Mol Biol 260(3):289–298. https://doi.org/10.1006/jmbi.1996.0399
19. Aziz RK, Pabst MJ, Jeng A et al (2004) Invasive M1T1 group a streptococcus undergoes a phase-shift in vivo to prevent proteolytic degradation of multiple virulence factors by SpeB. Mol Microbiol 51(1):123–134. https://doi.org/10.1046/j.1365-2958.2003.03797
20. van Opijnen T, Lazinski DW, Camilli A (2014) Genome-wide fitness and genetic interactions determined by Tn-seq, a high-throughput massively parallel sequencing method for microorganisms. Curr Protoc Mol Biol 106:7.16.1–17.16.24. https://doi.org/10.1002/0471142727.mb0716s106
21. Patel RK, Jain M (2012) NGS QC toolkit: a toolkit for quality control of next generation sequencing data. PLoS One 7(2):e30619. https://doi.org/10.1371/journal.pone.0030619
22. Stajich JE, Block D, Boulez K et al (2002) The Bioperl toolkit: Perl modules for the life sciences. Genome Res 12(10):1611–1618. https://doi.org/10.1101/gr.361602
23. Martin M (2011) Cutadapt removes adapter sequences from high-throughput sequencing reads. EMBnet J 17:10–12
24. Bolger AM, Lohse M, Usadel B (2014) Trimmomatic: a flexible trimmer for Illumina sequence data. Bioinformatics 30(15):2114–2120. https://doi.org/10.1093/bioinformatics/btu170
25. Langmead B (2010) Aligning short sequencing reads with bowtie. Curr Protoc Bioinformatics . Chapter 11:Unit 11.7. https://doi.org/10.1002/0471250953.bi1107s32
26. Langmead B, Salzberg SL (2012) Fast gapped-read alignment with bowtie 2. Nat Methods 9(4):357–359. https://doi.org/10.1038/nmeth.1923
27. Langmead B, Trapnell C, Pop M, Salzberg SL (2009) Ultrafast and memory-efficient alignment of short DNA sequences to the human genome. Genome Biol 10(3):R25. https://doi.org/10.1186/gb-2009-10-3-r25

28. Li H, Handsaker B, Wysoker A, Fennell T et al (2009) The sequence alignment/map format and SAMtools. Bioinformatics 25 (16):2078–2079. https://doi.org/10.1093/bioinformatics/btp352
29. Anders S, Huber W (2010) Differential expression analysis for sequence count data. Genome Biol 11(10):R106. https://doi.org/10.1186/gb-2010-11-10-r106
30. Burby PE, Nye TM, Schroeder JW, Simmons LA (2017) Implementation and data analysis of Tn-seq, whole-genome Resequencing, and single-molecule real-time sequencing for bacterial genetics. J Bacteriol 199(1): e00560–e00516. https://doi.org/10.1128/JB.00560-16
31. DeJesus MA, Ioerger TR (2013) A hidden Markov model for identifying essential and growth-defect regions in bacterial genomes from transposon insertion sequencing data. BMC Bioinformatics 14:303. https://doi.org/10.1186/1471-2105-1114-1303
32. DeJesus MA, Zhang YJ, Sassetti CM et al (2013) Bayesian analysis of gene essentiality based on sequencing of transposon insertion libraries. Bioinformatics 29(6):695–703. https://doi.org/10.1093/bioinformatics/btt043
33. McCoy KM, Antonio ML, van Opijnen T (2017) MAGenTA; a galaxy implemented tool for complete Tn-Seq analysis and data visualization. Bioinformatics 33(17):2781–2783. https://doi.org/10.1093/bioinformatics/btx320
34. Solaimanpour S, Sarmiento F, Mrázek J (2015) Tn-seq explorer: a tool for analysis of high-throughput sequencing data of transposon mutant libraries. PLoS One 10(5):e0126070. https://doi.org/10.1371/journal.pone.0126070
35. Zomer A, Burghout P, Bootsma HJ et al (2012) ESSENTIALS: software for rapid analysis of high throughput transposon insertion sequencing data. PLoS One 7(8):e43012. https://doi.org/10.1371/journal.pone.0043012

Chapter 5

Genetic Manipulation of Group A Streptococcus—Gene Deletion by Allelic Replacement

Timothy C. Barnett, Jessica N. Daw, Mark J. Walker, and Stephan Brouwer

Abstract

Genetic manipulation of *Streptococcus pyogenes* (Group A Streptococcus, GAS) has historically been a challenging process, with considerable variation in efficiency between different strains. Here, we outline an optimized, rapid method for creating markerless isogenic mutations that combines Gibson assembly cloning with a new temperature-sensitive plasmid, pLZts. This method is highly efficient and reduces the time needed to create GAS mutants to ~2–3 weeks, with the ability to prepare multiple mutants simultaneously.

Key words Group A Streptococcus, *Streptococcus pyogenes*, Allelic exchange, Isogenic mutant

1 Introduction

The ability to precisely manipulate the bacterial chromosome is central to studies of the genetic and molecular basis of bacterial disease and biochemistry. For bacterial pathogens, the proposal of a series of "molecular Koch's postulates" in 1988 has been universally accepted as the gold standard for studies of bacterial pathogenesis [1]. In the modern era of reverse genetics, transposon sequencing (*see* Chapter 4 on Tn-seq analyses), and other new approaches, the ability to validate gene function by creating precise isogenic strains that vary only in the gene of interest is central to the application of these postulates.

With the wealth of information on the epidemiology and gene content variation of global strains that has emerged in recent years [2–4], *Streptococcus pyogenes* (Group A Streptococcus, GAS) has undergone a renaissance of research into the molecular basis of disease [5, 6] and global initiatives into the development of a safe and effective vaccine [7]. Unfortunately, the tools for genetic manipulation of GAS have lagged considerably behind such

Thomas Proft and Jacelyn M. S. Loh (eds.), *Group A Streptococcus: Methods and Protocols*, Methods in Molecular Biology, vol. 2136, https://doi.org/10.1007/978-1-0716-0467-0_5,

initiatives. GAS strains vary considerably in both their ease of genetic manipulation and in their association with disease. Widely used laboratory strains (e.g., M49^{NZ131} and M6^{JRS4}) are relatively easy to manipulate genetically [8], but are not representative of contemporary disease isolates and show important phenotypic differences in disease models of GAS infection when compared with the most prevalent clinical strains (e.g., M1T1) [9, 10]. Furthermore, increasing antibiotic resistance (particularly due to *ermB* and *tetM*) associated with emerging strains limits the utility of many of the existing genetic tools that are based on these antibiotic resistance markers.

We have recently developed an optimized method for precise genetic manipulation of GAS that overcomes the limitations of existing methods. Our method combines Gibson assembly cloning with a new temperature-sensitive plasmid, pLZts [9, 11], a modified version of the *E. coli*-GAS shuttle vector pLZ12Spec that contains a temperature-sensitive *repA* gene from pJRS233 [12]. A 2-step temperature shift procedure is used to insert mutant alleles into the GAS genome, followed by excision of the wild-type allele. This highly efficient method reduces the time needed to create markerless GAS mutants to ~2–3 weeks. We have successfully used this method to manipulate multiple M types (including M12, M1, and M6) and routinely prepare multiple mutants simultaneously.

2 Materials

2.1 pLZts Vector Preparation

1. pLZts plasmid (*see* **Note 1**).
2. Restriction enzymes: *Bam*H1, *Eco*RI, and *Dpn*I (with supplied buffers).
3. Polymerase chain reaction (PCR) primers:

   pLZts-S 5'-AGATCTTCCTTCAGGTTATG-3'

   pLZts-A 5'-GGGCCCATCAGTCTGACG-3'
4. High-fidelity PCR reagents (e.g., KAPA HiFi, Kapa Biosystems).
5. 0.8% agarose gel.
6. DNA purification kits (Gel or PCR; *see* **Note 2**).

2.2 Insert Preparation and Cloning

1. DNA purification kit (e.g., Qiagen DNeasy Blood and Tissue Kit; *see* **Note 3**).
2. PCR primers (*see* **Note 4**).
3. PCR reagents (e.g., KAPA HiFi, Kapa Biosystems).
4. 0.8% agarose gel.

5. DNA purification kits (Gel or PCR; *see* **Note 2**).
6. Spectrophotometer (e.g., Nanodrop).
7. NEBuilder® HiFi DNA Assembly Cloning reagent (NEB).
8. Competent *E. coli* cells (any K12 laboratory strain, e.g., DH5α, TOP10).
9. SOC medium: 2% tryptone, 0.5% yeast extract, 10 mM NaCl, 2.5 mM KCl, 10 mM $MgCl_2$, 10 mM $MgSO_4$, and 20 mM glucose.
10. Luria-Bertani (LB) agar: 1% tryptone, 0.5% yeast extract, 85 mM NaCl, 1.2% w/v agar, containing 100 μg/mL spectinomycin (diluted from a 100 mg/mL stock solution in dH_2O, 1000×).
11. Plasmid purification kit (*see* **Note 5**).

2.3 GAS Transformation

1. GAS competent cells for the strain(s) of interest (*see* Chapter 4, Subheading 3.1.2 for the preparation of GAS competent cells).
2. 0.025-μm dialysis membranes (Millipore).
3. THYB: Todd-Hewitt broth containing 0.2% w/v yeast extract.
4. Electroporation apparatus and 2 mm electroporation cuvettes.
5. THYA: THY broth containing 1.5% w/v bacteriological agar.
6. THYA+Sp100: THY agar containing 100 μg/mL spectinomycin.

2.4 Allelic Exchange

1. THYA plates (*see* Subheading 2.3, **item 5**).
2. THYA+Sp100 plates (*see* Subheading 2.3 ,**item 6**).
3. THYB+Sp100: Todd-Hewitt broth containing 100 μg/mL spectinomycin.
4. THYB (*see* Subheading 2.3, **item 3**).
5. Incubators set at 28 °C and 37.5 °C.
6. Circulating water bath (*see* **Note 6**).

2.5 Mutant Screening

1. Self-locking PCR tubes (*see* **Note** 7).
2. Taq DNA polymerase and reaction buffer.
3. 0.8% agarose gel.
4. THYA plates (*see* Subheading 2.3, **item 5**).

3 Methods

The construction of GAS mutants requires a familiarity with general GAS bacteriological and molecular biology techniques. We recommend the excellent methods paper by Le Breton and McIver [8] for a detailed description of general GAS culture, DNA preparation, and transformation methods. The following protocol is limited to a precise description of our optimized cloning, allelic exchange, and colony screening methods using the pLZts vector. An overview of this process is given in Fig. 1.

3.1 pLZts Vector Preparation

1. Digest ~1 μg of pLZts with 10 U *Bam*H1 + 10 U *Eco*RI in a final volume of 30 μL for >3 h at 37 °C. Use buffers supplied with the restriction enzymes.
2. Following incubation, dilute the reaction to a final volume of 100 μL by adding 70 μL dH_2O.
3. Prepare serial tenfold dilutions (10^{-2} to 10^{-4}) and use as a template for PCR with primers pLZts-S and pLZts-A. We routinely use KAPA HiFi PCR enzyme, but any high-fidelity proofreading PCR enzyme is suitable. Follow manufacturer's instructions.
4. Analyze PCR products by gel electrophoresis on a 0.8% agarose gel.
5. To the reaction from the highest template dilution that gives a clean PCR product (3.7 kb), add 10 U *Dpn*I and incubate for 1 h at 37 °C to remove the PCR template DNA.
6. Purify the resulting pLZts vector fragment using a DNA purification kit according to manufacturer's instructions (*see* **Note 2**) and proceed to cloning (*see* **Note 8**).

3.2 Design and Preparation of Mutant Allele Insert DNA

1. Designing the desired mutation construct is most easily accomplished using the New England Biolabs online assembly tool for the NEBuilder cloning system (NEBuilder Assembly Tool, accessed at https://nebuilder.neb.com/#!/). For the sequence of the pLZts PCR fragment produced in Subheading 3.1 (*see* **Note 4**). The NEBuilder Assembly Tool will output the optimal primers for PCR amplifying the DNA fragments for the desired mutation construct.
2. Purify genomic DNA from the desired GAS strain(s) using a suitable DNA purification kit, according to manufacturer's instructions (*see* **Note 3**).
3. Prepare the insert DNA fragments by PCR using primers designed in **step 1** and GAS DNA prepared in **step 2** using a high-fidelity PCR enzyme (e.g., KAPA HiFi), according to manufacturer's instructions.
4. Analyze PCR products by gel electrophoresis on a 0.8% agarose gel and gel- or PCR-purify each fragment, as required (*see* **Note 2**).
5. Determine DNA concentration and purity for each fragment by measuring A_{260}/A_{280} absorbance values (pure double-stranded DNA will have an A_{260}/A_{280} ratio greater than 1.8, and an A_{260} of 1.0 is equivalent to a DNA concentration of 50 ng/μL).

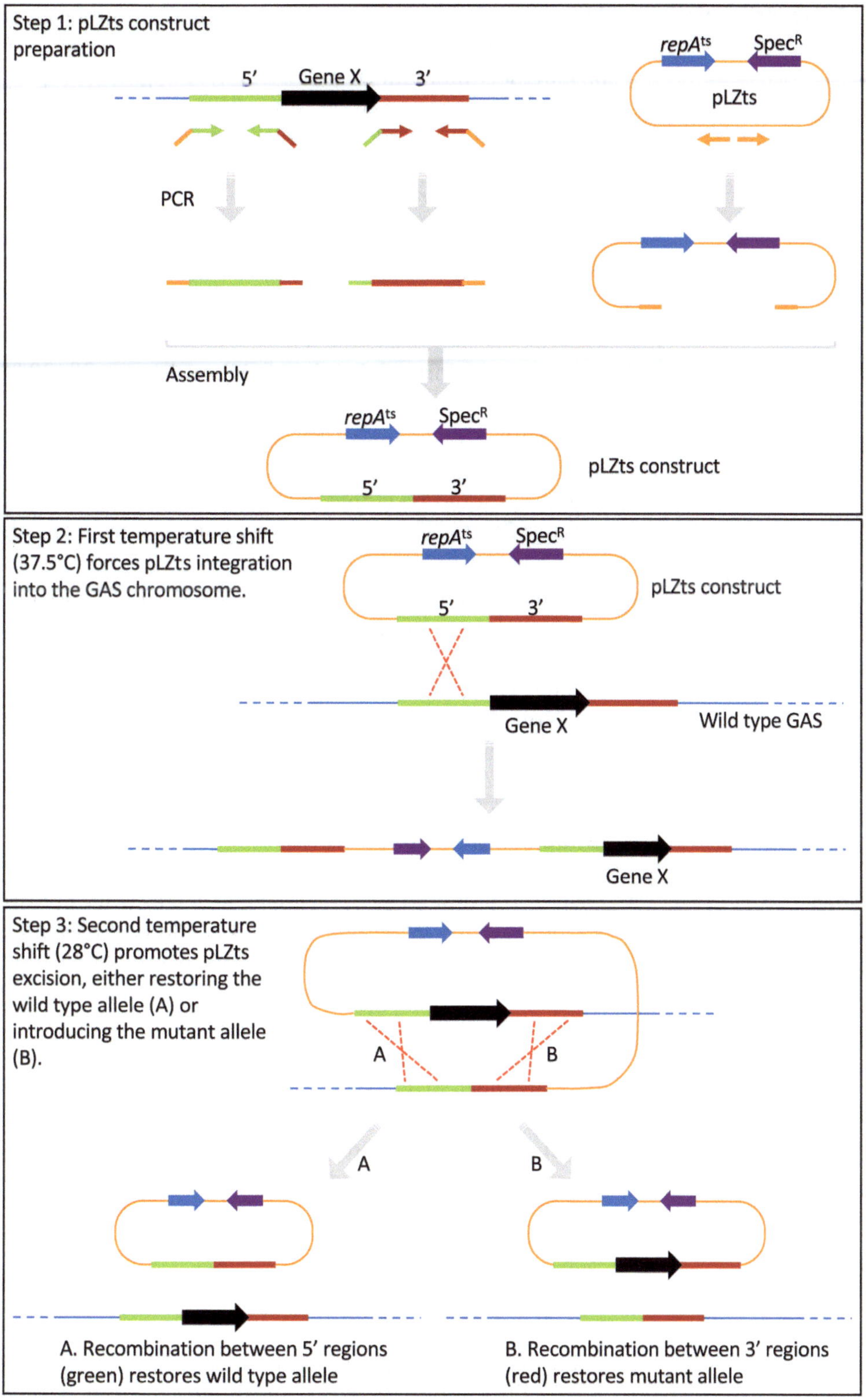

Fig. 1 Steps for introducing genome edits into GAS using pLZts. In **step 1**, regions of PCR primers (small arrows) that facilitate assembly are colored as follows: 5′ (green), 3′ (red), and pLZts (orange)

3.3 Gibson Assembly Cloning

1. Mix each of the PCR products prepared in Subheading 3.2 with 100 ng of the pLZts vector prepared in Subheading 3.1 to produce a twofold molar excess of insert fragments (optimized for 2–3 fragment assembly) in a final volume of 10 μL (*see* **Note 9**). Also include a parallel reaction that contains only the pLZts vector fragment; this will serve as a negative control for the assembly reaction.
2. Add 10 μL of NEBuilder® HiFi DNA Assembly Cloning reagent, mix gently, and incubate for 15–60 min at 50 °C.
3. Transform competent *E. coli* with 2 μL of the mix, according to manufacturer's instructions (*see* **Note 10**).
4. Incubate transformed *E. coli* in SOC medium at 28–30 °C for 2–3 h.
5. Plate 100 μL of 10^0 and 10^{-1} dilutions onto LB agar containing 100 μg/mL spectinomycin and incubate at 28 °C for 24 h. Colonies should be visible the next day when incubated overnight at 28 °C. The vector-only negative control plate should have very few or no colonies.
6. Pick and screen 6–10 colonies for the desired construct. We routinely culture individual colonies in 5 mL LB broth containing 100 μg/mL spectinomycin for 24–36 h at 28 °C, isolate plasmid DNA from 2 to 3 mL using a plasmid purification kit (*see* **Note 5**), and analyze by restriction enzyme digestion and DNA sequencing to confirm the desired construct (*see* **Note 11**). At this stage, we also recommend preparing a freezer stock from the leftover culture of the verified construct for future use (*see* **Note 12**).

3.4 GAS Transformation

1. Dialyze the desired pLZts mutagenesis construct prepared in Subheading 3.3 by placing a 20–30 μL drop of plasmid DNA onto the center of a 0.025 μm dialysis membrane floating in a petri dish of sterile dH_2O. After 15–30 min, carefully remove the dialyzed plasmid DNA into a sterile 1.5 mL tube.
2. Transform 10 μL of the desired mutagenesis construct into electrocompetent GAS for the target GAS strain, using standard GAS electrotransformation protocols (*see* Chapter 4, Subheading 3.1.2 for the preparation of GAS competent cells and the electroporation procedure). We routinely use the following electroporator settings (BioRad GenePulser):
   (a) Voltage = 1.75 kV.
   (b) Resistance = 400 Ω.
   (c) Capacitance = 25 μF.
3. Following electroporation, add the transformed GAS cells to 5 mL THY broth (without spectinomycin) and outgrow at 28 °C for 3 h.

4. Plate 100 μL of neat and 10^{-1} dilutions (prepared in THY broth) onto THYA+Sp100, and incubate at 28 °C for 36–48 h.
5. Streak-purify 2–3 isolated colonies overnight at 28 °C on THYA+Sp100.
6. Inoculate a single colony into 5 mL of THYB+Sp100 and culture overnight at 28 °C.

3.5 Allelic Exchange

The allelic exchange method is derived from the method described by Biswas et al. [13]. This method involves successive culture of pLZts-containing GAS strains at 37.5 °C and 28 °C to insert the plasmid into the GAS genome and excise the wild-type gene, respectively.

3.5.1 Plasmid Integration

1. Inoculate 10 mL of THYB+Sp100 with 100 μL of the transformed GAS strain (prepared in Subheading 3.4) and grow for 2 h at 28 °C (this step is ideally performed in a circulating water bath, but can be performed in a standard incubator if necessary; *see* **Note 6**).
2. Shift cultures to 37.5 °C in a circulating water bath and incubate for 3 h.
3. Plate 100 μL of serial tenfold dilutions (10^{-2}–10^{-7}) onto THYA+Sp100 plates prewarmed at 37.5 °C and incubate overnight at 37.5 °C to select for integrants. These will appear as larger colonies that grow out from a lawn of pinpoint colonies and should be visible by 24 h. It is important to use prewarmed plates for this step to minimize the growth of the lawn of pinpoint colonies.
4. Pick 3–4 large colonies onto a single loop and streak-purify onto THYA+Sp100 plates prewarmed at 37.5 °C. Incubate overnight at 37.5 °C.

3.5.2 Plasmid Excision

1. Pick several isolated colonies onto a single loop and inoculate into 10 mL of THYB+Sp100 and grow overnight at 37.5 °C.
2. Dilute 1/10,000 into 10 mL of THYB and culture at 28 °C for >30 h (i.e., 2 overnights).
3. Prepare serial tenfold dilutions in PBS and plate 100 μL each of the 10^{-3} to 10^{-7} dilutions onto THYA. Incubate overnight at 37.5 °C.
4. Patch individual colonies onto duplicate plates containing THYA+Sp100 and THYA. Incubate plates overnight at 37.5 °C.
5. Following overnight incubation, examine plates for clones with excised plasmids. These will be clones that grow on THYA, but not on THYA+Sp100. The resulting Sp-sensitive clones will be either the wild-type strain or the strain containing the desired

chromosomal mutation (*see* Fig. 1). These will need to be screened for the desired genotype by PCR (size difference for deletions, insertions, etc.; *see* Subheading 3.6) and/or DNA sequencing. We routinely screen 24 colonies by this process, with a desired mutant frequency of >30%. If the desired mutation is lethal, only clones containing the wild-type allele will be identified.

3.6 Mutant Screening

1. Pick part of a colony into a self-locking PCR tube (e.g., Eppendorf Safe-Lock, *see* **Note 7**) containing 10 μL of water using a sterile pipette tip (do not use wooden toothpicks, as these will absorb the water). Mark the location of the picked colony and number on the bottom of the plate, so you can refer back and pick the rest of the colony for outgrowth and for preparation of a glycerol stock, if it is correct.
2. Close the lid of each self-locking PCR tube and place tubes (without rack) on a microwave oven plate.
3. Set microwave to ~600 W (medium–high) and heat for 4 min. Briefly, spin down tubes using a microcentrifuge.
4. Use a multichannel pipette to transfer 0.5–1 μL of the supernatant to the amplification reaction mix as template for PCR analysis.
5. Analyze PCR products by gel electrophoresis on a 0.8% agarose gel. An example of results obtained using this method is shown in Fig. 2.
6. Streak-purify correct clone(s) onto THYA and prepare frozen stocks (*see* **Note 12**).

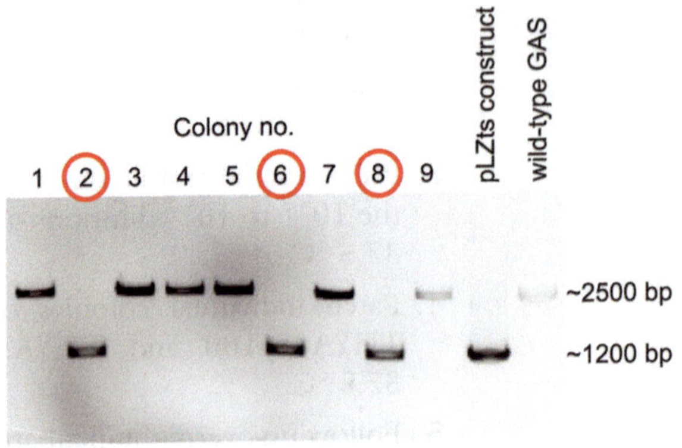

Fig. 2 Colony PCR screening for mutant GAS containing a gene deletion. Each lane represents an individual mutant colony. Red circles indicate strains with the expected band size following colony PCR

4 Notes

1. pLZts is a temperature-sensitive derivative of pLZ12Spec [14], produced by replacing the wild-type *repA* with the temperature-sensitive version from pJRS233 [12]. The sequence of pLZts can be obtained from GenBank (Accession No. MN073843). pLZts DNA can be obtained from Addgene (www.addgene.org; Addgene ID 128799).
2. The choice of DNA purification method depends on the relative purity of the 3709 bp pLZts PCR fragment. If the resulting PCR contains no visible contaminating DNA bands, then a routine PCR purification is sufficient (e.g., Qiagen QIAquick PCR or NEB Monarch PCR cleanup kit). If contaminating DNA bands are visible then we recommend gel-purifying the 3709 bp pLZts PCR fragment.
3. Isolation of GAS DNA from the parent strain can be accomplished by a number of methods. Our laboratory routinely employs the Qiagen DNeasy Blood and Tissue Kit, using the "Pretreatment for Gram-Positive Bacteria" protocol. For other methods, please refer to the article by Le Breton and McIver [8].
4. PCR primers are designed using the NEBuilder Assembly Tool. The mutant allele insert DNA requires ~400 bp (or greater) of homology on each side of the desired chromosomal mutation. For gene deletions, we routinely use the 600 bp upstream of the ATG start codon and 600 bp downstream of the stop codon. In practice, any chromosomal modification can be made, including missense and nonsense mutations, introduction of epitope tags, and introduction of gene cassettes (e.g., antibiotic markers, gene reporters). The sequence of the pLZts PCR fragment can be obtained by copying residues 25–3733 (5′-GGGCCCATCAGTCTGACG... CATAACCTGAAGGAAGATCT-3′) of the pLZts sequence (*see* **Note 1**). The forward (Fwd) primer for the upstream DNA fragment and the reverse (Rev) primer for the downstream DNA fragment can also be used to PCR a gene fragment for complementation using the pLZts procedure.
5. We routinely use a Qiagen QIAprep Miniprep kit for the screening of pLZts clones. As pLZts is very efficiently introduced into GAS, there is usually enough plasmid DNA of the correct clone left over for transformation into the desired GAS strain. However, larger amounts of plasmid DNA may be required for some difficult-to-transform GAS strains and should be prepared using a larger-scale plasmid isolation kit.
6. A circulating water bath is recommended for temperature shifts to integrate and excise pLZts constructs. If a circulating water bath is not available, we have had success by placing culture

tubes in a beaker of water preequilibrated to 28 °C or 37.5 °C and placed in a standard incubator for the duration of the temperature shift step.

7. The use of a self-locking PCR tube with a latched microwave-proof lid is extremely important, to remove any risk of the tubes exploding during this step. We routinely use Eppendorf Safe-Lock PCR tubes for this procedure.
8. The linear pLZts fragment produced in Subheading 3.1 can be stored at −20 °C and used for successive assembly reactions. We have successfully used stored pLZts fragments for at least 6 months.
9. Molar ratios will depend on the size of inserts to be assembled into the final pLZts construct. For 600 bp fragments, a 2:1 molar ratio of insert:pLZts is 32.4 ng of each insert for 100 ng of pLZts. For other insert sizes, the amount of insert can be calculated using an online calculator (e.g., nebiocalculator.neb.com).
10. *E. coli* competent cells are supplied with the NEBuilder HiFi DNA Assembly cloning kit (NEB #E5520S). Chemically- or electro-competent *E. coli* can also be obtained from several commercial sources or prepared by a variety of methods. In our laboratory, we recommend the method described by Hanahan [15].
11. Sanger DNA sequencing is recommended to confirm the desired pLZts clones are correct. We routinely sequence plasmids using a commercial service and with sequencing primers specific to each construct.
12. Freezer stocks (−80 °C) of GAS and *E. coli* strains can routinely be produced by adding 1 volume of sterile 80% glycerol (in dH_2O) to 3 volumes of an overnight bacterial culture, giving a final concentration of 20% glycerol.

Acknowledgments

The authors would like to thank Dr. Amanda Walker for technical assistance with preparing pLZts mutants. Timothy Barnett is supported by a Career Development Fellowship from the NHMRC-funded Improving Health Outcomes in the Tropical North: A multidisciplinary collaboration (Hot North; APP1131932). Mark Walker is supported by an NHMRC Principal Research Fellowship (APP1102621). The authors acknowledge support from the NHMRC research granting system and the Western Australia Department of Health.

References

1. Falkow S (2004) Molecular Koch's postulates applied to bacterial pathogenicity—A personal recollection 15 years later. Nat Rev Microbiol 2:67–72. https://doi.org/10.1038/nrmicro799
2. Carapetis JR, Steer AC, Mulholland EK, Weber M (2005) The global burden of group A streptococcal diseases. Lancet Infect Dis 5:685–694. https://doi.org/10.1016/S1473-3099(05)70267-X
3. Steer AC, Law I, Matatolu L et al (2009) Global *emm* type distribution of group A streptococci: systematic review and implications for vaccine development. Lancet Infect Dis 9:611–616. https://doi.org/10.1016/S1473-3099(09)70178-1
4. Davies MR, McIntyre L, Mutreja A et al (2019) Atlas of group A Streptococcal vaccine candidates compiled using large-scale comparative genomics. Nat Genet 51:1035–1043. https://doi.org/10.1038/s41588-019-0417-8
5. Walker MJ, Barnett TC, McArthur JD et al (2014) Disease manifestations and pathogenic mechanisms of group A *Streptococcus*. Clin Microbiol Rev 27:264–301. https://doi.org/10.1128/CMR.00101-13
6. Barnett TC, Bowen AC, Carapetis JR (2018) The fall and rise of group A *Streptococcus* diseases. Epidemiol Infect 147:e4. https://doi.org/10.1017/S0950268818002285
7. Schodel F, Moreland NJ, Wittes JT et al (2017) Clinical development strategy for a candidate group A Streptococcal vaccine. Vaccine 35:2007–2014. https://doi.org/10.1016/j.vaccine.2017.02.060
8. Le Breton Y, McIver KS (2013) Genetic manipulation of *Streptococcus pyogenes* (the group A *Streptococcus*, GAS). In: Current protocols in microbiology. John Wiley & Sons, Inc., Hoboken, NJ, pp 9D.3.1–9D.3.29
9. Soderholm AT, Barnett TC, Korn O et al (2018) Group A *Streptococcus* M1T1 intracellular infection of primary tonsil epithelial cells dampens levels of secreted IL-8 through the action of SpyCEP. Front Cell Infect Microbiol 8:160. https://doi.org/10.3389/fcimb.2018.00160
10. Barnett TC, Liebl D, Seymour LM et al (2013) The globally disseminated M1T1 clone of group A *Streptococcus* evades autophagy for intracellular replication. Cell Host Microbe 14:675–682. https://doi.org/10.1016/j.chom.2013.11.003
11. Turner AG, Djoko KY, Ong CY et al (2019) Group A *Streptococcus* coordinates manganese import and iron efflux in response to hydrogen peroxide stress. Biochem J 476(3):595–611. https://doi.org/10.1042/BCJ20180902
12. Perez-Casal J, Price JA, Maguin E, Scott JR (1993) An M protein with a single C repeat prevents phagocytosis of *Streptococcus pyogenes*: use of a temperature-sensitive shuttle vector to deliver homologous sequences to the chromosome of *S. pyogenes*. Mol Microbiol 8:809–819. https://doi.org/10.1111/j.1365-2958.1993.tb01628.x
13. Biswas I, Gruss A, Dusko Ehrlich S, Maguin E (1993) High-efficiency gene inactivation and replacement system for Gram-positive bacteria. J Bacteriol 175:3628–3635. https://doi.org/10.1021/ja01876a046
14. Husmann LK, Scott JR, Lindahl G, Stenberg L (1995) Expression of the Arp protein, a member of the M protein family, is not sufficient to inhibit phagocytosis of *Streptococcus pyogenes*. Infect Immun 63:345–348. https://doi.org/10.1042/BCJ20180902
15. Hanahan D (1983) Studies on transformation of *Escherichia coli* with plasmids. J Mol Biol 166:557–580

Chapter 6

Generation of Bioluminescent Group A Streptococcus for Biophotonic Imaging

Jacelyn M. S. Loh, Kar Yan Soh, and Thomas Proft

Abstract

Bioluminescence is a rapid and cost-saving technology that enables monitoring of bacteria in real time in animal infection models. Here, we report a method for labeling GAS with a set of plasmids we have developed and deposited with Addgene. These plasmids can be used to either integrate firefly luciferase (Ffluc) or red-shifted firefly luciferase (FflucRT) into the GAS genome or to introduce the reporters on plasmids which have been stabilized with a toxin–antitoxin system to prevent loss of plasmid in the absence of antibiotic selection.

Key words Bioluminescence, Toxin–antitoxin system, Biophotonic imaging, Firefly luciferase

1 Introduction

Biophotonic imaging is a nontoxic, highly sensitive, and noninvasive method that allows for real-time monitoring of microorganisms within whole living animals [1]. The technology is based on the detection of visible light produced by bioluminescent reactions (e.g., luciferase) or by excitation of fluorescent molecules (e.g., green fluorescent protein, GFP) using sensitive photon detectors, usually a charge-coupled device (CCD) camera. Multiple imaging of the same animal throughout an experiment allows disease progression to be followed with extreme accuracy, reducing the number of animals required to yield statistically meaningful data (Fig. 1) [2–5].

Bioluminescence is often preferred over fluorescence as autofluorescence is generally much brighter than autoluminescence and consequently produces a lower signal/noise ratio [6]. The most commonly used reporters for bioluminescence imaging are firefly luciferase (Ffluc) and Lux. Ffluc, derived from the firefly *Photinus pyralis*, generates light emission in the 550–575 nm range producing yellow–green light [7, 8]. There is also a red-shifted firefly luciferase derivative (FflucRT) with a light emission peak at

Thomas Proft and Jacelyn M. S. Loh (eds.), *Group A Streptococcus: Methods and Protocols*, Methods in Molecular Biology, vol. 2136, https://doi.org/10.1007/978-1-0716-0467-0_6, © Springer Science+Business Media, LLC, part of Springer Nature 2020

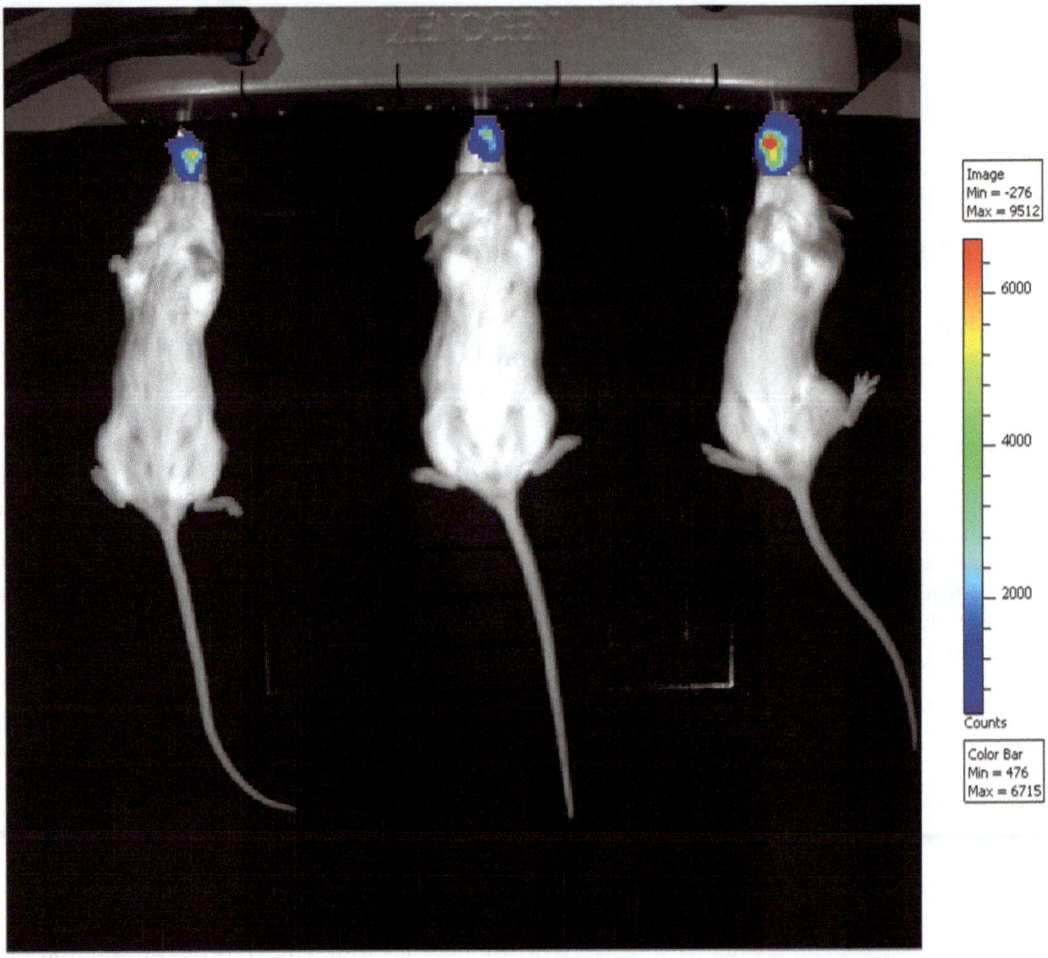

Fig. 1 Example of in vivo detection of bioluminescent GAS. FVB/n mice were intranasally inoculated with 10^5 CFU GAS strain 57G (M49-serotype clinical isolate) carrying pLZ12km2-P23:Ffluc. 100 μg/ml luciferin (15 μl/nare) was administered prior to imaging on an IVIS imaging system under isoflurane anesthesia

610 nm [9], which was generated by the introduction of six mutations into the *fflucRT* gene [10]. The bacterial luciferin–luciferase operon (lux) is encoded in the five-gene operon (*luxCDABE*) and emits a blue–green light with a maximum intensity at 490 nm [11]. Lux has the advantage that no exogenous luciferin substrate has to be administered. However, the large operon size (5.56 kb) generates a larger metabolic burden for the bacteria compared to the much smaller *ffluc* gene (1.65 kb).

There are two main strategies for introducing the reporter genes into bacteria, by integration into the bacterial genome or the use of plasmid-based systems. Integration into the genome has the advantage that selective antibiotics to maintain plasmid stability are not required. However, integrating genes into the genome is often far more technically challenging, and the single gene copy results in weaker signals. In contrast, plasmid-based systems provide multiple copies of the reporter gene, hence a stronger signal, but require administration of an antibiotic selection marker to prevent loss of the plasmid.

We have generated bioluminescence reporters for the biophotonic labeling of GAS using Ffluc and FflucRT [12]. The reporter genes were integrated onto the GAS genome by allelic replacement of the conserved *spy0535* gene, which encodes a nonessential sugar phosphate isomerase [13]. In addition, we have generated plasmid-based systems for both Ffluc and FflucRT using the *Streptococcus–Escherichia coli* shuttle vector pLZ12km [14] with the addition of the strong *Lactococcus lactis* promoter P23 [15]. To achieve segregational plasmid stability under nonselective conditions, we have also introduced the streptococcal ω–ε–ζ toxin–antitoxin (TA) cassette, originally found on the streptococcal plasmid pSM19035 [16]. The *zeta*-gene encodes a kinase that affects bacteria by inhibition of peptidoglycan synthesis and is neutralized by the labile epsilon-antitoxin protein.

2 Materials

2.1 Plasmids

1. pLZ12Km2-P23R:TA:ffluc (500 μg/ml).
2. pLZ12km2-P23:TA:FflucRT (500 μg/ml).
3. pFW11:Δspy0535:Ffluc (500 μg/ml).
4. pFW11:Δspy0535:FflucRT (500 μg/ml).

   All plasmids can be obtained free of charge from Addgene. Please refer to www.addgene.org.

2.2 Electro-competent GAS

1. Horse blood agar (HBA) plates.
2. Todd-Hewitt yeast (THY) medium: TH broth plus 0.2% yeast extract (*see* **Note 1**).
3. Refrigerated centrifuge.
4. Ice-cold filter-sterilized 0.5-M sucrose in MQ water.
5. Spectrophotometer (e.g., Nanodrop) (*see* **Note 2**).

2.3 Electroporation

1. Quartz cuvettes.
2. Electroporation cuvettes (2 mm).
3. Electroporator (*see* **Note 3**).
4. THY medium.
5. THY agar with 200 μg/ml kanamycin (*see* **Note 4**).
6. THY agar with 100 μg/ml spectinomycin (*see* **Note 4**).
7. Stationary incubator.

2.4 Biophotonic Imaging

1. General 96-well microtiter plates.
2. White opaque 96-well microtiter plates.
3. Luciferin (100 μg/ml in THY medium, freshly prepared).
4. Microtiter plate reader (*see* **Note 5**).

3 Methods

3.1 Preparation of Electrocompetent GAS

1. Streak a freezer stock of GAS onto HBA to obtain single colonies (*see* **Notes 6** and 7).
2. Select a single colony and grow in 1.5-ml THY medium O/N at 37 °C in a stationary incubator.
3. Inoculate a new culture in total volume of 15-ml THY medium (1:10) and grow to OD_{600} of 0.2–0.25.
4. Spin the culture in a refrigerated centrifuge for 10 min at 3600 × *g* at 4 °C.
5. Wash the bacterial pellet twice in 1 ml of ice-cold filter-sterilized 0.5-M sucrose and resuspend in 50-μl ice-cold filter-sterilized 0.5-M sucrose.

3.2 Electroporation of GAS

1. Add 2–4 μl plasmid DNA to the 50-μl bacterial culture and transfer to a prechilled 2-mm electroporation cuvette.
2. Electroporate the bacteria using the electroporator with a single pulse of 2100 V, capacitance at 25 μF, and resistance at 200 Ω (*see* **Note 8**).
3. Immediately add 1 ml of THY medium and incubate for 2 h at 37 °C in a stationary incubator (*see* **Note 9**).
4. Plate bacteria on THY/Kan plates (for pLZ12 plasmids) or THY/Spec plates (for pFW11 plasmids) and incubate ON at 37 °C.

3.3 Selection of Bioluminescent Bacteria

1. Inoculate single GAS colonies in 50 μl of THY medium in a white opaque 96-well microtiter plates and grow at 37 °C for 5 h (*see* **Note 10**).
2. Inoculate untreated GAS as negative control.
3. Add 50 μl of luciferin (100 μl/ml in THY).
4. Measure light emission in a plate reader.

3.4 Quantification of Bioluminescence

1. Grow an O/N culture of the single colonies that produced the highest light emission in Subheading 3.3 (*see* **Note 11**).
2. Inoculate 1-ml THY medium with O/N culture (1:20) and grow to mid-log phase (OD_{600} of 0.5–0.6).

3. Prepare a serial dilution of the GAS culture in a general 96-well microtiter plate.
4. Transfer 50 μl of the diluted bacteria into a white opaque 96-well microtiter plate and add 50 μl of luciferin (100 μl/ml in THY).
5. Measure light emission in a plate reader.
6. Drop plate 3 replicas (15 μl) from each well of the general 96-well microtiter plate onto a HBA plate and incubate O/N at 37 °C (*see* **Note 12**).
7. Count bacterial colonies for each drop. Average the triplicates and calculate the total number of bacteria in each of the 50-μl aliquots used for light detection. This will allow to generate a graph where bioluminescent units (BLUs) are plotted against viable cells, and this should show a linear correlation (*see* **Note 13**).

4 Notes

1. THY medium can be replaced with Brain Heart Infusion (BHI) medium.
2. We used a NanoDrop 2000 (Thermo Scientific).
3. We used a Gene Pulser Xcell^{TK} (Bio-Rad).
4. THY agar can be replaced with BHI agar.
5. We used an EnSpire™ 2300 Multilabel Reader (Perkin Elmer).
6. THY or BHI plates can also be used, but HBA has the advantage of differentiating for β-hemolytic bacteria to avoid contamination.
7. Any GAS strain can be used. However, we found it difficult with some strains to integrate the reporters onto the genome. The plasmid-based system can also be used with nongroup A Streptococcus, including *S. agalactiae*, *S. dysgalactiae*, *S. equi*, *S. iniae*, *S. salivarius*, and *S. vestibularis* [17]. However, the toxin–antitoxin system might not be functional in some of these species.
8. It is important that the bacterial culture is kept at 4 °C all the time.
9. Increase incubation time to 3 h for reporter plasmid integration into the genome.
10. White plates are recommended as they reflect light and will maximize the light output signal.
11. It is recommended to prepare a freezer stock of the bioluminescent bacteria for future use.

12. One drop will require an area of about 4 cm^2, so one HBA plate can be inoculated with multiple drops.
13. The background signal was established by measuring unlabeled GAS at a range between 10^2 and 10^4 CFU and resulted in 68 ± 6 BLU. This allows to determine the detection limit for each bacterial strain. In our hands, the detection limit for M1-FFluc (GAS serotype M1 strain SF370 transformed with pLZ12km-P23:Ffluc) was <50 CFU.

Acknowledgments

We thank Andreas Podbielski (University of Rostock, Germany) for sending us the pFW11 vector, Nobuhiko Okada (Kitasato University, Tokyo, Japan) for providing the pLZ12Km2 vector, and Urszula Zielenkiewicz (Polish Academy of Sciences, Warsaw, Poland) for sending the pBT286 plasmid containing the TA cassette. This project was supported by a research grant from the Health Research Council (HRC) New Zealand.

References

1. Andreu N, Zelmer A, Wiles S (2011) Noninvasive biophotonic imaging for studies of infectious disease. FEMS Microbiol Rev 35 (2):360–394. https://doi.org/10.1111/j.1574-6976.2010.00252.x
2. Avci P, Karimi M, Sadasivam M et al (2018) In-vivo monitoring of infectious diseases in living animals using bioluminescence imaging. Virulence 9(1):28–63. https://doi.org/10.1080/21505594.2017.1371897
3. Hoffman RM, Zhao M (2006) Whole-body imaging of bacterial infection and antibiotic response. Nat Protoc 1(6):2988–2994. https://doi.org/10.1038/nprot.2006.376
4. Hoppe AD, Seveau S, Swanson JA (2009) Live cell fluorescence microscopy to study microbial pathogenesis. Cell Microbiol 11(4):540–550. https://doi.org/10.1111/j.1462-5822.2009.01283.x
5. Hutchens M, Luker GD (2007) Applications of bioluminescence imaging to the study of infectious diseases. Cell Microbiol 9 (10):2315–2322. https://doi.org/10.1111/j.1462-5822.2007.00995.x
6. Troy T, Jekic-McMullen D, Sambucetti L, Rice B (2004) Quantitative comparison of the sensitivity of detection of fluorescent and bioluminescent reporters in animal models. Mol Imaging 3(1):9–23. https://doi.org/10.1162/15353500200403196
7. Niwa K, Ichino Y, Kumata S et al (2010) Quantum yields and kinetics of the firefly bioluminescence reaction of beetle luciferases. Photochem Photobiol 86(5):1046–1049. https://doi.org/10.1111/j.1751-1097.2010.00777.x
8. Zhao H, Doyle TC, Coquoz O et al (2005) Emission spectra of bioluminescent reporters and interaction with mammalian tissue determine the sensitivity of detection in vivo. J Biomed Opt 10(4):41210. https://doi.org/10.1117/1.2032388
9. Andreu N, Zelmer A, Sampson SL et al (2013) Rapid in vivo assessment of drug efficacy against Mycobacterium tuberculosis using an improved firefly luciferase. J Antimicrob Chemother. https://doi.org/10.1093/jac/dkt155
10. Branchini BR, Ablamsky DM, Murtiashaw MH et al (2007) Thermostable red and green light-producing firefly luciferase mutants for bioluminescent reporter applications. Anal Biochem 361(2):253–262. https://doi.org/10.1016/j.ab.2006.10.043
11. Meighen EA (1994) Genetics of bacterial bioluminescence. Annu Rev Genet 28:117–139. https://doi.org/10.1146/annurev.ge.28.120194.001001
12. Loh JM, Proft T (2013) Toxin-antitoxin-stabilized reporter plasmids for biophotonic

imaging of group A streptococcus. Appl Microbiol Biotechnol 97(22):9737–9745. https://doi.org/10.1007/s00253-013-5200-7

13. Park HS, Francis KP, Yu J, Cleary PP (2003) Membranous cells in nasal-associated lymphoid tissue: a portal of entry for the respiratory mucosal pathogen group A streptococcus. J Immunol 171(5):2532–2537
14. Okada N, Tatsuno I, Hanski E et al (1998) Streptococcus pyogenes protein F promotes invasion of HeLa cells. Microbiology 144 (11):3079–3086
15. Que YA, Haefliger JA, Francioli P, Moreillon P (2000) Expression of Staphylococcus aureus clumping factor A in Lactococcus lactis subsp. cremoris using a new shuttle vector. Infect Immun 68(6):3516–3522. https://doi.org/10.1128/iai.68.6.3516-3522.2000
16. Lioy VS, Pratto F, de la Hoz AB et al (2010) Plasmid pSM19035, a model to study stable maintenance in Firmicutes. Plasmid 64 (1):1–17. https://doi.org/10.1016/j.plasmid.2010.04.002
17. Loh JM, Proft T (2016) Stabilized plasmid-based system for bioluminescent labeling of multiple streptococcal species. Biotechnol Lett 38(1):139–143. https://doi.org/10.1007/s10529-015-1967-1

Part II

Application of Group A Streptococcus Omics

Chapter 7

Whole Genome Sequence Analysis and Population Genomics of Group A Streptococci

Jake A. Lacey, Taylah B. James, Steven Y. C. Tong, and Mark R. Davies

Abstract

Whole-genome sequencing (WGS) is used to determine the genetic composition of an organism. This fast-moving field is continually evolving through technical advancements and the development of new bioinformatic tools for analyzing genomic data; however, the basic principles and processes for defining and processing high-quality genome sequence information remain unchanged. Here, we introduce some considerations and describe some commonly used bioinformatic steps for processing raw genome sequence data to generate genome assemblies through to understanding basic population genomics.

Key words Population genomics, Comparative genomics, Genome sequencing, Next-generation sequencing, Reference genome, Group A Streptococcus, Streptococcus pyogenes

1 Introduction

Rapid technological advances in whole-genome sequencing (WGS) approaches over the last decade are transforming how research and clinical laboratories apply genomics for biological insight through to improved clinical and public health surveillance of infectious disease agents [1, 2]. The unprecedented resolution afforded by WGS, in addition to the diminishing costs and the ability to sequence multiple samples simultaneously, is driving the transition from classical single-gene-based molecular diagnostics to large-scale whole-genome-derived comparative genomics for understanding evolving pathogen populations.

Genotyping of Group A Streptococci (GAS) has changed and adapted to new sequencing technologies over the years. All classic GAS molecular typing markers can now be determined from WGS data [3]. For example, traditional Lancefield M-serotyping has moved to PCR-based *emm* genotyping (sanger sequencing of PCR products targeting the 5′ region of the *emm* gene) and multilocus sequencing typing of seven "housekeeping" genes, the first multigene sequence approach at defining GAS population

Thomas Proft and Jacelyn M. S. Loh (eds.), *Group A Streptococcus: Methods and Protocols*, Methods in Molecular Biology, vol. 2136, https://doi.org/10.1007/978-1-0716-0467-0_7,

structure, can also readily be determined by WGS. The reason why WGS has revolutionized the population biology field is that it provides the ability to compare relationships of strains at millions of nucleotides (comprising thousands of genes), a resolution only possible by unbiased genome-wide approaches. However, the utility of WGS for GAS genotyping relies on future development of reproducible strategies and bioinformatic approaches (especially within public health settings).

Sequencing a GAS genome, either using short-read sequencing to generate a draft genome (of multiple chromosomal fragments, termed contigs) or long-read technologies to generate a complete "single" chromosome, enables an unbiased approach to assess the genetic makeup of the organism. GAS is a global pathogen. GAS evolution is a dynamic process driven by ongoing genomic flux (genes entering and leaving a population), a process often associated with homologous (sequence replacement between two strains) and nonhomologous (incorporation of new sequence such as prophage, transposable and/or plasmid elements) recombination [4]. Recent approaches now provide the framework for defining GAS strain relationships within a global framework [4]. Importantly, ongoing sequencing programs, especially those encompassing previously unsampled genotypes and/or strains from unsampled geographical regions, will likely identify new GAS populations (lineages/phylogroups/clusters). The ability to interrogate existing genome databases enables users to address questions regarding the global distribution and variation within genes of interest. Collectively, genome sequencing is a powerful tool that will continue to revolutionize our ability to unravel the complex biology of this pathogen as well as enabling the "real-time" assessment of the emergence, transmission and evolution of GAS.

1.1 The Streptococcus pyogenes Genome

The average *S. pyogenes* genome, based on 192 complete *S. pyogenes* genomes in public databases (e.g., RefSeq—*see* Subheading 2.2, **item 1**), includes a median genome size of 1.84 Mbp (min 1.70 Mbp, max 1.98 Mbp), a guanine–cytosine (GC) content of 38.5% (min 38.2%, max 38.7%) and 1731 coding sequences (CDS) (min 1552, max 1882). A typical *S. pyogenes* genome encodes 5–6 16s rRNA operons, 57–67 tRNAs, 0–10 prophage elements and 0–1 plasmids (plasmid size ranges from 2.3 kb to 28.9 kb with a GC percentage of 31.02–37.33%).

While there is natural variation within the genomes of *S. pyogenes*, for example, there are 1122 MLST types and at least 220 *emm* types that are globally distributed. There are limiting features due to the host adapted nature of the species, such as genome size, number of 16s operons and GC content which all have a narrow range of variation. Therefore, if a GAS genome has more than 7 16s rRNA operons, a GC content higher than 39% or has a genome size larger than 2 Mbp, the researcher/analyst needs to consider the quality of the sequence data and/or bioinformatic processing steps to check for contamination. Like all experimental protocols, biological inferences rely on data quality.

1.2 Bioinformatic Workflow

The bioinformatic workflow for analyzing genome sequence information involves a series of algorithms that process raw sequence information generated by sequencing technologies into nucleotide "reads" with quality control information. The "fastq" file format is a consensus sequence format containing numeric quality score information on each nucleotide for sequencing read exchange between different sequencing technologies. The process of assessing genomic variation from raw sequence reads (fastq format) requires bioinformatics-based analysis platforms (Fig. 1). A number of plug-and-play software or cloud-based packages are available (such as Galaxy, PATRIC, Geneious, and CLC genomics workbench) (*see* **Note 1**). Like most software packages, these systems are currently restricted in their ability scale up to the thousands of genome sequences currently available or the ability to incorporate more bespoke analyses, yet this is likely to change as the software evolves. The following sections describe the general bioinformatic workflow for analyzing and interrogation of GAS sequence data.

Here we demonstrate some methods and tools to perform bioinformatic analyses of GAS genome sequence data. This is not intended to be extensive but rather shows a snippet of particular protocols. There are many different tools that can be utilized to perform the same task and consequently, there is no "one-size-fits-all" for bioinformatics (*see* **Note 2**). The dynamic nature of software development involves creation of new tools and constant updates to existing ones. For the protocols presented in this chapter, we are using tools that are either highly citied (*see* **Note 3**) and/or regularly maintained and updated by bioinformaticians (*see* **Note 4**). We find these tools are useful for GAS and are capable of producing consistent outputs. Most of the tools discussed are command-line tools, in which some have also been incorporated into more structured "clickable" graphical user interfaces (GUI).

Before someone considers producing and analyzing whole-genome sequencing data, there are important points to consider:

1. What is the "question" to be addressed? Appropriate study design and well-curated metadata are key for WGS analyses, as sequencing without a biological question in mind can result in a significant time commitment with no discernible output.
2. Are appropriate computational resources available?
3. Is collaboration with a sequencing facility and bioinformatic group required?
4. Is there a plan for the amount of sequencing and the type of data needed?
5. How familiar is the user with WGS pipelines?

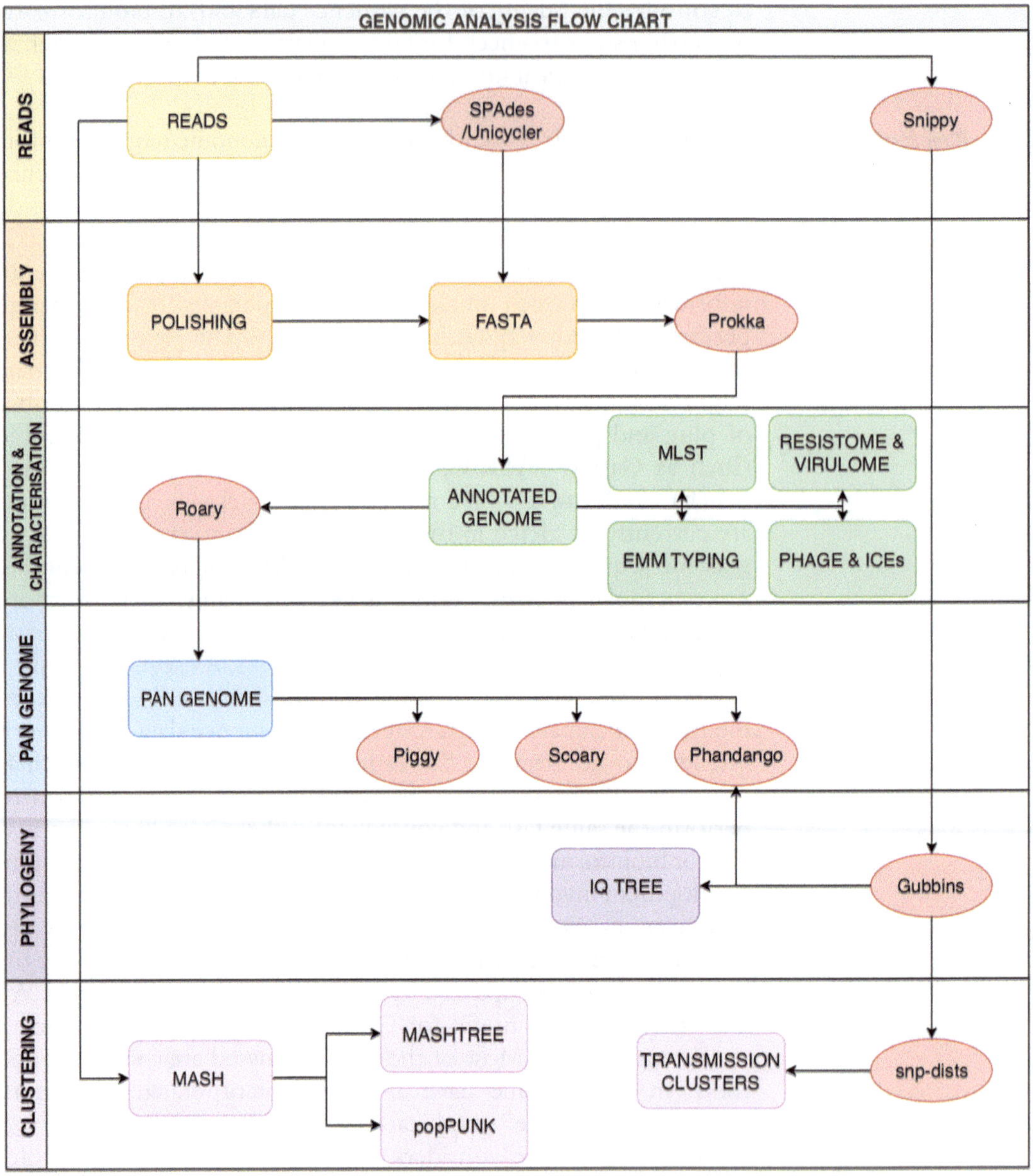

Fig. 1 Flow chart of the genomic processes used in the analysis of *Streptococcus pyogenes* WGS data

These points need to be taken into account before analyses are performed. A large amount of sequence data can become difficult to interpret without an informed study design. Also, the amount of time to perform the analyses can be exponential pending the number of genomes and outcomes envisaged. Being familiar with the WGS pipelines is important. Extensive help files/manuals are available for all the tools mentioned throughout this document and will provide detailed information for each of commands and options.

1.3 Quality Control (QC)

Low-quality sequence data can be a consequence of technical problems during the sequence run or as a result of contamination in the starting DNA extraction. Identifying such problems early is required to prevent issues in the downstream analysis and biological inference. The QC of a sample is made up of multiple steps, where samples that do not meet the requirements at each step are discarded from downstream analysis. Usually, the first stage of QC is to check the quality of the reads and then to perform "cleaning" and classification of sequence reads.

1.4 Genome Assembly

Genome assembly is a bioinformatic process where overlapping stretches of nucleotide sequence (reads) are stitched together (like a puzzle) to create a representation of the original chromosome from which the DNA originated. *De novo* genome assemblies assume no prior knowledge of the source DNA sequence length, layout, or composition. The goal of a sequence assembler is to produce long contiguous pieces of sequence (contigs) from sequence reads. The contigs are sometimes ordered and oriented in relation to one another to form scaffolds. There are many different assembly tools available; however, the most common type of assembler utilizes de Bruijn graph theory. The largest challenge in assembly is due to identical sequences present within a genome. For example, *S. pyogenes* genomes carry 5–6 identical 16s rRNA operons. This provides a challenge when assembling from short-read sequences (up to 150 bp) as the assembly algorithms are unable to determine the correct location of the operons, so the assembly "stops" or "breaks", resulting in a fragmented genome containing multiple contigs. This challenge has largely been overcome with long-read sequencing (Oxford Nanopore and PacBio), but long-read sequencing requires higher quality DNA, increased costs and more computational resources (*see* **Note 5**).

1.5 Genome Annotation and Curation

Annotation is one of the most useful steps in interpreting biological significance of genome sequence data. Genome annotation and characterization is the process of identifying the locations of genes and all of the coding regions in a genome assembly and inferring functional significance to those elements. The biological accuracy of this process relies on a curated database of "known" gene function. There are two groups of annotation: structural annotation and functional annotation. Structural annotation consists of identifying the coding regions, such as open-reading frame identification, gene structure, characterization of coding regions and examining location of regulatory motifs. Functional annotation consists of attaching biological, such as biochemical function, expression and regulation to those elements. There is one tool that is widely adopted as the standard for bacterial genome annotation termed Prokka [5] (*see* **Note 6**).

1.6 Recombination Detection

Genetic recombination in bacteria is the exchange of genetic material between different species or different strains of the same species, which can lead to the production of strains with a combination of traits that differ from those found in the progenitor (parent) strain. Recombination can occur by multiple methods with the classically understood processes being transformation, transduction, conjugation and genomic rearrangements. Recombination is believed to play a large role in the evolution of *S. pyogenes.* Therefore, being able to characterize and flag regions of the genome that are potentially recombinogenic can be useful for examining evolutionary relationships and adaptation. There are multiple tools for detecting genomic regions with potential recombinogenic signatures, one example is Gubbins [6] (Genealogies unbiased by recombination in nucleotide sequences). Other gene-specific examples include FastGEAR [7] and ClonalFrameML [8] (*see* **Note** 7).

1.7 Phylogenetic Analysis

A phylogenetic tree is a diagram that represents evolutionary relationships between populations of strains (genome sequences in this instance). Phylogenetics are estimations of how a species or subpopulations of strains have evolved from a series of common ancestors. While there are many methods for building and estimating phylogenetic trees, one of the most common methods is using maximum-likelihood (ML). ML is a statistical method for estimating unknown parameters of a probability model that is very well understood and is appropriate for data like DNA sequences, where we can reasonably model the stochastic process. ML is robust towards many of the assumptions in the evolutionary model. Even with very short sequences, it generally outperforms alternative methods such as parsimony or distance methods. Tree topologies can also be evaluated and all sequence data can be used to infer tree topologies and branch lengths.

In bacterial population genomics, phylogenetic trees are built based on single nucleotide polymorphisms (SNPs), which are substitutions of single nucleotides occurring within a genome relative to a reference genome sequence. In bacterial genomes, SNPs may fall within coding sequence of genes or noncoding region of genes. SNPs that change the amino acid sequence of a protein are known as nonsynonymous SNPs ("may" change function) and those that do not change the amino acid sequence are known as synonymous SNPs (no change in function). Indels are a term that summarizes insertions and deletions of bases in a genome relative to another genome(s). Indels can range from large genomic deletions or insertions of several hundred kB to a single base event. When Indels occur within a gene, it can result in a frameshift mutation (if the codon triad is affected) or in some cases a nonsense mutation (introduction of a premature stop codon in a gene). SNPs are not just important from a biological perspective but from an evolutionary perspective as SNPs can accumulate over time in bacterial populations. We can use SNP data to investigate many facets of bacterial populations such as evolution and transmission. One key assumption when building phylogenetic trees is that evolution has occurred vertically (i.e., through ancestry, such as nucleotide mispairing during DNA polymerase replication). As indels and recombination (Subheading 1.6) violate this assumption (as SNPs are acquired "horizontally"), these regions of the genome should be excluded when building phylogeny. However, recombination is an important biological property, so the information should be considered when studying the evolutionary biology/population genetics of the organism.

1.8 Population Structure

The ability to determine the genetic relationship of a population of organisms plays a key role in tracking the spread of ever evolving pathogens. Such information may be important in defining whether an outbreak has occurred through to global transmission patterns [1, 2]. GAS is both phenotypically diverse in addition to being a genetically diverse pathogen with globally disseminated lineages [4]. Bioinformatic tools have recently been developed that incorporate both relationships in core and accessory genes to define clusters of related strains [9]. We have recently applied this approach to a global GAS genome database, providing a framework to rapidly identify clusters of related GAS [4]. This iterative platform facilitates new genomes to be populated into this framework enabling rapid comparison of GAS populations across the globe, a key component to future global surveillance.

1.9 Commonly Used Terminology

1. *Alignment*: Similarity-based arrangement of DNA, RNA, or proteins sequences. In this context, subject and query sequence should be orthologous and reflect evolutionary, not functional or structural relationships.
2. *Annotation*: Associating structural or functional information to genome sequence data.
3. *Assembly*: Computational reconstruction of the "parent" genome sequence from genome sequence reads.
4. *Barcode*: Short-sequence identifier for individual labeling (barcoding) of sequence libraries enabling multiple genomes to be sequenced simultaneously.
5. *BLAST*: Algorithm for comparing sequence information. Commonly used for comparing a query "unknown" gene/protein sequence against a library/database with "known" information to infer possible function/structural characteristics.
6. *Contig*: A contiguous linear stretch of DNA or RNA consensus sequence. Constructed from a number of smaller, partially overlapping, sequence fragments (reads).

7. *Coverage*: Also known as “sequencing depth". Coverage refers to the average number of sequence reads per locus.
8. *de novo assembly*: Refers to the reconstruction of contiguous sequences without making use of any reference sequence.
9. *GC content*: The relative proportion (percentage) of guanine (“G”) and cytosine (“C”) bases in a DNA/RNA sequence.
10. *Gene ontology*: Structured terminology for the functional classification of gene and gene product across species and research areas.
11. *Indel*: Insertion/deletion of nucleotide sequence relative to another sequence.
12. *K-mer*: Length of DNA or protein sequence of specified length “K". Used by many assembly algorithms and clustering techniques.
13. *Mapping*: Term used to describe alignment of genome sequence reads to a comparator “reference” sequence.
14. *Library*: Collection of DNA fragments modified in a way that is appropriate for downstream analyses, such as high-throughput sequencing.
15. *Masking*: Converting a DNA sequence “ACTG” to the uninformative character “N". Often used to highlight regions of a genome alignment excluded from downstream analyses.
16. *N50*: Statistic used to describe a set of assembly contigs or scaffolds. It is defined as the length for which the collection of all contigs of that length or longer contains at least half of the total lengths of the contigs.
17. *Read*: Often referred to as a stretch of nucleotide sequence synthesized during a genome sequence experiment (often in fastq format).
18. *Paired-end*: In short-read sequencing protocols (e.g., Illumina), this specifies that sequencing of DNA is determined from both ends of the DNA fragment. Gold standard for generating high-quality short-read sequence data, enabling more accurate read alignment and identification of insertion-deletion (indel) variants, as well as facilitating better chances of detecting genomic rearrangements and repetitive sequence elements.
19. *Single-end*: In short-read sequencing protocols (e.g., Illumina), sequencing of DNA from only one end. Produces half as much information from a genome library as paired-end sequencing approaches. Best suited for gene-targeted methods such as RNA sequencing, Tn-seq or ChIP-seq.

20. *Short-read sequencing*: General term for sequencing technologies (e.g., Illumina) that generates sequence read fragments of 75–300 bp in order to preserve high-quality data.
21. *Long-read sequencing*: General term for sequencing technologies (e.g., Pacific Biosciences, Oxford Nanopore) that generates sequence read fragments (average of >10,000 bp).
22. *Scaffold*: Two or more contigs joined together using read-pair information.
23. *Transcriptome*: Complete RNA profile of a genome sequence.

2 Materials

In this section, we provide a list of representative bioinformatic software. It is recommended that the latest versions be used. We only discuss those that we have first-hand experience in, are routinely maintained and are freely available on public repositories such as Github (links listed). Installing such software can be problematic as all software will require certain dependencies to be installed for successful operation. It should also be noted that operating system updates can cause version control problems.

2.1 Computational Requirements

The computational requirements for bioinformatic analysis of bacterial genomes vary greatly depending on how many isolates are being examined and the nature of the analyses being performed. Typically, bioinformatic tools are command-line based, without a graphic user interface (GUI), which can provide limitations within windows operation systems. Therefore, most bioinformaticians use a Unix-based operating system such as used by macOS or Ubuntu. Ideally, users will have access to a high-performance computing cluster (HPC) through commercial or institutional providers. Alternatively, cloud computing is a cost-efficient way of accessing high-performance computational resources.

A typical laptop with a quad core CPUs (8 threads) and 32 GB of ram can be used for many bioinformatic tools. The major limiting factor is scalability as basic laptop hardware is only capable of working with "tens" of bacterial genomes. Now with hundreds of genomes being used in single projects, a HPC is typically required to handle datasets of that size in a reasonable period of time.

2.2 File Formats

The most common file formats used in analysis of whole-genome sequencing data include the following:

1. *FASTA* (nucleotide sequence(s), file extension .fas, .fa, .fasta).
2. *FASTQ* (nucleotide sequence including quality scores, file extension .fastq, .fq; often supplied in a compressed file format such as gzip [.gz]).

3. *SAM* (sequence alignment, file extension .sam).
4. *BAM* (binary version of SAM, file extension .bam).
5. *GFF* (annotation file format, file extension .gff).
6. *GBK* (GenBank sequence database annotation format, file extension .gbk, gb).
7. *BED* (coordinates file used for annotation, file extension .bed).
8. *VCF* (variant call format file extension .vcf).

2.3 Public Genome Sequence Repositories

There are two main public repositories where raw sequence data (often in fastq format) and complete reference genomes are stored. All published manuscripts that generate or have used genome sequence should provide the unique identifier relating to each strain and/or study identifier. These identifiers enable the user to use software to rapidly download genome sequence reads (*see* **Note 8**).

2.3.1 Reference Sequence (RefSeq) Collection

The Reference Sequence (RefSeq) collection provides a comprehensive, integrated, nonreductant, well-annotated set of sequences, including genomic DNA, transcripts and proteins. They provide a stable reference for genome annotation, gene identification and characterization of mutations. RefSeq is a curated database of selected assembled genomes available in GenBank (National Institutes of Health (NIH) supported National Center for Biotechnology Information, NCBI). RefSeq transcript and protein records are generated by several processes, including manual curation; and propagation from annotated genomes that are submitted to members of the International Nucleotide Sequence Database Collaboration (INSDC). The main features of RefSeq include nonredundancy, explicitly linked nucleotide and protein sequences, updates to reflect current knowledge of sequence data and biology, data validation and format consistency. RefSeq is accessible via BLAST, Entrez and NCBI FTP site (*see* **Note 9**).

2.3.2 The Sequence Read Archive (SRA)

The sequence read archive (SRA) is the NIH primary archive of high-throughput sequencing data and part of the INSDC, which includes SRA along with the European Bioinformatics Institute (EBI) and the DNA Database of Japan (DDBJ). Data submitted to any of the three organizations are shared. SRA receives, processes and archives millions of sequencing files and currently stores over 10 petabases (10×10^{15}) of open access data that continues to expand exponentially. The aim of SRA is to archive raw sequencing data and alignment information from high-throughput sequencing platforms and makes sequence data available to the research community to enhance reproducibility and allow for new discoveries by comparing datasets (*see* **Note 10**).

2.4 Quality Control

2.4.1 Software Required

1. FastQC v0.11.8 (https://github.com/s-andrews/FastQC).
2. Trimmomatic v0.39 (https://github.com/timflutre/trimmomatic).
3. FastP v0.20.0 (https://github.com/OpenGene/fastp).
4. Kraken2 v2.0.8-beta (https://github.com/DerrickWood/kraken2).

2.4.2 Inputs Required

Sequence reads for each isolate (generally in fastq format). Can be compressed (such as .gz format).

2.5 Genome Assembly

2.5.1 Software Required

1. Spades v3.13.0 (https://github.com/ablab/spades).
2. Pilon v1.23 (https://github.com/braodinsitures/pilon).
3. Bandage (https://github.com/rrwick/Bandage).
4. Unicycler v0.4.8 (https://github.com/rrwick/Unicycler).
5. Poretools (https://github.com/arq5x/poretools).
6. Deepbinner (https://github.com/rrwick/Deepbinner).

2.5.2 Inputs Required

Quality controlled screened sequence reads for each isolate (generally in fastq format).

2.6 Genome Annotation and Curation

2.6.1 Software Required

1. prokka v1.13.7 (https://github.com/tseemann/prokka).
2. mlst v2.16.2 (https://github.com/tseemann/mlst).
3. emm-typing-tool (https://github.com/phe-bioinformatics/emm-typing-tool).
4. PHASTER (http://phaster.ca/).
5. ICEberg v2.0 (http://db-mml.sjtu.edu.cn/ICEberg/).
6. ABRicate v0.8.13 (https://github.com/tseemann/abricate).
7. ARIBA v2.14.1 (https://github.com/sanger-pathogens/ariba).
8. Artemis or another genome visualization/editing tool (https://github.com/sanger-pathogens/Artemis).

2.6.2 Inputs Required

1. Genome assembly (fasta format).
2. Sequence reads (fastq format).

2.7 Pangenome and Accessory Genome

2.7.1 Software Required

1. prokka v1.13.7 (https://github.com/tseemann/prokka).
2. roary v3.12.0 (https://github.com/sanger-pathogens/Roary).
3. scoary v1.6.16 (https://github.com/AdmiralenOla/Scoary).
4. piggy v1.0 (https://github.com/harry-thorpe/piggy).

2.7.2 Input Required

The recommended input for Scoary, Roary and Piggy is the GFF format output produced using Prokka v1.13.7. Each of these tools requires greater than three genomes and is typically run on tens to hundreds of genomes in a single run.

2.8 Phylogenetic Analysis

2.8.1 Software Required

1. snippy v4.4.1 (https://github.com/tseemann/snippy).
2. gubbins v2.3.4 (https://github.com/sanger-pathogens/gubbins).
3. snp-dists v1.0 (https://github.com/tseemann/snp-dists).
4. snp-sites v1.0 (https://github.com/sanger-pathogens/snp-sites).
5. artemis (or genome visualization tool) (https://github.com/sanger-pathogens/Artemis).
6. IQ-TREE v1.6.11 (https://github.com/Cibiv/IQ-TREE).

2.8.2 Inputs Required

1. Reference Genome (ideally a closely related closed reference genome)
2. Read sets (reads in fastq format for each isolate to be included in the phylogenetic tree).

2.9 Population Structure

2.9.1 Software Required

1. mash v2.1.1 (https://github.com/marbl/Mash).
2. mashtree v0.56 (https://github.com/lskatz/mashtree).
3. PopPUNK (https://github.com/johnlees/PopPUNK).

2.9.2 Inputs Required

1. Genome assembly (fasta format).
2. Preexisting global cluster database (PopPUNK optional).

3 Methods

Command-line examples provided are written in generic form where "Strain_A" is a general term used for the name of the "identifier" associated with the genome sequence of the strain.

3.1 Accessing Public Genome Sequence Data

While a number of users will generate their own genome sequence data (fastq format) from a service provider, collaborator, or in-house sequencing facility, many study designs will often include the need to download raw genome sequence data from public databases.

Downloading short reads can be done from multiple sources and many different tools. One of the simplest ways is to download a collection of reads based on a Bioproject accession ID from the European Nucleotide Archive (ENA); ENA can be searched by Bioproject ID or Sample ID and using the ena-ftp-downloader in a GUI, where a list of Biosample and SRR accession numbers are listed. Reads based on these identifiers can then be selected and downloaded for a command-line option to downloading genomes from GenBank (*see* **Note 11**).

3.2 Quality Control (QC) of Genome Sequence Data

3.2.1 Check Reads

FastQC is a tool that can import sequence reads in fastq format and provide a quick overview so that problematic areas can be investigated.

fastqc Strain_A_1.fq.gz Strain_A_2.fq.gz -o Strain_A_fastqc_summary_out -t 18

3.2.2 Investigate Read Statistics

FastQC produces multiple summary graphs and tables with statistics on the reads, including base and sequence quality scores, GC content, N content, flag sequence duplication and overrepresented sequences and providing information on adapter and k-mer content. FastQC can be used to assess whether re-sequencing of the sample needs to occur, or if the reads are sufficient quality to continue with downstream analysis.

3.2.3 Clean Reads

Cleaning reads involve the removal of commercial adaptor sequences, low-quality bases and low-quality reads from the set of reads belonging to a single sample. Examples of tools that can perform this task is Trimmomatic and FastP, which cut adaptor and other specific sequence from fastq reads.

trimmomatic-0.39.jar PE –threads 8 --phred33 Strain_A_1.fq.gz Strain_A_2.fq.gz Strain_A_1P.fq.gz Strain_A_1U.fq.gz Strain_A_2P.fq.gz Strain_A_2U.fq.gz

--phred specifies the base quality encoding. Output files 1P = paired forward reads, 2P = paired reverse reads, 1U = unpaired forward reads, 2U = unpaired reverse reads.

fastp -i Strain_A_1.fq.gz -I Strain_A_2.fq.gz -o Strain_A_output_1.fq.gz -O Strain_A_output_2.fq.gz

Inputs for paired-end compressed reads are flagged by the -i and -I and outputs are designated by the -o -O flags.

3.2.4 Screen for Contamination

The simplest way to remove contaminating reads is to use a k -mer-based analysis against a well-curated genome classification database. Kraken2 [10] is a taxonomic classification system using exact k-mer matches to achieve high accuracy and fast classification speeds the classifier matches for a k-mer within a query sequence to the lowest common ancestor of all genomes containing the giving k-mer. Kraken2 is a command-line tool that can be run as below on raw sequence reads in fastq format (*see* **Note 12**).

kraken2 --db database_genomes --threads 8 --output Strain_A_output --use-names --paired --gzip-compressed Strain_A_1.fq.gz Strain_A_2.fq.gz

Both classified and unclassified reads can be sent to another file for later processing, using the --classified-out and --unclassified-out options. Kraken2 can also be run on fasta files:

kraken2 --db database_genomes --threads 8 --output Strain_A_output --use-names Strain_A.fasta

3.2.5 Investigation of Contamination

Examine output for contaminated reads/sequence:

1. Percentage of fragments covered by the clade rooted at this taxon.
2. Number of fragments covered by the clade rooted at this taxon.
3. Number of fragments assigned directly to this taxon.
4. A rank code, indicating (U)nclassified, (R)oot, (D)omain, (K)ingdom, (P)hylum, (C)lass, (O)rder, (F)amily, (G)enus, or (S)pecies. Taxa that are not at any of these 10 ranks have a rank code that is formed by using the rank code of the closest ancestor with a number indicating the distance from that rank. For example, "G2" is a rank code indicating a taxon is between genus and species and the grandparent taxon is at the genus rank.
5. NCBI taxonomic ID number.
6. Indented scientific name.

3.2.6 Other QC Metrics

There are multiple other metrics used for QC of a sample, such as being within the expected genome size (after genome assembly), having appropriate number of coding sequences (after genome annotation), correct number and sequence of 16s rRNA operons and the presence of species-specific markers. For example, *emm* and the MLST genes in *Streptococcus pyogenes* (*see* Subheading 3.4).

3.3 Building Draft and Reference Genomes by Genome Assembly

Many tools can be used to investigate the genome sequence of an isolate; however, many rely on a genome assembly where reads are "assembled" into contiguous sequences. There are diverse tools for genome assembly; here, we discuss two methods: assembly from short-read-only data and assembly from hybrid data (long-read and short-read data). The outcomes from genome assembly are either a draft genome or a complete genome. For *S. pyogenes*, a complete genome is represented by a single chromosome that is circular and potentially a single circular plasmid. Any genome assembly that is not a single chromosome is referred to as a draft genome.

3.3.1 Draft Genome Assembly from Short-Read Sequences

Too high- and too low-read coverage can result in issues with assembly algorithms. Aim for ~100× estimated coverage of the genome (*see* **Note 13**). This can be done using seqtk on each read file where "100,000" reads are subsampled.

Subsample Reads

seqtk Strain_A_1.fastq.gz "100,000" | pigz > Strain_A_1.fq.gz

Assemble the Genome

There are many tools that can be used for assembly of short-read sequence data. Spades [11] is one of the most commonly used assemblers. Subsampled paired-end reads are used as an input into spades to produce contigs.

spades.py -1 Strain_A_1.fq.gz -2 Strain_A_2.fq.gz -t 18 -m 250 Strain_A_spades_assembly

Spades produces several outputs, a FASTA file of contigs and scaffolds, as well as an assembly graph, which can be used to visualize the graph to troubleshoot and improve the genome assembly.

Visualize the Assembly Graph (Optional)

Bandage is a tool used to visualize the assembly graph [12]. The assembly.gfa can be loaded into Bandage. This will visualize where the contradictions and assembly breaks in the genome occur. Bandage can also utilize BLAST, so the user can find a locus of interest, for example, plasmid or antimicrobial resistance sequences.

Polishing the Genome (Error Correction)

Pilon (version 1.23) is a software tool that takes a draft genome assembly (fasta format) and through mapping of the original sequence reads (fastq files), it corrects errors (SNPs) in the genome.

pilon --genome Strain_A_spades _assembly.fasta --bam StrainA_reads_mapped.bam --output StrainA_polished --outdir Strain_A_pilon_out --changes --vcf

The "--changes" (if specified) produces a file listing the changes in the output files and --vcf (if specified) will generate a vcf file of the changes.

Check Quality

You have now generated a high-quality "draft" genome. For *S. pyogenes* using paired-end short reads, we would expect the number of contigs in a high-quality genome assembly to be between 10 and 100 contigs. To get a lower number of contigs or to close a genome, long-read sequencing is required and a hybrid assembly should be used.

3.3.2 Reference Genome Assembly from Oxford Nanopore Long-Read and Short-Read Genome Assembly Using Unicycler [13]

When performing long-read sequencing (current long-read sequencing chemistry has a natively high error rate relative to current short-read chemistry), short-read sequencing from the exact same DNA preparation is required for optimal genome assembly with reduced errors. While different DNA preparations could be used, it is not recommended as mutations and genome rearrangements can occur between sample preparations.

Convert Nanopore fast5 into fastq Format (*See* Note 14)

poretools fastq –type best –min-length 1000 Strain_A.fast5 > Strain_A_nanopore_reads.fastq

Assemble the Genome

unicycler -1 Strain_A_1.fq.gz -2 Strain_A_2.fq.gz -l Strain_A_nanopore_reads.fastq -o Strain_A_unicycler_hybrid_output -t 26 --keep 3 --vcf

The -1 and -2 options are the short-read inputs and the -l is the long-read input -o is the output directory and --keep 3 will keep all intermediate files and --vcf will produce a vcf output from reads mapped back to the genome to look at heterozygous sites. There are also multiple modes for the assembler to bridge contig breaks (--conservative, --normal, --bold). Unicycler [13] also has the pilon polishing step (see Subheading 3.4.1) as part of the software. This is used as default -but can be ignored using the --no_pilon flag (not recommended).

Visualize the Assembly Graph (Optional)

As discussed for short-read draft genome assemblies, the assembly.gfa can be loaded into Bandage [12] to inspect the Unicycler assembly graph.

3.4 Annotation and Curation of Genome Assemblies

3.4.1 Gene and Feature Annotation

QC Genome Assembly

An appropriately assembled genome can be a single contig (for a long-read assembly) or multiple contigs (ideally <100 contigs) in fasta format. Check genome statistics to see if it is the appropriate length and GC content. For *S. pyogenes*, a genome size of 1.6–2 Mbp and a GC content of ~38% is expected.

Genome Annotation

Annotate the genome (fasta format) using Prokka.

prokka Strain_A.fa --outdir Strain_A_annotation –prefix Strain_A --locustag Strain_A --compliant --addgenes --genus Streptococcus --species pyogenes --strain Strain_A --cpus 16

The example run above will take a fasta file "Strain_A.fasta" and annotate the contigs and put all outputs into a directory "Strain_A_annotation", all files will be prefixed with "Strain_A", and annotation will be run compliant to NCBI's rules of annotations --compliant. Gene names will be included in the annotation using --addgenes and the organisms' details are included using the --genus "Streptococcus", --species "pyogenes" and --strain "Strain_A" flags. This will output multiple files including standardized output files such as GenBank and GFF3 formats, which will be utilized in downstream analysis.

Check for Errors

Enter the prokka output folder and examine the error file (.err). This will contain any errors and misannotated genes that will need to be manually investigated.

Check Annotation Statistics

Check the genome statistics file (.txt) and compare to the original genome assembly. Running –compliant forces the annotation of a genome to comply with GenBank/ENA/DDJB standard for genome assemblies, including no contigs smaller than 200 bp in length and annotation of both genes and CDS.

Relocate GenBank and FASTA for Further Analysis

Copy the GenBank (.gbk) the GFF3 Format (.gff) and nucleotide FASTA (.fna) from prokka output into a new directory as they will be used for further genome characterization of the genome including MLST and *emm* typing and investigation of mobile genetic elements such bacteriophage and integrative conjugative elements.

3.4.2 Assign Multilocus Sequence Type (MLST) Alleles

Within the directory, use mlst to scan contigs and assign MLST alleles based on pubMLST database.

mlst Strain_A.gbk --db Streptococcus pyogenes --threads 8 > Strain_A_mlst.tab

This will produce a tab-delimited output Strain_A_mlst.tab with the isolate name, the MLST scheme used and the assigned MLST profile, followed the allele assignment for each of the seven alleles. A tilde symbol will be used for alleles that have a similar, but not exact, database match. These should be investigated and submitted to database curators for assignment.

3.4.3 Assign emm Type

The *emm*-typing-tool [14] uses sequence reads to assign *emm* type (*see* **Note 15**).

emm_typing.py -1 Strain_A_1.fq.gz -2 Strain_A_2.fq.gz -o Strain_A_emm_output

The *emm*-typing-tool produces multiple output files, which include the following:

1. EMM_log.txt—contains variants for allele sequences and coverage statistics.
2. Result_XML file—contains the ID; NGS sample ID, Version, Final_EMM-type, EMM validated (*emm*1-124) and EMM-Non-validated (*emm*125+).

3.4.4 Screen for Bacteriophage

The same GenBank file can be used to characterize the bacteriophage using the webtool PHASTER [15]. PHASTER is used via a web interface where the user uploads a FASTA format file or a GenBank format file. PHASTER produces summary table of all predicted bacteriophage sequences with a predicted score. PHASTER also produces a detailed file for each of the predicted regions. An important part of the PHASTER output is the region position, which provides the nucleotide coordinates corresponding to the location of the putative bacteriophage within the query GenBank file. Take note that PHASTER attempts to identify possible chromosomal prophage attachment (*att*) sites when defining prophage boundaries (GAS prophage are generally no larger than 45kbp). Using the coordinates of the predicted bacteriophage, these can be input as a new feature into the GFF3 (.gff) format annotation file from prokka using Artemis or another genome visualization/editing tool.

3.4.5 Screen for Integrative Conjugative Elements (ICE)

The same GenBank file can be used to identify ICE, or transposon-like elements using the web-based tool ICEfinder and ICEberg2 [16]. The outputs of ICEfinder are similar to what is observed in PHASTER, where the length and location of the predicted ICE are characterized.

Using the coordinates of the predicted ICE, these can be input as a new feature into the GFF3 (.gff) format annotation file from prokka using Artemis or another genome visualization/editing tool.

3.4.6 Screen for Known Virulence Factors

There are multiple tools that can be used to screen for known virulence factors and antimicrobial resistance genes. One tool that can do both is ABRicate. The software accesses several databases that are regularly being updated: Resfinder, CARD, ARG-ANNOT, NCBI BARRGD, NCBI, EcOH, PlasmidFinder, Ecoli_VF and VFDB.

abricate Strain_A.fna --db vfdb --mincov 70 --minid 70 > Strain_A_virulome.tab.

The example above screens the virulence factor database using a minimum identify and coverage of 70% to assign the presence of the gene in virulence database.

If multiple isolates are screened using the same threshold against the same database, they can be summarized into a single table using the abricate –summary option, where all the .tab outputs are located within the same directory.

*abricate –summary *.tab > virulome_summary.tab*

3.4.7 Screen for Known Antimicrobial Resistance Genes

1. ABRicate can be used to screen for the presence of antimicrobial resistance genes.

   abricate Strain_A.fna --db ncbi --mincov 90 --minid 90 > Strain_A_resistome.tab

   The example above screens the NCBI database of antimicrobial resistance genes using a minimum identity and coverage length of 90% to assign the presence of the gene in virulence database. If multiple isolates are screened using the same threshold against the same database, they can be summarized into a single table using the abricate --summary option where all the .tab outputs are located within the same directory.

2. ARIBA [17] utilizes reads to assign the presence of antimicrobial resistance genes including resistance produced due to point mutations in conserved genes. ARIBA can also be used for assigning MLST alleles.

   ariba run reference_database Strain_A_1.fq Strain_A_2.fq Strain_A_ariba_output_dir

You *should* now have a list of specific outputs for each isolate. This information includes an annotated genome in multiple formats, MLST and *emm* type designation, location of bacteriophage, ICE and a profile of virulence factors and antimicrobial resistance.

3.4.8 BLAST Screening of Genomes for Genes of Interest

BLAST-based tools are the most commonly used approach for screening genome assemblies. In many instances, a user may want to screen for genes that are not represented in public databases. We have recently released a BLAST-based script for screening genome assemblies with a user-supplied set of genes [4]. The user can make their own query set of genes (in single or multifasta format) and a set of genome assemblies (e.g., one or more whole genomes).

screen_assembly3.py -q GeneA.fa -i folder_of_genomes -t 16 -p 95 -l 95

The example above uses the tool screen_assembly3.py [4] where a folder containing fasta files of genomes is being screened using a fasta file containing a nucleotide sequence(s) for a query gene(s) "GeneA". For hits to be reported, a minimum of 95% gene identify (−p 95) and 95% of the gene length (−l 95) is required. This tool outputs any hits within the specified thresholds and produced sequence alignments, which can be used to look at genetic variants of the gene and produce phylogenetic trees.

3.5 Defining the Pangenome and Investigating Accessory Genome Variation in a Population

3.5.1 Check Timestamps of Prokka GFF3 Files

First you need to check the GFF3 format (.gff) files produced from prokka. This is important as mixing isolates run at different times could result in unexpected errors due to updates or changes in the database used for the annotation. This may require reannotating genomes (*see* Subheading 3.5).

3.5.2 Copy GFF3 (.gff) Files to New Directory

Roary, Scoary and Piggy each require a single directory containing a single GFF3 file for each isolate to be included.

3.5.3 Calculate the Pangenome

Roary is a high-speed pangenome pipeline that calculates the pangenome and is saleable to thousands of samples. Roary clusters all the coding regions based on specified thresholds and assigns the presence and absence of each cluster to an isolate. Roary requires some thresholds to be set. Including the minimum identify for a BlastP hit (-I default 95%) and whether or not gene synteny or gene order should be accounted for when assigning a sequence to a cluster (−s) which in turn can have a large effect on the total number of genes in the pangenome.

*roary *.gff -p 16 -f roary_i90_s_output -I 90 -s -r -y*

The example command above is run on all .gff files in the directory from which the command is run (∗.gff), the blastp identity (−i) to assign clusters is set at 90% and gene synteny is not account for (−s) -r produces R plots of the pangenome statistics and -y adds the genes inference data to the outputs (*see* **Note 16**).

roary ∗*.gff -p 16 -f roary_i90_s_output -i 90 -e -n*

3.5.4 Pangenome Summary

One of the outputs from Roary is a text file summary_stastistics.txt. The file shows the number of genes in the core and accessory as an overview and how frequently they occur in the input isolates. If the number of core genes is 0 or the total number of genes is very high, then it is likely that there is contamination.

3.5.5 Investigate the Presence and Absence of Genes

One of the most useful outputs from Roary is the gene_presence_absence.csv that lists each gene and which samples it is in. Each gene has a unique identifier for the protein but also has the locus_-tag identifier for each isolate (if annotated with prokka typically, the locus tag will be Strain_name_gene_number, e.g., StrainA_1010) (*see* **Note 17**).

3.5.6 Pangenome-Wide Association

Scoary is the microbial pangenome-wide association tool [18]. Scoary is designed to take the gene_presence_absence.csv from Roary as well as traits file created by the user to calculate the associations between all genes in the pangenome and the traits. It reports a list of genes sorted by strength of association per trait. The trait's file is a binary (1 or 0) .csv file where columns are the traits, for example, invasive, pharyngitis, skin sores and row names are the Strain names (that need to match the corresponding file used in Roary) and 1 is where the trait is observed and 0 is where the trait is absent.

scoary -g gene_presence_absence.csv -t traits.csv -o scoary_output_directory –threads 8

Scoary outputs a single .csv file per trait investigated from the traits file. The results consist of genes that were found to be associated with the trait, sorted according to significance. By default, Scoary reports all genes with a naive *p*-value <0.05, but the user can change the cutoff value and use adjusted *p*-values instead (*see* **Note 18**).

3.5.7 Intergenic Region Investigation (IGR)

Piggy is a tool used for analyzing the intergenic component of bacterial genomes [19]. Like Roary, Piggy requires bacterial genome assemblies in GFF3 format (ideally produced from Prokka). Piggy also requires the output folder from Roary and is recommended to have run Roary with the -s option.

piggy --indir ∗*.gff --outdir piggy_output --roary_dir roary_i90_s_output*

Piggy produces a number of output files including an output similar to the presence/absence matrix produced by Roary

IGR_presence_absence.csv - An IGR presence/absence matrix with the same structure as that produced by Roary. The IGRs in this file are named according to the genome and gene neighborhood of the IGR. Piggy also provides information on whether the IGR have switched or rearranged to different part of the genome potentially playing a role in bacterial metabolism or virulence (*see* **Note 19**). From these analyses, a user should have a comprehensive set of outputs to investigate the genomic relatedness of the pangenome of a set of isolates. This will provide a deep understanding of the current status of a set of isolates, such as changes in the accessory genome content of perhaps selective pressures and genome rearrangements. Combined with phylogenomic and ML trees, this forms the basis for strain investigations and genome comparisons.

3.6 Phylogenetic Analysis

3.6.1 Select a Reference Genome

Choose an appropriate reference genome for evolutionary relationships (*see* **Note 20**).

Haploid Variant Calling

There are many tools that can be used for investigating and calling SNPs in bacterial genomes. One of the most widely used is Snippy- - A rapid haploid variant calling and core genome alignment tool. Snippy finds SNPs between a haploid reference genome and the sequence reads of a sample. It will find both substitutions (SNPs) and insertions/deletions. Snippy produces a consistent set of output files in a single folder. Snippy requires a reference genome in fasta, EMBL or GenBank format and sequence read files (in fastq format). This step will need to be repeated for each isolated to be included in the phylogenetic tree.

snippy --cpus 16 --outdir Strain_A_SNPs --reference Reference_genome.gbk --R1 Strain_A_R1.fq.gz --R2 Strain_A_R2.fq.gz

Calculating the Core Genome

If SNPs are called from multiple isolates against the same reference strain, Snippy can produce an alignment of "core SNPs", which can be used to build a phylogenetic tree. For example, running snippy-core on three isolates Strain_A, Strain_B, and Strain_C would include the following:

snippy-core --ref Reference_genome.gbk Strain_A_SNPs/ Strain_B_SNPs/ Strain_C_SNPs/

Snippy-core produces several outputs including a core SNP alignment (.aln), a whole-genome SNP alignment (.full.aln) that includes invariant sites and multiple .txt files. SNPs residing within phage and ICE sequences are normally excluded from phylogenetic analyses as these can be masked from the alignment at this stage (*see* **Note 21**).

Make a Quick Tree

To speed up recombination detection, a tree is required as a prior. We use the -fast option within iqtree to produce a rapid tree. This is a good indicator of preliminary tree structure but should not be used as a final tree. This is also an opportunity to examine if major outliers are present in the dataset (*see* **Note 22**).

iqtree -s core.aln -m GTR + F + G4 -fast > quick_core.tree

Removal of Recombination

One genome alignment recombination detection tool, Gubbins, takes the core.full.aln from snippy and the quick_core.tree from the fast run of IQ-TREE and calculates SNP densities in a sliding window to estimate whether SNPs are within the clonal frame or recombinogenic (*see* **Note 23**).

run_gubbins.py core.full.aln -s quick_core.tree -m 5 -c 24 -r GTRGAMMA -p gubbins_screen

Gubbins produces multiple outputs including EMBL and GFF3 format recombination predictions that can be loaded into Artemis or Phandango for visualization and the snp-distribution summaries in .csv and .vcf format (*see* **Note 24**).

The output used for further phylogenetic inference is gubbins_screen.filtered.polymorphic.sites.fasta an alignment of the filtered polymorphic sites.

Remove Ambiguous Characters

The gubbins_screen.filtered.polymorphic.sites.fasta includes some ambiguous characters in the sequence alignment. These are uninformative for producing ML phylogenies and can be removed using snp-sites [20].

snp-sites gubbins_screen.filtered.polymorphic.sites.fasta -c -o final_core.aln

The output from snp-sites final_core.aln is the final core alignment input used for producing ML phylogenies.

Calculating the Proportion of SNPs That Are Influenced by Recombination

This will be done by running Snp-dists on two files - the alignment before and the one after recombination detection.

snp-dists final_core.aln > final_core.snp-dists.tsv

This is the pairwise SNP calculations for all strain comparisons with recombinogenic SNP sites removed, so we also need to investigate the pairwise SNPs in the original alignment (before recombination was removed).

snp-dists core.aln > core.snp-dists.tsv

The pairwise comparisons can be compared to see the influence of recombination in each strain and how it compares to all other strains.

Examine Recombinogenic Regions

Recombinogenic regions can be investigated by overlaying some of the Gubbins outputs over the annotated reference genome. Artemis is a genome visualization tool where the Gubbins .gff and .embl outputs can be loaded over the .gff or .gbk of the reference genome. This will highlight all SNP regions or recombinogenic blocks flagged as potential recombinogenic regions relative to the annotated reference genome. This is useful in examining which genes and sets of genes may have arisen through a horizontal gene transfer event or a genome rearrangement (*see* **Note 24**).

Calculate Constant Sites

Producing a phylogenetic tree using only the variant sites results in significant reductions in computing time; however, branch lengths are not a true representation of the population structure. IQtree can account for this by including the constant sites -fconts. To calculate the constant sites from a "core.full.aln", the following is performed:

snp-sites -C core.full.aln

The outputs from snp-sites are put into -fconts parameter in IQtree and alternatively, this can be manually calculated by examining GC% content and averaging over the genome size.

Produce Maximum-Likelihood (ML) Phylogeny

IQ-TREE can be used to produce a ML phylogeny [21]. The core alignment with recombinogenic SNPs removed is used as the alignment input (−s), the constant sites -fconst as defined above are also included. The evolutionary model used (in most cases) for *Streptococcus pyogenes* nucleotide alignments is the General-Time-Reversible gamma 4 model (GTR+F+G4). 1000 bootstraps (−bb 1000) and approximate likelihood ratio test (−alrt 1000) are performed to provide support for all tree branches.

iqtree -s final_core.aln -m GTR+F+G4 -fconst 558000, 342000, 342000, 558000 -bb 1000 -alrt 1000 -v -nt 16

From IQ-TREE, several outputs are produced. The final_core.aln.iqtree file is one of the log files with all the parameters and options included. The .iqtree file also provides the summary statistics for the ML phylogeny. Two tree files are produced: the .contree and the .treefile. The treefile is the ML phylogeny (the best tree that was found). The .contree is the consensus tree (average of all the possible trees, not the best tree).

Visualize the Tree

Import the final_core.treefile into tree visualization software (such as iTOL - Interactive Tree of Life, Figtree, or ggtree in R). From this protocol, the user should have a high-quality ML tree that conforms to the clonal frame from the most recent common ancestor.

3.7 Analysis of Population Structure

In order to examine the global relationships of GAS genomes, k -mer-based analysis can be used to provide a fast determination of the population structure. Example tools for this include mash [22], mashtree and PopPUNK [9]. In the case for mash and mashtree, these tools provide a rapid clustering to a set of isolates, which can be used to determine clusters of relatedness (such as determining which isolates should be investigated further as potential transmission or outbreaks from outliers or background isolates).

3.7.1 Mash and Mashtree

Mash is the fast genome and metagenome distance estimation using minhash [22]. Mash extends the MinHash dimensionality-reduction technique to include a pairwise mutation distance and *P* value significance test, enabling the efficient clustering and search of massive sequence collections. Mash reduces large sequences and sets into small representative sketches, from which divergence or variance between samples can be estimated. Mash takes a set of reads or fasta files and estimates distances between them and stores the distances in files called sketches. Mash involves a multiple step process.

1. *mash sketch StrainA.fasta StrainB.fasta*
2. *mash paste StrainB.msh > StrainB .mash* * repeat with Strain C
3. *mash dist -t StrainB.mash StrainC. mash > output_table_mash_distances.*

This will output a pairwise table of Strain A, B and C, where Strain A is the reference strain and B and C are sketched against A. Each of these steps are very fast and this process can be applied to thousands of genomes with very low computational capacity (*see* **Note 25**). Mashtree is used to get an estimate of the clustering observed in a set of isolates. While Mashtree produces a tree, it is not a phylogenetic tree, but rather a clustering of isolates based on mash distances. Mashtree like mash can take multiple inputs including reads, fasta and GenBank or precomputed mash distances.

mashtree StrainA.fasta StrainB.fasta StrainC.fasta --genome-size 1800000 --mindepth 0 --mincpus 16 > mashtree.dnd

This will produce a "tree" from three isolates. Where the genome size of *S. pyogenes* is 1,800,000 bp in length, --mindepth 0 ignores rare k- mers and --mincpus is number of cpus to use for mashtree (*see* **Note 26**).

3.7.2 PopPUNK

PopPUNK is a k -mer-based genome assembly clustering tool [9]. This system calculates distances of both the core (common) and accessory (variable) genome of an assembly relative to a database of isolates. This database was established to be iterative in that new clusters can be assigned as they are identified and new strains are added to existing clusters. It is after this step that refined microevolution studies should be undertaken using (or by making) a closely related reference genome within that assigned cluster/lineage/phylogroup. PopPUNK requires a folder of isolates as the input file. PopPUNK will use mash to create mash sketches and calculate core and accessory distances between samples that is used to fit a two-component Gaussian mixture model to the distances to find within-strain distances and then construct a network to define clusters. This is an effective way to examine the population structure of a large set of isolates.

poppunk --easy-run --r-files Strep_reference_list.txt --output Strep_output --threads 8 --plot-fit 5 --min-k 13 --full-db

Other flags can be added to PopPUNK that can be used to create output files for visualization, including --microreact, --cytoscape, --phandango, and --grapetree (*see* **Note 27**). To assign new GAS genomes to the existing GAS database [4], PopPUNK will use the existing database (comprising at least 360 representative genomes) and assign to a preexisting cluster or provide a "new" cluster identifier:

1. Download and extract the representative *S. pyogenes* cluster database (https://figshare.com/articles/Streptococcus_pyogenes_PopPUNK_database/6931439/3):
   tar xf GAS_query_db.tar.bz2
2. Assign new genomes (where query.list is a list of the assemblies to assign):
   poppunk --assign-query --ref-db GAS_query --distances GAS_query/GAS_query.dists --model-dir GAS_query --q-files query.list --output GAS_assignment --threads 8

4 Notes

1. There are multiple plug-and-play options for whole-genome bioinformatic analyses. These packages often contain a suite of tools many of which are discussed in this chapter. These plug-and-play options typically have a major limiting factor; the computational requirements. Command-line tools reduce computational requirements that can be taken up by using graphic user interfaces (GUIs). Most plug-and-play GUIs are limited to analyzing tens of genomes.
2. Not all bioinformatic analysis is equal. Some tools that work effectively for bacterial species such as *E. coli* may not be appropriate for *S. pyogenes.* It is important to understand the biology of the pathogen, in particular, the genome structure, so the correct tools can be used.
3. If a tool/software is published in a research article, citations can be an indication of its wider applicability. Typically, if a tool is highly citied, it is likely to be easy to install and will create reproducible outputs. If a publication has a low citation rate and has been published for some time (a few years) without updates, this can be used as an indication of a problematic software tool.
4. Some tools are not published but typically are hosted on websites such as GitHub. A tool can be evaluated for its effectiveness by the number of "downloads", "updates", and "commits". Be wary of tools that have a low number of any of the above metrics.

5. While long-read sequencing has dramatically improved genome assemblies, there is still some challenges. In particular, the presence of isoforms (structural variants of the bacterial chromosome) where bacterial strain can be switching between two different structural states. As a colony (which is used for sequencing) contains millions of cells, one chromosomal structure is difficult to resolve. Furthermore, the sequencing error rates associated with some long-read platforms are generally higher than in short-read technologies. For accuracy, we recommend a combination of long- and short-read sequencing approaches for reference genome generation.
6. Prokka: rapid prokaryotic genome annotation, at the time of writing this has received 3080 citations. Another tool used frequently of annotation is the RAST server: rapid annotations subsystems technology, which has received 6340 citations. For RAST, users submit a genome to an online server for annotation, while Prokka can be done in the command-line in a secure server or computer. The problem with both of these approaches is neither are currently an accepted form for integration into the central reference genome portals of NCBI and RefSeq, that uses its own annotation pipeline (such as PGAP—Prokaryotic Genome Annotation Pipeline) with internal metrics for consistency. Irrespective of the annotation pipeline used, a user should always be wary of annotation terminology as they are inferred by homology and/or related protein domain structure. Chains of annotation can lead to functional associations that require laboratory confirmation and as such are often annotated with vague terms such as "hypothetical", "putative", or "probable" gene products.
7. ClonalFrameML and Gubbins determine recombinogenic lineages by investigating relative SNP distribution across a genome alignment, where unusually high clusters of SNPs within a sliding window (of defined length) are flagged as regions potentially arisen through recombination. These tools are best suited to microevolution studies of evolutionary related strains (less than a couple of thousand SNPs) and not for single gene(s) analyses. FastGEAR uses a Hidden Markov Model (HMM) approach and is performed on a gene-by-gene basis. The relationships of individual genes are compared to a ML phylogeny to examine whether the genes conform to the phylogeny or are distinct. This enables ancestry of each gene to be examined and whether recombination of the gene has occurred between different phylogenetic lineages.
8. Downloading sequence reads can be time and space (storage) consuming. Paired-end *S. pyogenes* reads for a single isolate (~1.8 million base pairs) at an estimated 100× coverage are approximately 1 GB in size. A balance between capacity and usability is a constant challenge.

9. RefSeq stores validated closed genomes for each species. It is important to note that not all closed reference genomes are of the same quality. Originally, it was very time consuming, expensive and required a lot of people to close a genome to a high standard, with a significant amount of validation. We are now able to close a genome rapidly using only bioinformatic tools with long- and short-read technology; however, this speed does not always mean accurate genomes. We are still unsure about bias in genome assemblies using different tools and/or sequence technology, as these have not been benchmarked for all species. This is an ongoing research area.
10. It is highly recommended to start working from raw sequence reads (fastq files) when conducting a genomic-based study. Reads are the raw sequence data containing base quality score information. Different bioinformatic tools (and different versions therein) apply different rules when interpreting this information, which is lost after processing the data into assemblies/alignments. Users should, where possible, base the analysis from raw data rather than preinterpreted data (contigs/assemblies). A user also can screen for contamination (*see* Subheading 3.1), which may have been unknown to the submitter of the sequence.
11. ena-download-project.pl is one way to download a lot of genomes from the same resource. There are similar tools for NCBI https://github.com/kblin/ncbi-genome-download where all isolates from the same species can be downloaded from NCBI. For example, all complete genomes of *S. pyogenes.*
12. Screening for contamination is the most important step in any sequence analysis. It is important to remember that laboratory reagents are often contaminated, so it is likely that the final sequence reads are also contaminated. Consequently, best practice is to control and remove any contamination within a sample, thus enabling accurate inferences from sequence data and downstream analyses.
13. "Coverage" of a genome refers to the sequence depth or how many times each base pair of a genome has been sequenced. If a genome is sequenced to an estimated 100× coverage, that means that the genome has been sequenced 100 times. As a rule of thumb, a minimum coverage of 30× is required for accurate base pair assignment. For genome assembly, too little or much data for an assembly algorithm can lead to errors. It is best practice to "subsample" the sequence reads to a defined estimated coverage (e.g., 100×) to ensure reproducible interpretation.
14. Extensive QC is required on long-read data for the best genome assemblies, while not mentioned in this protocol, deepbinner, minionQC and pycoQC are all valuable tools for investigating the QC of nanopore long-reads.

15. The *emm* type database housed at the CDC is constantly being updated as more strains from around the globe are being sequenced. The accuracy of the *emm* database is being refined to remove M-like *enn* and *mrp* genes (also located in the Mga regulon) that can confound *emm* sequence type designations. It should also be noted that some GAS strains can harbor more the one "*emm*" gene and the Mga regulon itself is evolutionarily dynamic and can recombine. Alternative assembly-based *emm* typing tools are available (e.g., https://github.com/sanger-pathogens/mlst_check/blob/master/bin/get_emm_sequence_type), yet they have not been benchmarked for accuracy.
16. Roary can also be used to create multifasta alignments of core genes. This can be done by adding the -e and -n flags.
17. Roary includes tRNAs as part of the pangenome calculations. tRNAs are not genes and should not be included. The user can interrogate the .csv to calculate how many tRNAs are present in the pangenome. Within the .csv file, roary also has a QC column. Users should look at the column to screen for genes that are marked as "investigate" – these are genes that could not be assigned to a cluster (and are likely pseudogenes (gene fragments) or due to sequence artefacts). If there are too many of these in the outputs, the users should re-evaluate the QC metrics that were used, as a low-quality genome/genome annotation was likely used in the analysis.
18. Scoary can also be used to do association studies with SNPs using a VCF file as an input: vcf2scoary.
19. The outputs from Piggy can also be added to Scoary to look at genome-wide associations of intergenic regions.
20. The choice of reference is an important step. Not only is a complete high-quality reference genome preferred, but also the reference should be as closely related to the set of isolates that are being compared. The more related the reference is to the population being analyzed, the higher the resolution and the higher accuracy of the tree to dissect recent evolutionary relationships. As cost-effective long-read sequencing is available, best practice involves generating a reference strain from one of the isolates in the sample set.
21. Snippy can also mask certain regions of the genome by providing a BED file with the *--mask* parameter. Any SNPs in those regions will be excluded. This is common for masking phage regions and known recombinogenic hotspots.
22. Not all phylogenetic trees are equal. In many cases, speed and accuracy are inversely related. While the –fast option in IQtree is a way to quickly test for outlier isolates (to be removed for further analysis), it should not replace an accurate run. ML

algorithms require a model of nucleotide evolution. The choice of nucleotide substitution models for bacterial phylogenetics is a chapter in its own right. Generally speaking, the GTR+F+G4 has been used in a number of GAS phylogenetic studies. The General Time Reversible (GTR) model accounts for unequal rates of base substitutions and base frequency (as nucleotide transitions are more common than transversions). "F" represents empirical base frequencies (indicating base frequencies are unequal—often matched to GC content). The discrete gamma model (G4) indicates that substitution rates are estimated across four nucleotide sites and that they have an equal probability. IQtree also has a model finding option to find which model best fits the data (-m).

23. Gubbins cannot distinguish elevated densities of SNPs arising through recombination or from other processes such as selection pressure. Therefore, it is important that high-quality sequences are used and mutational hotspots are known. This is also highly dependent on the reference genome used. A highly divergent reference genome will result in a larger amount of recombinogenic regions being flagged. The -m option in Gubbins is the minimum number of SNPs in a sliding window for a region to be classified as a recombination (default is 3, but for *S. pyogenes*, a value of 5 is recommended).

24. The .per_brach_statistics.csv output has a lot of information that can be interrogated further. Where each recombinogenic region and SNP site is shown in terms of length and how many isolates carry those events. The node labels in this output correspond to the node in the. final_tree.tree output file. The per_brach_statistics.csv files show the node position, total SNPs at that position, number of SNPs inside recombinations, number of SNPs outside recombinations, number of recombination blocks, bases in recombination, the recombination/mutations ratio (r/m), the ratio of recombination events to point mutations (rho/theta) and genome length that is maintained in the clonal frame (conserved since most recent common ancestor).

25. mash dist can be set to output a table file -t and thresholds of maximum p value -v and maximum distance -d can be used when attempting to assign clusters. To examine all relationships, leaving these default (1.0) will report all results/combinations/comparisons. However, if a user is only interested in isolates that closely match, then setting -d and -v can be used. This can be used for a rapid rule or rule out for specific isolates (say in a surveillance study searching for specific MLST or *emm* types).

26. You can get a more accurate tree with the minimum abundance finder. Simply give *--mindepth 0*. This step helps ignore very unique kmers that are more likely read errors. Without this step, mashtree will run more rapidly but will produce less accurate clusters.

    Mashtree can add confidence values using jack knifing. For each jack knife tree, 50% of hashes are used. Confidence values are calculated from the jack knife trees using BioPerl.

    *mashtree_jacknife.pl --reps 100 --numcpus 12 *.fastq.gz --mindepth 0 > mashtree_jacknife.dnd*

    Boostrapping can also be performed using mashree_bootstrap

    *mashtree_bootstrap.pl --reps 100 --numcpus 12 *.fastq.gz --min depth 0 > mashtree.bootstrap.dnd*

    However, this is not a substitute for a phylogenetic tree, but rather this can be used as a first look into the structure of the isolates after QC, as a dataset may have unexpected outliers which may not want to be included. Mashtree gives a rapid method for detecting outliers before significant computational time is invested. Even if bootstrapping or jack-knifing is performed, it is still not a substitute for a ML or Bayesian phylogenetic analyses.

27. This is a simplified example. PopPUNK requires quite a few steps of validation and examining the data. There is an extensive documentation provided on the PopPUNK website which should be investigated if running PopPUNK. At the moment, new cluster identifiers are not automatically incorporated into the preexisting database.

Acknowledgments

This work was supported by NHMRC project grants (#1130455, #1165876 and #1098319). S.Y.C.T. is an Australian National Health and Medical Research Council (NHMRC) Career Development Fellow (#1145033). M.R.D is an University of Melbourne C.R. Roper Fellow.

References

1. Gardy JL, Loman NJ (2018) Towards a genomics-informed, real-time, global pathogen surveillance system. Nat Rev Genet 19 (1):9–20. https://doi.org/10.1038/nrg.2017.88
2. Klemm E, Dougan G (2016) Advances in understanding bacterial pathogenesis gained from whole-genome sequencing and phylogenetics. Cell Host Microbe 19(5):599–610. https://doi.org/10.1016/j.chom.2016.04.015
3. Bessen DE, Smeesters PR, Beall BW (2018) Molecular epidemiology, ecology, and evolution of group a streptococci. Microbiol Spectr 6(5). https://doi.org/10.1128/microbiolspec.CPP3-0009-2018
4. Davies MR, McIntyre L, Mutreja A et al (2019) Atlas of group a streptococcal vaccine

candidates compiled using large-scale comparative genomics. Nat Genet 51(6):1035–1043. https://doi.org/10.1038/s41588-019-0417-8

5. Seemann T (2014) Prokka: rapid prokaryotic genome annotation. Bioinformatics 30 (14):2068–2069. https://doi.org/10.1093/bioinformatics/btu153
6. Croucher NJ, Page AJ, Connor TR et al (2015) Rapid phylogenetic analysis of large samples of recombinant bacterial whole genome sequences using Gubbins. Nucleic Acids Res 43(3):e15. https://doi.org/10.1093/nar/gku1196
7. Mostowy R, Croucher NJ, Andam CP et al (2017) Efficient inference of recent and ancestral recombination within bacterial populations. Mol Biol Evol 34(5):1167–1182. https://doi.org/10.1093/molbev/msx066
8. Didelot X, Wilson DJ (2015) ClonalFrameML: efficient inference of recombination in whole bacterial genomes. PLoS Comput Biol 11(2): e1004041. https://doi.org/10.1371/journal.pcbi.1004041
9. Lees JA, Harris SR, Tonkin-Hill G et al (2019) Fast and flexible bacterial genomic epidemiology with PopPUNK. Genome Res 29 (2):304–316. https://doi.org/10.1101/gr.241455.118
10. Wood DE, Salzberg SL (2014) Kraken: ultrafast metagenomic sequence classification using exact alignments. Genome Biol 15(3):R46. https://doi.org/10.1186/gb-2014-15-3-r46
11. Bankevich A, Nurk S, Antipov D et al (2012) SPAdes: a new genome assembly algorithm and its applications to single-cell sequencing. J Comput Biol 19(5):455–477. https://doi.org/10.1089/cmb.2012.0021
12. Wick RR, Schultz MB, Zobel J et al (2015) Bandage: interactive visualization of de novo genome assemblies. Bioinformatics 31 (20):3350–3352. https://doi.org/10.1093/bioinformatics/btv383
13. Wick RR, Judd LM, Gorrie CL et al (2017) Unicycler: resolving bacterial genome assemblies from short and long sequencing reads. PLoS Comput Biol 13(6):e1005595. https://doi.org/10.1371/journal.pcbi.1005595
14. Kapatai G, Coelho J, Platt S et al (2017) Whole genome sequencing of group a streptococcus: development and evaluation of an automated pipeline for emmgene typing. PeerJ 5:e3226. https://doi.org/10.7717/peerj.3226
15. Arndt D, Grant JR, Marcu A et al (2016) PHASTER: a better, faster version of the PHAST phage search tool. Nucleic Acids Res 44(W1):W16–W21. https://doi.org/10.1093/nar/gkw387
16. Liu M, Li X, Xie Y et al (2019) ICEberg 2.0: an updated database of bacterial integrative and conjugative elements. Nucleic Acids Res 47 (D1):D660–D665. https://doi.org/10.1093/nar/gky1123
17. Hunt M, Mather AE, Sanchez-Buso L et al (2017) ARIBA: rapid antimicrobial resistance genotyping directly from sequencing reads. Microb Genom 3(10):e000131. https://doi.org/10.1099/mgen.0.000131
18. Brynildsrud O, Bohlin J, Scheffer L et al (2016) Rapid scoring of genes in microbial pan-genome-wide association studies with Scoary. Genome Biol 17(1):238. https://doi.org/10.1186/s13059-016-1108-8
19. Thorpe HA, Bayliss SC, Sheppard SK et al (2018) Piggy: a rapid, large-scale pan-genome analysis tool for intergenic regions in bacteria. Gigascience 7(4):1–11. https://doi.org/10.1093/gigascience/giy015
20. Page AJ, Taylor B, Delaney AJ et al (2016) SNP-sites: rapid efficient extraction of SNPs from multi-FASTA alignments. Microb Genom 2(4):e000056. https://doi.org/10.1099/mgen.0.000056
21. Nguyen LT, Schmidt HA, von Haeseler A et al (2015) IQ-TREE: a fast and effective stochastic algorithm for estimating maximum-likelihood phylogenies. Mol Biol Evol 32 (1):268–274. https://doi.org/10.1093/molbev/msu300
22. Ondov BD, Treangen TJ, Melsted P et al (2016) Mash: fast genome and metagenome distance estimation using MinHash. Genome Biol 17(1):132. https://doi.org/10.1186/s13059-016-0997-x

Chapter 8

Group A Streptococcus Transcriptome Analysis

Misú Sanson and Anthony R. Flores

Abstract

Global gene expression analyses in bacteria have undergone a dramatic transformation. Prior to the development of high-throughput sequencing technologies, real-time PCR or microarray studies were the mainstay of assessing differences in gene expression in bacteria. Real-time PCR remains a critical tool for targeted gene expression analyses. However, microarray studies have given way to the plethora of advantages in RNA sequencing (RNA-seq) for the determination of global gene expression (i.e., transcriptome). Increased accessibility to high-throughput sequencing and user-friendly bioinformatics data analysis software have made RNA-seq technology use more widespread. Here, we provide comprehensive methods to perform RNA sequencing of *Streptococcus pyogenes* strains grown in vitro in standard laboratory media, including cell growth, RNA extraction, ribosomal RNA depletion, and library construction. Considerations for library sequencing and data analysis are also provided.

Key words Group A Streptococcus, Differential gene expression, Transcriptome, Global gene expression, RNA sequencing, *Streptococcus pyogenes*, RNA-seq

1 Introduction

Streptococcus pyogenes (Group A Streptococcus; GAS) is an obligate human pathogen responsible for a variety of diseases. GAS disease phenotype can, at least partially, be attributed to specific gene regulation imparted by 13 two-component systems and several stand-alone transcriptional regulators. The ability to generate isogenic mutants in a variety of GAS strain backgrounds has led to the need to critically assess global gene transcription in an effort to understand the relationships between gene expression and elaboration of disease phenotype. Historically, transcriptional differences between strains have been either targeted using reverse transcriptase, real-time PCR, or global using microarray (e.g., Affymetrix). While microarray has yielded useful information regarding GAS global gene regulation [1–4], the laborious chip design for use on a limited strain set and relatively decreased quantitative sensitivity has driven the need for more accurate interrogation of the GAS transcriptome. Simultaneously, high-throughput, next-generation

Thomas Proft and Jacelyn M. S. Loh (eds.), *Group A Streptococcus: Methods and Protocols*, Methods in Molecular Biology, vol. 2136, https://doi.org/10.1007/978-1-0716-0467-0_8,

sequencing (NGS) technologies have become increasingly affordable and accessible to researchers. The result has been an explosion of RNA sequencing (RNA-seq)—a technique that takes advantage of the ability to make complementary DNA (cDNA) from messenger RNA (mRNA) and subsequently apply high-throughput sequencing technology—providing an unprecedented examination of global gene transcription in GAS and other organisms.

The scale in which genomes and transcriptomes can now be analyzed has been revolutionized by NGS. For example, RNA-seq of nearly 500 GAS strains was combined with classical virulence assessment in 50 strains to provide an intimate look into the virulence of *emm28* GAS, demonstrating the truly high-throughput ability of RNA-seq [5]. However, the majority of investigators are interested in smaller scale transcriptome comparisons between different growth conditions or isogenic mutant strains. Several modifications based on the goals of the investigative team are possible with RNA-seq, and protocols can be tailored to yield strand-specific information, transcriptional start sites, and small noncoding or regulatory RNAs among other applications. The amount of data generated from a single RNA-seq experiment can be daunting for the bench scientist, leading many to seek fee-for-service experimental design, data generation, and analysis or collaboration with expert teams in RNA-seq.

The goal of this chapter is to provide a detailed description of the procedures necessary to produce high-quality RNA-seq data for GAS transcriptome analysis (Fig. 1). From the outset, it should be noted that the techniques described below may not be suitable for every laboratory. In fact, many core facilities at large academic centers and institutes and private sequencing companies provide RNA-seq as a fee-for-service (Fig. 1) and are a consideration for small projects or infrequent users of RNA-seq. However, as RNA-seq becomes a staple in a number of laboratories, the techniques described in this chapter will hopefully serve as a foundation for efficient and evermore affordable RNA-seq data generation and analysis in GAS experimentation. Figure 1 shows the general workflow for any RNA-seq experiment and should serve as a roadmap for the reader of this chapter. As with any experimental design, proper attention to strain selection (including strain background and mutant selection) and growth conditions is a critical first step in RNA-seq (essential to eliminate “garbage in, garbage out”). We next describe adaptation of commercially available kits for the extraction of RNA from GAS cells followed by the imperative step of ribosomal RNA depletion and enrichment of messenger RNA. RNA-seq library construction and sequencing are described nearly unmodified from commercial manufacturer instructions. Finally, we describe common bioinformatics pipelines (commercial software-based and freeware) used in analysis. Along the way, we

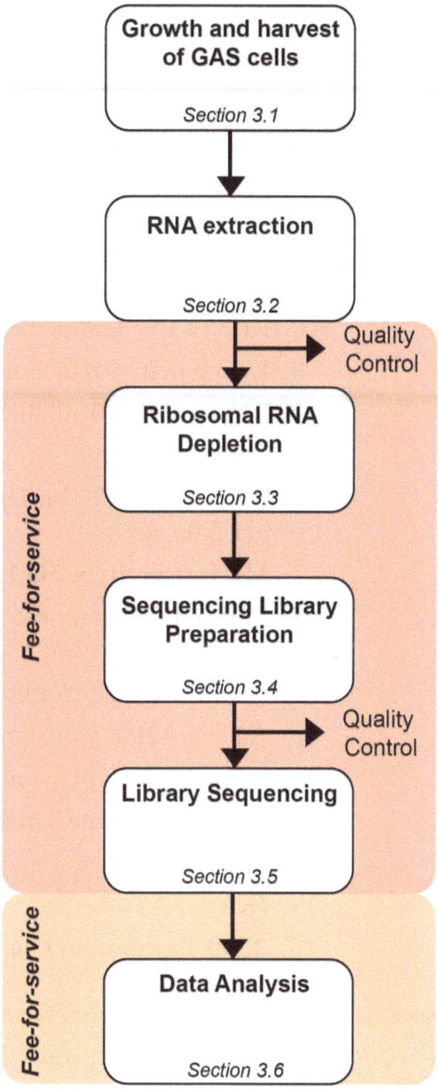

Fig. 1 Workflow in a typical GAS RNA-sequencing experiment. Shaded regions (salmon and tan) indicate steps that are typically provided as a fee-for-service by institutional core facilities or private companies

provide extensive notes on potential pitfalls and their avoidance based on our experience.

2 Materials

2.1 Cell Growth and Harvest

1. Blood agar: Tryptic Soy Agar (TSA) + 5% sheep blood agar plates.
2. THY broth: Todd Hewitt broth supplemented with 0.2% yeast extract.

3. Screw cap conical polypropylene 15-ml and 50-ml tubes.
4. Sterile inoculating loops.
5. Spectrophotometer (visible spectrum).
6. RNAprotect Bacteria Reagent [Qiagen].
7. Liquid nitrogen.

2.2 RNA Extraction and Processing (Quantification and Evaluation of Integrity)

1. β-mercaptoethanol (βme).
2. RNeasy Mini Kit [Qiagen]: RLT (lysis) buffer, RWI (buffer) RPE (wash) buffer.
3. Lysing matrix B in 2-ml tubes [MP Biomedicals].
4. RNAse-free 1.5-ml microfuge tubes.
5. Absolute ethanol.
6. RNAse-free TE buffer: 10 mM Tris–HCL, 1 mM EDTA pH 7.0.
7. MP Fast-prep-24 5G Homogenizer.
8. Turbo DNA-free Kit [Invitrogen]: DNAse, DNAse buffer, DNAse inactivation reagent.
9. NanoDrop Spectrophotometer.
10. Agilent 4200 TapeStation System.
11. RNA ScreenTape [Agilent].
12. RNA ScreenTape Sample Buffer [Agilent].
13. RNA ScreenTape Ladder [Agilent].
14. IKA MS3 Vortexer.
15. 2200 TapeStation Controller Software.

2.3 rRNA Removal and Sequencing Library Preparation

1. Magnetic rack or stand for 1.5-ml tubes.
2. 0.2-ml, 0.5-ml, and 1.5-ml RNAse-free microfuge tubes.
3. Thermocycler.
4. Water bath or heating block.
5. Vortex mixer.
6. ScriptSeq Complete Kit (Bacteria) [Epicentre].
7. Ribo-Zero Kit (for Bacteria) [Epicentre (Illumina)].
8. RNEasy MinElute Cleanup Kit [Qiagen].
9. FailSafe PCR enzyme Mix [Lucigen].
10. MinElute PCR purification columns [Qiagen].
11. ScriptSeq Index Primers 1–48 (dependent on number of samples and sequencing instrument used) [Illumina].
12. Agentcourt AMPure XP System [Beckman Coulter].
13. Qubit Fluorometer [Invitrogen].

14. Qubit dsDNA HS Assay Kit [Invitrogen].
15. 500-μl thin-walled PCR tubes.
16. High Sensitivity D5000 ScreenTape [Agilent].
17. High Sensitivity D5000 Sample Reagents (Buffer and Ladder) [Agilent].

3 Methods

3.1 Cell Growth and Harvest

1. Grow strain of interest on a blood-agar plate for ~12 h.
2. Using a sterile, disposable inoculating loop pick several colonies and inoculate 10 ml of THY broth (or other standard laboratory media) (*see* **Note 1**).
3. Grow for 9–12 h (*see* **Note 2**) at 37 °C in an incubator with 5% CO_2.
4. Inoculate fresh prewarmed media using a 1:50 dilution with the culture that was grown overnight (*see* **step 3**).
5. Measure the absorbance ($O.D._{600nm}$) every hour until the cells are at mid-logarithmic phase of growth (~$O.D._{600}$: 0.5–1.0).
6. Transfer 1 ml of sample to a 15-ml conical tube containing 2 ml of RNAprotect cell reagent. (The ratio of RNAprotect to sample should be 2:1. If larger volumes of cells are to be collected, the volume of RNAprotect should be increased accordingly).
7. Allow samples to incubate with RNAprotect for at least 10 min at room temperature (*see* **Note 3**).
8. Pellet samples using high-speed centrifugation (>4000 × *g*) for 10 min.
9. Discard supernatant without disturbing the cell pellet and making sure to eliminate all the liquid residue.
10. Flash freeze the cell pellets using liquid nitrogen and store in a −80 °C freezer until ready to proceed with RNA extraction.

3.2 RNA Extraction and Processing (Quantification and Evaluation of RNA Integrity)

RNA extraction described here is a column-based format adapted from the manufacturer's protocol for the Qiagen RNeasy extraction kit. For a large number of samples, the protocol can be carried out using a 96-well plate format (*see* **Note 4**).

3.2.1 RNA Extraction

1. Remove the frozen cell pellets from the −80 °C and thaw on ice.
2. Prepare 2-ml screw cap tubes by adding approximately ~290 mg of lysing matrix B to each tube (one tube per sample to be used in **step 6**).

3. Prepare individual 1.7-ml microfuge tubes by adding 900 μl of 100% ethanol to each tube. (One tube per sample to be used in **step 12**.)
4. Prepare the lysis buffer by adding 10 μl β-mercaptoethanol (βme) to 1 ml of RLT buffer (*see* **Note 5**). You will need 650 μl of RLT (with βme) per sample (e.g., for 12 samples add 80 μl of βme to 8 ml of RLT).
5. Resuspend the cell pellet in 100 μl of RNase-free TE pH 7.0.
6. Transfer the suspended cell pellets to a fast prep tube containing the lysing matrix B and place on ice until all samples have been transferred to fast prep tubes.
7. Secure tubes with lysing matrix beads and suspended bacterial pellets in the FastPrep at 6 m/s for 20 s (*see* **Note 6**).
8. Briefly place the tubes on ice after the fast prep run.
9. Centrifuge the tubes at >13,000 × *g* for 10 s to bring sample to the bottom of the tube.
10. Add 650 μl of RLT (with βme) to the sample tubes.
11. Repeat FastPrep (**step** 7) at 6 m/s for 20 s.
12. Centrifuge samples at >13,000 × *g* for 1 min.
13. Transfer 600 μl of supernatant from the samples to the microcentrifuge tubes containing 900 μl of ethanol. Mix by gently pipetting up and down.
14. Transfer 750 μl of lysate/ethanol mixture to the RNeasy purification columns.
15. Spin at >8000 × *g* for 30 s and discard flow-through.
16. Transfer the remaining 750 μl of lysate for each sample to the purification columns.
17. Spin at >8000 × *g* for 30 s and discard flow-through and collection tubes.
18. Place purification columns in a new collection tube (provided in kit).
19. Add 700 μl of buffer RWI to each column and spin at >8000 × *g* for 30 s. Discard flow-through and collection tubes.
20. Place purification columns in a new collection tube (provided in kit).
21. Add 500 μl of RPE buffer to each individual column.
22. Spin at >8000 × *g* for 30 s. Discard flow-through.
23. Repeat wash by adding 500 μl of buffer RPE to each individual column (total of 2 washes).
24. Discard flow-through and spin at >8000 × *g* for 30 s to remove residual ethanol.

25. Elute using 50 μl of RNAse-free water (*see* **Note 7**).
26. Discard purification columns and save flow-through (contains the total RNA).
27. 25. Add 6 μl of Turbo-free DNAse buffer and 2 μl of Turbo-free DNAse I to each of the eluted samples.
28. Incubate for 40 min at 37 °C.
29. Add 2 μl of Turbo-free DNAse I to each sample and incubate at 37 °C for 40 min.
30. Add 2 μl of Turbo-free DNAse I to each sample and incubate at 37 °C for 40 min (samples will be treated with DNAse I a total of three times to remove all traces of DNA) (*see* **Note 8**).
31. Add 12 μl of DNAse inactivation agent to each sample and mix by gently pipetting.
32. Centrifuge at >8000 × *g* for 3 min.
33. Remove 40 μl of supernatant from each sample into a new labeled RNAse-free microfuge tube and keep samples on ice.
34. Proceed with RNA quantification and quality check.

3.2.2 RNA Quantification and Quality Control

RNA quantification described below is performed using a NanoDrop spectrophotometer. RNA size and integrity are evaluated using high-resolution automated electrophoresis with a 4200 TapeStation instrument. If the 4200 TapeStation is not available, suitable alternatives to check quality of RNA include the Agilent 4150 Tapestation or Agilent 2100 Bioanalyzer (*see* **Note 9**).

1. Quantify RNA by using 1 μl on a NanoDrop spectrophotometer. Yield generally ranges between 100 and 350 ng/μl depending on input cells. Our experience demonstrates that using a NanoDrop spectrophotometer gives a more accurate quantity than automated gel electrophoresis systems.
2. Proceed to RNA quality check by allowing TapeStation reagents to equilibrate at room temperature for 30 min prior to use.
3. Vortex mix all reagents before using.
4. Prepare the ladder by adding 5 μl of RNA sample buffer to 1 μl of RNA Ladder.
5. Prepare samples by mixing 5 μl of RNA sample buffer with 1 μl of RNA for each sample.
6. Spin down your samples and vortex using IKA MS3 vortexer at 2000 rpm for 1 min.
7. Spin down tubes to position samples at the bottom.
8. Heat samples and ladder at 72 °C for 3 min.
9. Place samples and ladder on ice for 2 min.

10. Spin samples once again to position at the bottom of the tube.
11. Load samples on the 4200 TapeStation instrument.
12. Use the 2200 TapeStation Controller Software to specify sample and file names and to begin sample evaluation.

3.3 Ribosomal RNA Depletion

The removal of ribosomal RNA (rRNA) and subsequent cDNA library preparation described is based on Epicentre ScriptSeq Complete Kit (Bacteria) user's protocol with very few adaptations (https://www.illumina.com/products/scriptseq-bacteria.html) (*see* **Note 10**). Removal of rRNA is a crucial step in the preparation of cDNA libraries for RNA-Seq. The Ribo-Zero magnetic kit is the most effective product in removing ribosomal RNA from total extracted bacterial RNA [14].

3.3.1 Bead Washing Procedure

1. Allow all of the Magnetic Core Kit components to equilibrate at room temperature.
2. Mix the magnetic beads by pipetting gently but thoroughly.
3. Calculate the amount of magnetic beads required for the total number of reactions you are setting up. (Each reaction requires 225 μl of magnetic beads).
4. Dispense a maximum of 1350 μl of magnetic beads into a 1.5-ml microfuge tube. (This will be enough for six reactions.) If setting up a greater number of rRNA removal reactions, you can wash multiple aliquots of 1350 μl or a larger amount using a 15-ml tube if a 15-ml magnetic stand is available.
5. Place the 1.5-ml microfuge tubes that contain the aliquoted beads on the magnetic stand for 1 min or until the solution is clear.
6. Keep the microfuge tube on the magnetic stand and remove and discard the supernatant. Be cautious, the supernatant contains 0.1% sodium azide which is toxic.
7. Remove the tubes from the magnetic stand and add an equal volume of RNAse-free water. Mix by pipetting.
8. Place the tubes on the magnetic stand once again and let them stand until the solution is clear.
9. Keep the microfuge tube on the magnetic stand and remove and discard the supernatant.
10. Repeat **steps 8–10** until you have washed the beads a total number of two times with RNAse-free water.
11. Remove the tubes from the magnetic stand and add the volume of magnetic bead resuspension solution equal to the number of reactions times 60 μl (for six reactions add 360 μl). Mix thoroughly by pipetting up and down.

12. The volume of beads and resuspension solution will be additive, so each reaction will use 65 μl of resuspended beads.
13. Aliquot 65 μl of the washed magnetic beads into 1.5 ml RNAse-free microfuge tubes and add 1 μl of RiboGuard RNAase inhibitor to each tube.
14. Store tubes at room temperature until required (after treatment of RNA with Ribo-Zero removal solution; *see* Subheading 3.3.2, **step 4**).

3.3.2 Treatment of Total RNA Samples with Ribo-Zero rRNA Removal Solution

As per the manufacturer's protocol (https://www.illumina.com/products/scriptseq-bacteria.html), the volume of RNA sample and the volume of the Ribo-Zero rRNA removal solution used per reaction will depend on the amount of total RNA (calculated above using NanoDrop). Refer to Table 1.

1. In a 0.2-ml RNAse-free PCR tube combine in the following order:

   x μl RNAse-free water.

   4 μl RiboZero reaction buffer.

   1–5 μg Total RNA sample.

   y μl Ribo-Zero removal solution.

   40 μl total volume.
2. Mix reactions together by gently pipetting and incubate at 68 °C for 10 min.
3. Remove reactions tubes from 68 °C and incubate at room temperature for 5 min.
4. Using a pipette, add the treated RNA to the 1.5-ml microfuge tube that contains the washed magnetic beads, and without changing the pipet tip, immediately mix thoroughly the contents of the tube by pipetting at least ten times. Immediately vortex tube at medium speed and keep on a rack at room temperature.
5. Repeat the same process for each sample. It is important to always add the treated RNA sample to the magnetic beads and pipette immediately, to avoid clumps forming that can impact the efficiency of rRNA removal.

Table 1
Volume of Ribo-Zero rRNA removal solution to use based on the amount of input RNA

Amount of input total RNA (μg)	Maximum volume of total RNA added to each reaction (μl)	Volume of Ribo-Zero rRNA removal solution used per reaction (μl)
1–2.5	28	8
>2.5–5	26	10

6. Incubate the 1.5-ml microfuge tubes at room temperature for 5 min.
7. Vortex reactions at medium speed for 10 s and place at 50 °C for 5 min on a heating block.
8. After the incubation, remove the tubes from heat block and immediately place them on the magnetic stand until the solution appears clear (1–2 min).
9. Remove and transfer 85–90 μl of supernatant (contains the rRNA-depleted RNA) from each sample into a clean RNAse-free microfuge tube.
10. Place samples on ice and proceed to purification. Alternatively, samples can be stored in the −80 °C.

3.3.3 Purification of the rRNA-Depleted Sample

Purification of the rRNA-depleted sample is performed using a modified Qiagen RNeasy MinElute protocol which allows recovery of small RNAs. Alternative clean-up methods include ethanol precipitation or the use of AmpPure beads. For the use of alternative methods, refer to Illumina's user manual (https://www.illumina.com/products/scriptseq-bacteria.html).

1. Adjust the sample to a volume of 100 μl using RNase-Free water.
2. Add 350 μl of Buffer RLT and mix by pipetting.
3. Add 550 μl of 100% ethanol to the RNA and mix well by pipetting.
4. Transfer 500 μl of the sample to the RNEasy MinElute spin columns placed on the 2-ml collection tube.
5. Close the lid and centrifuge for 15 s at 8000 × *g*. Discard flow-through.
6. Transfer the remaining sample to the column and repeat the centrifugation at 8000 × *g* for 15 s.
7. Discard flow-through and collection tube. Place the columns in new collection tubes.
8. Add 500 μl of buffer RPE to the spin columns.
9. Centrifuge for 15 s at 8000 × *g*.
10. Add 500 μl of 80% ethanol to the RNEasy MinElute spin column.
11. Close lid and centrifuge for 2 min at 8000 × *g*. Discard flow-through and collection tube.
12. Place the columns in new 2-ml collection tubes.
13. Centrifuge at >10,000 × *g* for 5 min with the lids of the column open to ensure the spin-column membranes will dry and no residual ethanol will be carried over during the elution step.

14. After centrifugation, place the column in new RNAse-free microfuge tube.
15. Add 12 μl of RNAse-free water directly to the center of the spin column membrane.
16. Centrifuge at full speed for 1 min to elute the clean, rRNA-depleted RNA.
17. It is typical to obtain 10 μl after elution.
18. Place samples on ice for immediate use, or alternatively, it can be stored in the −80 °C.

3.4 Sequencing Library Preparation (See Note 11)

The volume of clean rRNA-free RNA used in the ScriptSeq V2 library preparation kit depends on the amount of total RNA that was treated. As per manufacturer suggestions, follow Table 2 to determine the input volume of treated RNA.

3.4.1 Fragmentation of RNA and cDNA Synthesis Primer Anneal

1. In a 0.2-ml PCR tube, set up the following reaction:

   x μl Nuclease-free water.

   y μl rRNA-depleted RNA (Table 2).

   1 μl RNA fragmentation solution.

   2 μl cDNA synthesis primer.

   12 μl total volume.
2. Incubate reactions at 85 °C for 5 min.
3. Stop fragmentation by placing tubes on ice.
4. Prepare the cDNA synthesis master mix by combining for each sample:

   3 μl cDNA synthesis premix.

   0.5 μl 100 mM DTT.

   0.5 μl StarScript Reverse transcriptase.

   4 μl total volume.
5. Add 4 μl of the cDNA master mix to each fragmentation reaction, and mix by pipetting.
6. Incubate at 25 °C for 5 min followed by 42 °C for 20 min.

Table 2
Volume of Ribo-Zero-treated RNA to use in ScriptSeq v2 library preparation procedure

Amount of total RNA used in the Ribo-Zero procedure (μg)	Maximum volume of purified Ribo-Zero-treated RNA used for the ScriptSeq kit (μl)
1	9
2.5	4
5	1

7. Cool reactions to 37 °C and pause the thermocycler.
8. Add 1 μl of Finishing Solution to each of the sample tubes and mix by pipetting.
9. Incubate at 37 °C for 10 min, followed by 95 °C for 3 min.
10. Cool the reactions to 25 °C and pause the thermocycler.
11. Prepare the Terminal tagging master mix by combining for each reaction on ice:

    7.5 μl Terminal Tagging Solution.

    0.5 μl DNA polymerase.

    8.0 μl total volume.
12. Add 8.0 μl of the Terminal tagging master mix to each reaction and mix well by pipetting.
13. Incubate each reaction at 25 °C for 15 min, followed by 95 °C for 3 min.
14. Cool reactions to 4 °C.

3.4.2 Purification of di-Tagged cDNA

Purification of the cDNA is necessary before amplification of the library. Described below is purification using Qiagen MinElute purification kit, with few modifications of the manufacturer's protocol.

1. Add 5 volumes of buffer PB to 1 volume of the PCR reaction and mix well by pipetting.
2. With the MinElute column in a 2-ml collection tube, transfer the sample to the column and centrifuge at for 1 min.
3. Discard flow-through and place column back into the same collection tube.
4. Wash by adding 750 μl of buffer PE to each column and centrifuge at for 1 min.
5. Discard flow-through and centrifuge for 1 min at maximum speed to completely remove ethanol.
6. Elute DNA by adding 25 μl of Elution Buffer to the center of the membrane. Let the column stand for 1 min before centrifugation at high speed for 1 min.
7. The eluate volume is typically 22.5 μl, which will be used entirely in the PCR amplification reaction (*see* Subheading 3.4.3, **step 1**). Keep samples on ice, or alternatively, they can be stored at −20 until ready for use.

3.4.3 Amplification of the RNA-seq Library

To amplify the library, each reaction will be set up using a distinct barcode (enables multiplexing or the ability to process multiple samples simultaneously when sequencing) instead of a reverse primer. The number of barcodes needed will depend on the number of samples and the instrument used for sequencing.

1. In a 0.2-ml PCR tube, mix the following reactions:

   22.5 μl Clean di-tagged cDNA.

   25 μl FailSafe PCR PreMix E.

   1 μl Forward PCR Primer.

   1 μl Index oligo.

   0.5 μl FailSafe PCR Enzyme (1.25 U).

   50 μl total volume.
2. Denature the DNA at 95 °C for 1 min.
3. Follow denaturation by 15 cycles of: 95 °C for 30 s, 55 °C for 30 s, and 68 °C for 3 min. It is important to not over amplify the library (*see* Subheading 3.4.5). Thus, we recommend a maximum of 10 cycles.
4. Incubate at 68 °C for 7 min.
5. Proceed immediately to library purification.

3.4.4 Purification of the RNA-Seq Library

Purification of the amplified library will eliminate primer dimers and other minor contaminants that may alter sequencing. Described below is purification performed using the Beckman Coulter AmPure XP system. For alternative purification methods please refer to the manufacturer's user manual (https://www.illumina.com/products/scriptseq-bacteria.html).

1. Allow the AMPure XP beads to equilibrate to room temperature.
2. Prepare 400 μl of fresh 80% ethanol at room temperature for each sample.
3. Vortex the beads well until they are homogenous.
4. Add 50 μl of the beads to each PCR tube.
5. Mix thoroughly by pipetting the entire volume of each tube at least ten times.
6. Transfer each 100 μl of the bead and library mixture to a clean 1.5 ml.
7. Incubate the tubes at room temperature for 15 min.
8. Place the tubes on in a magnetic stand at room temperature for at least 5 min, or until the liquid appears clear.
9. Remove and discard the supernatant from each tube by using a pipette, without disturbing the beads.
10. Keep the tubes on the magnetic stand and add 200 μl of 80% ethanol to each tube, without disturbing the beads.
11. Incubate the tubes at room temperature for 30 s, then remove and discard the supernatant.

12. Repeat **steps 10** and **11** until two washes with 80% ethanol are completed.
13. Allow the plate to air dry on the magnetic stand for 15 min at room temperature.
14. Add 20 μl of nuclease-free water to each tube and remove tubes from the magnetic stand.
15. Resuspend the beads by pipetting at least ten times.
16. Incubate the tubes at room temperature for 2 min.
17. Place the tubes on the magnetic stand and incubate for 5 min or until the solution appears clear.
18. Transfer the supernatant (contains the clean RNA-seq library) to a clean microcentrifuge tube.
19. Proceed to quantity and quality check.

3.4.5 RNA-Seq Library Quantification and Quality Assessment

To quantify the library, we recommend using the Qubit Fluorometric Quantitation.

1. Set up the required number of 500-μl thin-walled tubes for your samples and for two standards.
2. Prepare enough Qubit working solution by diluting the Qubit dsDNA HS Reagent 1:200 in Qubit dsDNA HS Buffer. You will need 190 μl of working solution for each standard and 199 μl of working solution for each sample.
3. Transfer 190 μl of the working solution to each of the tubes for the standards and add 10 μl of the Qubit standard to the appropriate tube, then mix by vortexing 2–3 s.
4. Transfer 199 μl of working solution to individual assay tubes (1 for each sample).
5. Add 1 μl of each RNA-seq library sample to one of the tubes containing 199 μl of working solution. Vortex 2–3 s.
6. Allow the tubes with sample and working solution to incubate for 2 min at room temperature.
7. Read each of the standards to calibrate the Qubit before reading the samples.
8. Read each individual sample and record the concentration (ng/μl) of the library.

3.4.6 RNA-seq Library Quality and Size Distribution Assessment

Quality assessment of the RNA-seq library and adequate size distribution can be achieved using the DNA High Sensitivity kit for the Agilent TapeStation (*see* **Note 12**).

1. Allow the TapeStation reagents (High Sensitivity D5000) to equilibrate at room temperature for 30 min.
2. Prepare the ladder by mixing 2 μl of sample buffer with 2 μl of ladder.

3. Prepare the samples by mixing 2 μl of sample buffer with 2 μl of DNA.
4. Spin down and vortex using the IKA MS3 vortexer at 2000 rpm for 1 min.
5. Spin down tubes to bring sample to bottom of the tubes.
6. Load samples into the TapeStation instrument and run software to analyze your samples.

3.5 Library Sequencing

Several factors must be considered when determining library sequencing. For example, the sequencing instruments used (e.g., Illumina's MiSeq or HiSeq) affect read length and the number of samples (discussed in Subheading 3.5.2). We routinely use the described procedures to prepare libraries to run on a HiSeq 4000 instrument. The overwhelming majority of users of the described procedures for library preparation will not directly run samples on the sequencing instrument. Rather, samples are generally given to a sequencing core facility. We recommend consultation with sequencing support staff prior to library generation to ensure compatibility and adequate read coverage for the proposed experiment. Below are the major considerations to discuss with technical support or managers of sequencing facilities.

3.5.1 Single-Versus Paired-End

In the design of the RNA-seq experiment, a decision must be made for single-end (SE) or paired-end (PE) sequencing. As the names imply, SE indicates sequencing from one end of the cDNA fragment, while PE yields sequence from both ends (or directions). PE sequencing provides twice the number of reads and is almost always more expensive. However, compared to SE, PE sequencing aids in mapping to repetitive regions such as mobile genetic elements (e.g., prophage and integrative conjugative elements). Also, as PE sequencing has become increasingly more common for many sequencing applications, advanced planning is necessary for most sequencing facilities to accommodate SE sequencing requests. Our experience has been that the advantages of PE sequencing outweigh the differences in cost, and thus, we recommend PE. Ultimately, the decision will be up to individual investigators and should be discussed with sequencing facilities far in advance.

3.5.2 Higher Coverage (Reads) Versus More Replicates

The number of replicates used in a RNA-seq experiment (*see* **Note 11**) deserves special consideration. In general, increasing the number of replicates reduces the influence of technical noise generated during RNA extraction and library preparation, while biological variation is more difficult to control. Most analyses require a minimum of three biological replicates to perform statistical analyses. However, the individual needs and goals of the experiment may dictate more or less replicates. Large experiments involving many strains can very rapidly increase in cost with additional replicates. In general, we aim for four biological replicates when comparing individual strains (e.g., parental versus isogenic mutant) or growth conditions (e.g., in vitro rich medium versus ex vivo human saliva). For a more in-depth discussion of experimental design including replicate considerations, the reader is directed to Conesa et al. [6].

Read coverage (*see* **Note 13**) is another important aspect to account for in experimental design. In contrast to whole-genome sequencing projects where read coverage is calculated on a per nucleotide basis, for RNA-seq experiments, the total number of reads mapped to a given gene (or annotated coding sequence) is used for quantification and in downstream calculations. Increasing read length (range usually 50–300 bp) and use of paired reads improves mapping to the reference genome. Studies in *E. coli* suggest that as few as 1 million reads is sufficient to detect nearly 100% of open reading frames [7]. In our experience, a minimum of 5 million reads (regardless of length or PE versus SE) yields sufficient power to detect differences between genes with an average expression level [8, 9]. For a typical RNA-seq experiment, we target 15 million reads per sample/replicate to account for variability during library generation. Exceeding 25 million reads per sample/replicate, except under special circumstances, is not recommended given the substantial increase in computational resources necessary to process and analyze data (*see* Subheading 3.6).

3.6 RNA-seq Data Analysis

Generation of high-quality libraries and raw RNA-seq data are essential for downstream analyses. The in-depth discussion necessary to guide the user through analyses precludes extensive discussion in this chapter. For an overview of the general principles and considerations, the reader is directed to Chatterjee et al. [10]. Of note, the discussion in Chatterjee et al. is in specific reference to eukaryotic transcriptomic analyses, but the concepts are roughly similar for prokaryotic analyses.

The amount of raw data generated and analyses required can be challenging for the bench scientist. In addition, some tools may require knowledge of less commonly used operating systems (e.g., Linux) or command-line interface (CLI) that may dissuade investigators using RNA-seq infrequently to invest significant time. Thus, as indicated in Fig. 1, analysis can also be performed as a fee-for-service and is often in conjunction with RNA-seq library preparation and sequencing services. Figure 2 provides an overview of RNA-seq analysis. In general, all steps in data analysis can be achieved on a desktop computer (Windows, MacOS, or Linux based). Newer, multiprocessor laptops are also quite capable for basic analyses. Many of the tools indicated in Fig. 2 require some knowledge or comfortability using a CLI. The advantages of using the indicated (or similar) tools are (1) ability to customize analyses based on user need(s) and (2) all are available for download free of charge. If RNA-seq is to become a staple of your GAS research, we

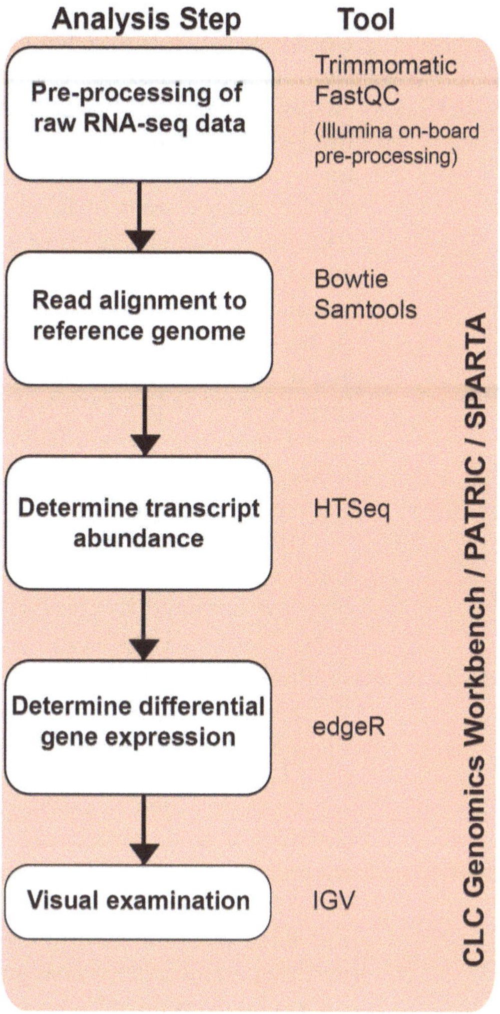

Fig. 2 Flowchart detailing steps in RNA-seq data analysis. Shading (salmon) indicates software packages combining the referenced (or similar) tools. References for tools are as follows: Trimmomatic [15], FastQC (https:/www.bioinformatics.babraham.ac.uk/projects/fastqc/), Bowtie [16], Samtools [17], HTSeq [18], edgeR [19], and IGV (integrated genome viewer) [20]

recommend the time investment or working directly with bioinformaticians to assemble custom analysis pipelines. SPARTA (simple program for automated reference-based bacterial RNA-seq transcriptome analysis) [11] is a free downloadable program that integrates the packages outlined in Fig. 2. For users interested in gaining more experience using CLI and building their own custom RNA-seq analysis pipeline, SPARTA is an excellent resource toward this end. Recognizing that many users are not comfortable using a CLI or installing software (some with multiple additional dependences), a few all-in-one packages have been developed (ranging from free to very expensive).

CLC Genomics Workbench (Qiagen Bioinformatics) and Geneious (https://www.geneious.com/) are stand-alone, multiplatform (e.g., Windows, MacOS, Linux) programs with RNA-seq analysis capabilities. Both utilize a windows interface facilitating use by inexperienced users. In addition, technical supports including written and online video tutorials are excellent resources for the novice. All the steps of RNA-seq analysis are performed including trimming; mapping reads to a reference, quantification, and differential gene expression. Both software options provide excellent visual display of read mappings and allow export of data for figure or table generation. The major downside is the significant cost to acquire and maintain either program but may be available for use by investigators at large academic centers or institutes. Several additional packages are available in both software packages beyond RNA-seq that may be of use to investigators potentially offsetting the cost of acquisition.

Alternatively, PATRIC (Pathosystems Resource Integration Center) consists of an online package (https://www.patricbrc.org/) maintained by the National Institute of Allergy and Infectious Diseases supporting integrated data and analysis tools for biomedical research on bacterial infectious diseases. The PATRIC RNA-seq analysis service offers two independent analysis pipelines for RNA-seq (the tuxedo suite of analysis tools [12] and Rockhopper [13]) and has an excellent online tutorial for beginners. The service is provided to investigators free of charge and is generally suitable for noncomplex analyses. The main limitations are the need to upload potentially large datasets (data are maintained in an individual user account and are not publicly shared but may be restricted by some institutions) and lack of ability to substantially modify parameters for analyses. PATRIC can be accessed through the online web portal or through a command-line interface. Outputs can be exported/downloaded for offline analyses.

4 Notes

1. The number of biological replicates and type of media to be used are to be determined prior to initiation of the protocol. In general, for most experiments involving parental/isogenic mutant comparisons, we aim for a minimum of three biological replicates. It should be noted that multiple environmental factors contribute to replicate differences including batches of media and subtle differences between individual incubators. In addition, if media other than THY are to be used (such as C-medium), differences in cell density at given phases of growth must be accounted for and determined prior to initiation of the RNA-seq protocol to achieve the ideal samples for processing.

2. Prolonged (e.g., >12 h) overnight cell growth should be avoided. Our experience with several GAS *emm* types including 1, 3, 4, 87, and isogenic mutant backgrounds demonstrates that prolonged (i.e., >12 h) overnight growth results in erratic growth (e.g., prolonged lag phase leading up to logarithmic growth) upon subculture. Presumably, the erratic growth is due to the majority of cells from long overnight incubations being in late stationary growth phase. Thus, we aim to subculture when cells are close to peak density in rich medium but not yet in stationary phase—a practice particularly useful when multiple strains are to be grown in biological replicates.
3. The manufacturer's protocol states a minimum of 10-min incubation prior to continuation of the protocol. In our hands, samples are stable at room temperature for several hours and up to 7 days at 4 °C with minimal RNA degradation facilitating collection and processing of multiple samples (replicates or mutant strains) at the same time.
4. Nearly identical kits and reagents for RNA extraction can be obtained from Qiagen for 96-well plate extractions. Details are not provided in the protocol given the relative rarity that large-scale extractions would be performed. However, the principles described in the protocol for RNA extraction and quality control remain the same.
5. β-mercaptoethanol is a hazardous reagent and should be aliquoted in a manner to protect the user from harmful fumes.
6. Settings are provided for the FastPrep 24 instrument. If using the FastPrep 5G, settings for GAS (under "*Streptococcus pyogenes*") are preprogrammed but may also be entered manually. Ballistic homogenization using other machines has proven to give lower RNA yields or is untested in our hands.
7. The manufacturer's protocol suggests re-loading the eluate on the same column to maximize RNA elution. However, for standard procedures where cells are not limiting, we have not observed significantly greater yields using this procedure. It may be considered when cells are harvested at phases of growth (e.g., prior to OD_{600} 0.5) where RNA yield may be lower.
8. DNA removal using DNAse is a critical step prior to ribosomal RNA removal and subsequent cDNA synthesis. Alternative (and significantly less expensive) on-column DNAse treatment is given in the manufacturer protocol. However, we recommend an additional quality control step of qRT-PCR using a standard housekeeping gene such as *proS* or *tufA* to demonstrate DNA is not detectable prior to proceeding if using on-column DNAse treatments.
9. RNA quality using the Agilent Bioanalyzer or TapeStation is given by the RNA integrity number (RIN). The RIN is defined

by an algorithm that takes into account peaks associated with bacterial 16S and 23S ribosomal RNA and tell-tale signals of RNA degradation. The original RIN numbers were defined on the Agilent Bioanalyzer 2100 and ranged from 1 (totally degraded RNA) to 10 (completely intact RNA). The TapeStation cannot directly calculate the RIN but instead estimates the value (RINe). Moreover, due to the estimation, RINe values of 10 are not achievable using bacterial total RNA on the TapeStation. Rather, values >8.0 are comparable in overall quality to samples with RIN values >9.5 on the Bioanalyzer. (Multiple discussions with Agilent technical support have not resulted in a direct ratio or conversion from the Bioanalyzer RIN at the time of writing this chapter).

10. Unfortunately, during the preparation of this chapter, the Script-Seq Complete Kit [Epicentre (Illumina) Cat. No. BB1206/BB1224], which includes the Ribo-Zero Kit, Magnetic Core Kit, and v2 RNA-Seq Library preparation kit, was discontinued. A new kit by Illumina using similar technology is set to be commercially available by January 2020. We advise following the manufacturer's instructions for the new kit when it becomes available. Alternative rRNA depletion kits are available including Ribominus (ThermoFisher) and MicrobExpress (Invitrogen). However, neither kit has been tested in our laboratory. Based on Petrova et al. [14], Ribominus performs nearly as well as Ribo-Zero.

11. The protocol described here is for paired-end sequencing (most common). However, it is worth considering the advantages/disadvantages of paired versus single-end libraries in the experimental design and is discussed in Subheading 3.5.

12. Our experience demonstrates that Qubit is accurate in quantification of the cDNA library. Typically, our concentrations average 20 ng/μl. Size distribution is assessed using the HS DNA Tapestation kit. The average size distribution is between 200 and 300 bp and should be observed as a single peak on the electrophoretogram.

13. Read coverage = (number of reads) × (read length)/genome size. For example, 250,000 paired-end reads (total of 500,000 reads) of 150-bp length yields approximately 40× coverage for most GAS genomes (average size of 1.9 Mb). For RNA-seq experiments, only coding regions (annotated genes or coding sequence) and total reads mapped to a given coding region are used in quantification in downstream analyses.

References

1. Sitkiewicz I, Green NM, Guo N et al (2009) Transcriptome adaptation of group B *Streptococcus* to growth in human amniotic fluid. PLoS One 4(7):e6114. https://doi.org/10.1371/journal.pone.0006114
2. Sitkiewicz I, Green NM, Guo N et al (2010) Adaptation of group A *Streptococcus* to human amniotic fluid. PLoS One 5(3):e9785. https://doi.org/10.1371/journal.pone.0009785
3. Sitkiewicz I, Musser JM (2006) Expression microarray and mouse virulence analysis of four conserved two-component gene regulatory systems in group A *Streptococcus*. Infect Immun 74(2):1339–1351. https://doi.org/10.1128/IAI.74.2.1339-1351.2006
4. Virtaneva K, Porcella SF, Graham MR et al (2005) Longitudinal analysis of the group A *Streptococcus* transcriptome in experimental pharyngitis in cynomolgus macaques. Proc Natl Acad Sci U S A 102(25):9014–9019
5. Kachroo P, Eraso JM, Beres SB et al (2019) Integrated analysis of population genomics, transcriptomics and virulence provides novel insights into Streptococcus pyogenes pathogenesis. Nat Genet 51(3):548–559. https://doi.org/10.1038/s41588-018-0343-1
6. Conesa A, Madrigal P, Tarazona S et al (2016) A survey of best practices for RNA-seq data analysis. Genome Biol 17:13. https://doi.org/10.1186/s13059-016-0881-8
7. Haas BJ, Chin M, Nusbaum C et al (2012) How deep is deep enough for RNA-Seq profiling of bacterial transcriptomes? BMC Genomics 13:734. https://doi.org/10.1186/1471-2164-13-734
8. Flores AR, Jewell BE, Yelamanchili D et al (2015) A single amino acid replacement in the sensor kinase LiaS contributes to a carrier phenotype in group A *Streptococcus*. Infect Immun 83(11):4237–4246. https://doi.org/10.1128/IAI.00656-15
9. Flores AR, Olsen RJ, Cantu C et al (2017) Increased pilus production conferred by a naturally occurring mutation alters host-pathogen interaction in favor of carriage in Streptococcus pyogenes. Infect Immun 85(5):e00949–e00916. https://doi.org/10.1128/IAI.00949-16
10. Chatterjee A, Ahn A, Rodger EJ et al (2018) A guide for designing and analyzing RNA-Seq data. Methods Mol Biol 1783:35–80. https://doi.org/10.1007/978-1-4939-7834-2_3
11. Johnson BK, Scholz MB, Teal TK, Abramovitch RB (2016) SPARTA: simple program for automated reference-based bacterial RNA-seq Transcriptome analysis. BMC Bioinformatics 17:66. https://doi.org/10.1186/s12859-016-0923-y
12. Trapnell C, Roberts A, Goff L et al (2012) Differential gene and transcript expression analysis of RNA-seq experiments with TopHat and cufflinks. Nat Protoc 7(3):562–578. https://doi.org/10.1038/nprot.2012.016
13. McClure R, Balasubramanian D, Sun Y et al (2013) Computational analysis of bacterial RNA-Seq data. Nucleic Acids Res 41(14): e140. https://doi.org/10.1093/nar/gkt444
14. Petrova OE, Garcia-Alcalde F, Zampaloni C, Sauer K (2017) Comparative evaluation of rRNA depletion procedures for the improved analysis of bacterial biofilm and mixed pathogen culture transcriptomes. Sci Rep 7:41114. https://doi.org/10.1038/srep41114
15. Bolger AM, Lohse M, Usadel B (2014) Trimmomatic: a flexible trimmer for Illumina sequence data. Bioinformatics 30 (15):2114–2120. https://doi.org/10.1093/bioinformatics/btu170
16. Langmead B, Trapnell C, Pop M, Salzberg SL (2009) Ultrafast and memory-efficient alignment of short DNA sequences to the human genome. Genome Biol 10(3):R25. https://doi.org/10.1186/gb-2009-10-3-r25
17. Li H, Handsaker B, Wysoker A et al (2009) The sequence alignment/map format and SAMtools. Bioinformatics 25 (16):2078–2079. https://doi.org/10.1093/bioinformatics/btp352
18. Anders S, Pyl PT, Huber W (2015) HTSeq--a python framework to work with high-throughput sequencing data. Bioinformatics 31(2):166–169. https://doi.org/10.1093/bioinformatics/btu638
19. Robinson MD, McCarthy DJ, Smyth GK (2010) edgeR: a bioconductor package for differential expression analysis of digital gene expression data. Bioinformatics 26 (1):139–140. https://doi.org/10.1093/bioinformatics/btp616
20. Robinson JT, Thorvaldsdottir H, Winckler W et al (2011) Integrative genomics viewer. Nat Biotechnol 29(1):24–26. https://doi.org/10.1038/nbt.1754

Chapter 9

Protocol for Proteome Analysis of Group A Streptococcus

Laura Wilk

Abstract

This chapter describes a protocol for the separation of different cell fractions from *Streptococcus pyogenes* and sample preparation for analysis by mass spectrometry. The presented approach can be used for the analysis of all subcellular proteomes from *S. pyogenes* and enables identification of drug or vaccine targets within a certain cellular fraction. Here, proteins integral to the plasma membrane are of particular interest for the development of new antimicrobial therapies.

Key words *Streptococcus pyogenes*, Gram-positive bacteria, Bacterial proteome, Cellular fractionation, secretome, Cell wall, Membrane, Mass spectrometry

1 Introduction

Streptococcus pyogenes (Group A Streptococcus, GAS) is a Gram-positive bacterium, which consists of three cellular compartments: the cytoplasm, the plasma membrane, and a cell wall consisting mainly of peptidoglycan [1]. A protective outer membrane such as found in Gram-negative bacteria is missing. Although this makes the cells exposed to the environment, extracellular proteins are still present in Gram-positive bacteria [2, 3]. Extracellular proteins are synthesized in the cytoplasm and following translocation across the plasma membrane, they are secreted into the surroundings or stay associated with the cell surface [2, 3]. Surface-associated proteins are either adhered to the cell wall or the plasma membrane. Cell wall proteins can stay attached to the peptidoglycan by covalent anchoring or noncovalent interactions in Gram-positive bacteria [2, 3]. A prominent example of cell wall anchoring in GAS is the M- and M-like proteins [1]. Particularly interesting is the presence of proteins on the GAS surface, which were previously thought to be confined to the cytoplasm. These proteins, such as glyceraldehyde-3-phosphate dehydrogenase (GAPDH), are termed "moonlighting proteins," since they appear to serve more than one cellular function [3, 4]. They are translocated via an unknown mechanism

Thomas Proft and Jacelyn M. S. Loh (eds.), *Group A Streptococcus: Methods and Protocols*, Methods in Molecular Biology, vol. 2136, https://doi.org/10.1007/978-1-0716-0467-0_9,

and likely retained at the surface by charge or hydrophobic interactions [4]. Another class of proteins located at the cellular surface are the lipoproteins, which are tethered to the cytoplasmic membrane and might represent anchored equivalents to their counterparts in the Gram-negative bacterial periplasm [3]. The periplasmic space is located in between the cell wall and the plasma membrane in Gram-negative bacteria, and the existence of a similar separating space has been suggested also for Gram-positive bacteria [1]. Additional soluble proteins could be present in such a putative periplasmic space in GAS and be retained by membrane proteins. Integral membrane proteins are also synthesized in the cytoplasm and then inserted into the plasma membrane [5]. These proteins are of special interest, since they are involved in many important cellular processes, such as signaling and transport, and they are, therefore, interesting new targets for antimicrobial research.

So far, different techniques for analysis of the GAS proteome have been employed, such as extraction of cell wall-associated proteins with mutanolysin or lysozyme [6], "surface shaving" aided by proteases [7], and fractionation of whole bacterial cells following breakage, yielding mixtures of the cellular compartments, respectively [8, 9]. The method presented herein now comprises a fractionation protocol that yields separated cellular fractions, especially with respect to the membrane proteome.

For analysis of the GAS secretome, this protocol follows the common approach of growing the bacteria in protein-reduced medium in order to reduce the number of nonstreptococcal proteins [10]. The growth medium containing secreted proteins is then harvested. The preparation of cell wall extracts is based on a method using the muralytic enzyme mutanolysin [11]. The remaining protoplasts are then disrupted by sonication. Unbroken protoplasts and cellular debris are removed by centrifugation. The supernatant is subjected to ultracentrifugation in order to separate cytoplasmic proteins from crude membranes. Membrane-associated peripheral proteins are nonintegral to the plasma membrane and are isolated by treating the crude membranes with high salt containing buffer followed by ultracentrifugation. This step yields both the extracted peripheral proteins and purified membranes. The preparation of samples from these subcellular fractions for mass spectrometric analysis is described. For subsequent performance of liquid chromatography-mass spectrometry, protein identification, and data analysis, the reader is referred elsewhere [12, 13].

In order to study novel drug or vaccine targets, it is important to gain a better knowledge of the proteins produced by GAS and their cellular location. This protocol provides a cellular fractionation method that allows complete GAS proteome analysis on the basis of separated fractions. Importantly, the entire isolated membrane proteome can be investigated.

2 Materials

Prepare all solutions with ultrapure water (Milli-Q® water), unless indicated otherwise. We do not add sodium azide to solutions. Prepare and store all reagents at RT, unless stated otherwise. Please review and follow the manufacturers' safety data sheets for all reagents used. *S. pyogenes* is a human pathogen and should be handled with the required safety precautions.

2.1 Bacterial Culture

1. Group A Streptococcus (*S. pyogenes*) strain (*see* **Note 1**).
2. Blood agar plates.
3. 0.3% THY medium: 30 g/l Todd-Hewitt (TH) broth, 3 g/l yeast extract. Prepare 1 × 110 ml in small Erlenmeyer flask and 2 × 1 l in two large Erlenmeyer flasks. Autoclave prior to use.
4. Protein-reduced TH broth: 30 g/l TH broth; passed through a 10-kDa molecular weight cutoff ultrafiltration membrane disc in a stirred cell prior to autoclavation. Discard rententate. Prepare 1 × 100 ml in small Erlenmeyer flask (*see* **Note 2**).
5. Incubator.
6. Photometer and disposable polystyrene cuvettes (10 × 4 × 45 mm).

2.2 Preparation of Secreted Extracellular Proteins

1. 0.22-μm pore size filter unit.
2. Phosphate buffered saline (PBS): 137-mM NaCl, 2.7-mM KCl, 10-mM Na_2HPO_4, 1.8-mM KH_2PO_4, pH 7.4. Filter sterile or autoclave and prechill on ice before use.
3. 10-kDa molecular weight cut off centrifugal filtration unit.

2.3 Subcellular Fractionation: Preparation of Cell Wall Proteins, Cytoplasmic Proteins, and Crude Membranes

1. PBS (prechilled on ice). *See* Subheading 2.2 **item 2**.
2. TE buffer: 50-mM Tris–HCl, 1-mM EDTA, pH 8. Filter sterile or autoclave and store at 4 °C. Prechill on ice before use.
3. TES buffer: 20% (w/v) sucrose in TE buffer. Filter sterile or autoclave and store at 4 °C. Prechill on ice before use.
4. 100-mM 4-(2-aminoethyl)benzenesulfonyl fluoride hydrochloride (AEBSF). Prepare stock solution in water and store at −20 °C. Add to buffers just before use.
5. 5000-U/ml Mutanolysin solution: Resuspend mutanolysin from *Streptomyces globisporus* ATCC 21553 in chilled and sterile filtered K_2HPO_4, pH 6.2 (1 ml is sufficient for 2-l bacterial culture). Prepare fresh each time and store on ice before use.
6. 100 mg/ml lysozyme solution: Resuspend chicken egg white lysozyme in chilled TES buffer (0.2 g is sufficient for 2-l bacterial culture). Prepare fresh each time and store on ice before use.

7. Mutanolysin/lysozyme mix: Mix 20-ml TES buffer with 2-ml lysozyme solution and 1-ml mutanolysin solution. Prepare immediately before use and place on ice.
8. Protease and Phosphatase Inhibitor Mini Tablets, EDTA-free (Pierce™, Thermo Scientific). Use 1 tablet per 10-ml buffer.
9. Small autoclaved Erlenmeyer flask (empty).
10. 50-ml conical centrifugation tube (e.g., Falcon® tube).
11. Sterile microcentrifuge tubes.
12. Probe sonicator (we use a Branson Sonifier 150).
13. Ultracentrifuge, rotor, and centrifugation tubes capable of centrifugation at 200,000 × g (we use a Type 50.2 Ti tube rotor in an Optima Beckman ultracentrifuge).

2.4 Isolation of Membrane-Associated Peripheral Proteins

1. TE buffer (*see* Subheading 2.3).
2. TEN buffer: TE buffer, 1-M NaCl. Filter sterile or autoclave, store at 4 °C. Prechill on ice before use.
3. Handheld glass Dounce homogenizer (for small volumes, e.g., 2 ml). Prechill on ice before use.
4. 3-kDa molecular weight cutoff centrifugal filtration unit.
5. Ultracentrifuge, rotor, and small volume centrifugation tubes capable of centrifugation at 200,000 × g (we use a TLA 120.2 rotor in an Optima Beckman ultracentrifuge).

2.5 Sample Preparation for Mass Spectrometry

1. ProteaseMAX™ (Promega).
2. 50-mM ammonium bicarbonate in liquid chromatography (LC)-grade water.
3. Bicinchoninic Acid (BCA) Protein Assay (BCA Kit, Pierce™, Thermo Scientific).
4. Microplate reader and 96-well microtiter plates.
5. 4–20% acrylamide gradient gel (BioRad).
6. Loading buffer: 4× Laemmli Sample Buffer complemented with β-mercaptoethanol (BioRad) according to manufacturer's protocol, freshly prepared.
7. Molecular weight marker: Precision Plus Protein™ Dual Xtra Prestained Protein Standard, 2–250 kDa range (BioRad).
8. Coomassie® Blue staining solution (GelCode™ Blue Safe Stain Reagent, Thermo Scientific). Store at RT.
9. Destain solution: MilliQ®-water.
10. Urea-buffer: 8-M urea, 0.1-M ammonium bicarbonate dissolved in LC-grade water (*see* **Note 3**).
11. 0.1-M ammonium bicarbonate in LC-grade water (*see* **Note 3**).

12. 0.5-M Tris(2-carboxyethyl)phosphine (TCEP).
13. 0.5-M iodoacetamide in LC-grade water (*see* **Note 3**).
14. Trypsin (sequencing grade, Promega).
15. 10% formic acid in LC-grade water (*see* **Note 4**).
16. C18 reverse-phase spin columns (Macro SpinColumns™, Harvard Apparatus).
17. LC-grade methanol.
18. LC-grade acetonitrile.
19. LC-grade water.
20. Buffer A: 2% acetonitrile, 0.2% formic acid in LC-grade water (*see* **Note 4**).
21. Buffer B: 50% acetonitrile, 0.2% formic acid in LC-grade water (*see* **Note 4**).
22. Thermo block or incubator.
23. Vacuum concentrator.
24. Ultrasonic water bath.
25. Mass spectrometric sample vials.

3 Methods

3.1 Bacterial Culture

1. Streak *S. pyogenes* onto a blood agar plate and incubate over night at 37 °C/5% CO_2 (*see* **Note 5**).
2. Inoculate a single, hemolytic colony (*see* **Note 6**) into 100-ml 0.3% THY. Incubate over night at 37 °C/5% CO_2 without shaking and grow cells to stationary phase for 16–18 h.
3. The next day, determine the optical density at 620 nm (OD_{620}) of the 100-ml preculture (*see* **Note 7**).
4. Add 50-ml preculture to 1-l 0.3% THY in a large Erlenmeyer flask (use two flasks in total to yield approx. 2-l final culture volume) and determine OD_{620} again.
5. In addition, take 5 ml from the same 100-ml preculture from **step 3** and add to 100-ml protein-reduced TH in a small Erlenmeyer flask (*see* **Note 8**). Determine the OD_{620} again. This 100-ml culture will be used only for collecting the secreted proteins.
6. Incubate all cultures at 37 °C/5% CO_2 without shaking and grow cells until mid-log phase is reached (*see* **Note 9**).

3.2 Preparation of Secreted Extracellular Proteins

1. Harvest the cells from the 100-ml culture with protein-reduced TH (*see* Subheading 3.1) at 12,000 × *g* for 15 min at 4 °C. Carefully take off the supernatant.

2. Sterile filter the supernatant through a 0.22-µm pore size filter unit and keep it on ice.
3. Concentrate 15 ml of the supernatant by ultrafiltration using a 10-kDa molecular weight cutoff centrifugal filtration unit at 4000 × *g* for 15 min at 4 °C.
4. Wash three times with 5-ml ice-cold PBS; the final volume of concentrated supernatant should be 1 ml.
5. Store supernatant containing secreted extracellular proteins at −80 °C in a sterile reaction tube until further use.

3.3 Subcellular Fractionation: Preparation of Cell Wall Proteins, Cytoplasmic Proteins, and Crude Membranes

1. Harvest the cells from both 1-l cultures (*see* Subheading 3.1) at 12,000 × *g* for 15 min at 4 °C.
2. Discard the supernatant and keep both cell pellets on ice.
3. Resuspend cell pellets in a total of 35 ml of ice-cold PBS by pipetting up and down.
4. Pellet the cells again at 12,000 × *g* for 15 min at 4 °C and discard the supernatant.
5. Store the PBS-washed pellet at −20 °C until further use (*see* **Note 10**).
6. Thaw the PBS-washed cell pellet and resuspend in 35 ml of ice-cold TE buffer containing 1-mM AEBSF (*see* **Note 11**) by pipetting up and down. No bacterial clumps should be visible after resuspension.
7. Pellet the cells again at 12,000 × *g* for 15 min at 4 °C and discard the supernatant.
8. Repeat **steps 6** and 7.
9. Resuspend cell pellet in 23 ml of ice-cold mutanolysin/lysozyme mix by pipetting up and down (avoid foaming and bacterial clumps).
10. Transfer to a small autoclaved Erlenmeyer flask and incubate for 2 h at 37 °C with shaking at 100 rpm.
11. Centrifuge for 10 min at 14,000 × *g* at RT.
12. Collect the supernatant containing the solubilized cell wall-associated proteins and store at −80 °C until further use.
13. Wash the pellet (remaining protoplasts) twice with 20-ml TES buffer containing 1-mM AEBSF by centrifugation for 5 min at 14,000 × *g* at RT to remove traces of solubilized cell wall-associated proteins.
14. Perform all further steps on ice with ice-cold buffers.
15. Resuspend pellet in 20-ml TE buffer containing two protease inhibitor tablets until no more bacterial clumps are visible. Transfer to 50-ml conical centrifugation tube and precool on ice.

16. Disrupt protoplasts by sonication with a tip sonicator (*see* **Note 12**): 6 × 10 s cycles at output 6, followed by 6 × 10 s cycles at output 10. Cool for 1.5 min on ice in between cycles.
17. Transfer cell suspension into a centrifugation tube and centrifuge at 10,000 × *g* for 15 min at 4 °C to remove any remaining cellular debris or unbroken protoplasts.
18. Transfer supernatant to an ultracentrifugation tube.
19. Balance tubes carefully and centrifuge sample at 200,000 × *g* for 1 h at 4 °C.
20. Carefully remove the supernatant containing the cytosolic proteins from the crude membrane pellet and keep on ice.
21. Rinse the membrane pellet three times with 2-ml TE buffer containing 1-mM AEBSF and take it up in 500 μl of the same buffer (*see* **Note 13**).
22. Store cytosolic proteins and crude membrane suspension in sterile reaction tubes at −80 °C until further use.

3.4 Isolation of Membrane-Associated Peripheral Proteins

1. Add chilled 700 μl of TE-buffer, 1-M NaCl to 300-μl crude membrane suspension.
2. Homogenize the membrane suspension with a small handheld glass Dounce homogenizer and incubated on ice for 30 min.
3. Transfer the membrane suspension to an ultracentrifugation tube.
4. Balance tubes carefully and centrifuge at 200,000 × *g* for 1 h at 4 °C.
5. Carefully remove the supernatant containing the membrane-associated peripheral proteins and keep the remaining membrane pellet on ice.
6. Concentrate the supernatant (~1 ml) to 200 μl by ultrafiltration with a 3-kDa molecular weight cutoff centrifugal filtration unit at 14,000 × *g* at 4 °C.
7. Wash the supernatant once with 1-ml TE-buffer containing 1-mM AEBSF (*see* **Note 14**) and concentrate again to 200 μl by ultrafiltration. Transfer to a sterile microcentrifuge tube.
8. Rinse the remaining purified membrane pellet three times with 1-ml TE-buffer containing 1-mM AEBSF and take it up in 300 μl of the same buffer.
9. Store membrane-associated peripheral proteins and purified membrane suspension in sterile reaction tubes at −80 °C until further use.

3.5 Sample Preparation for Mass Spectrometry

1. Prepare 1% ProteaseMAX™ stock solution in 50-mM ammonium bicarbonate (*see* **Note 3**).
2. Solubilize 200 μl of each membrane fraction (crude or purified) with 0.05% ProteaseMAX™ at RT for 20 min with a vortexer.
3. Inactivate the ProteaseMAX™ reagent by incubation at 95 °C for 5 min.
4. Centrifuge ProteaseMAX™ treated fractions at 16,000 × *g* for 10 min at 4 °C and collect the supernatants containing the solubilized membrane proteins.
5. Determine the protein concentration in all subcellular fractions using the BCA protein assay (*see* **Note 15**).
6. Carry out BCA protein assay in a microplate reader format: mix 25 μl of prediluted (*see* **Note 16**) sample with 200 μl of BCA working reagent (sample to BCA working reagent ratio of 1:8 (v/v)). Prepare bovine serum albumin (BSA) standards according to the manufacturer's instructions. Measure the absorbance at 562 nm and use the BSA standard curve to determine protein concentration of the samples.
7. Load 10 μg of protein per lane (determined by BCA assay) onto a 4–20% acrylamide gradient gel (BioRad) and run according to manufacturer's instructions. Estimate the protein content by eye.
8. Based on both estimates from **steps 5** and 7, transfer 20 μg of each subcellular fraction into a microcentrifuge tube for tryptic digest.
9. Add 8-M urea and 0.1-M ammonium bicarbonate to a final sample volume of 100 μl.
10. Reduce cysteine bonds by adding TCEP to a final concentration of 5 mM and incubation for 30–60 min at 37 °C.
11. Add iodoacetamide to a final concentration of 10 mM for alkylation and incubate for 30–60 min at 22 °C.
12. Dilute samples with 0.1-M ammonium bicarbonate to a final urea concentration of 1.5 M and add 1-μg sequencing grade trypsin to digest per 10 μg of protein. Incubate for 18 h at 37 °C in a thermo block or incubator.
13. Acidify samples with 10% formic acid to a final pH of 3.0 (add approximately 150 μl).
14. Purify the peptides with C18 reverse phase spin columns according to the manufacturer's instructions.
15. Dry peptides in a vacuum concentrator to complete dryness (2–4 h) and reconstitute the peptides in 40-μl 2% acetonitrile, 0.2% formic acid.

16. Sonicate samples in a water bath for 5 min at RT and transfer to a mass spectrometric sample vial for mass spectrometric analyses.

4 Notes

1. We used *S. pyogenes* strain AP1 (40/58; covS truncated) of the M1 serotype, which was obtained from the WHO Collaborating Centre for Reference and Research on Streptococci, Prague, Czech Republic.
2. Filtration of TH medium will take up to a few hours. The obtained protein-reduced TH growth medium will lack proteins with a molecular mass greater than 10 kDa.
3. Prepare fresh and use within the same day.
4. Do not use plastic laboratory equipment when diluting formic acid for the working solution.
5. We store *S. pyogenes* as glycerol stocks at −80 °C and steak the bacteria out onto fresh blood agar plates for each experiment.
6. Only select a clearly isolated and hemolytic colony for inoculation of growth medium. We use a colony counter for this purpose.
7. For *S. pyogenes* strain AP1, the expected OD_{620nm} is ~0.8.
8. The same preculture is used herein in order to obtain the proteome from one bacterial clone.
9. Cells will need approx. 5.5–6 h to reach mid-log phase when growing in 1-l culture flasks. For *S. pyogenes* strain AP1, the mid-log phase is reached at an OD_{620nm} of ~0.4.
10. The protocol was stopped at this stage due to lengthy culture growth and additional preparation of secreted proteins.
11. We used the water-soluble and nontoxic AEBSF as an alternative to the highly toxic PMSF. Prepare AEBSF-containing buffers immediately before use.
12. The reader is advised that this protocol aims at the preparation of cellular fractions for mass spectrometric analysis. Upscaling the whole preparation might require optimization of the protoplast disruption step.
13. Depending on centrifugation tube shape, it might be useful to cut the tube open for this step. The slightly sticky membrane pellet can be loosened with a small spatula and homogenized with a chilled small handheld glass Dounce homogenizer on ice.
14. This washing step is required to reduce the salt content for later mass spectrometric analyses.

15. It was not possible to determine the protein content for secreted proteins using the BCA protein assay, due to contaminating proteins from the media.
16. Some samples such as membranes might require dilution (e.g., by 1:10) for protein concentration determination.

Acknowledgments

The author thanks Lotta Happonen and Heiko Herwald (both Lund University, Lund, Sweden) for critical input.

References

1. Navarre WW, Schneewind O (1999) Surface proteins of gram-positive bacteria and mechanisms of their targeting to the cell wall envelope. Microbiol Mol Biol Rev 63(1):174–229
2. Silhavy TJ, Kahne D, Walker S (2010) The bacterial cell envelope. Cold Spring Harb Perspect Biol 2(5):a000414
3. Scott JR, Barnett TC (2006) Surface proteins of gram-positive bacteria and how they get there. Annu Rev Microbiol 60:397–423
4. Fischetti VA (2016) M protein and other surface proteins on streptococci. In: Ferretti JJ, Stevens DL, Fischetti VA (eds) Streptococcus pyogenes: basic biology to clinical manifestations. University of Oklahoma Health Sciences Center, Oklahoma City, OK. https://www.ncbi.nlm.nih.gov/books/NBK333431/. Accessed 1 June 2017
5. Xie K, Dalbey RE (2008) Inserting proteins into the bacterial cytoplasmic membrane using the sec and YidC translocases. Nat Rev Microbiol 6(3):234–244
6. Callegari EA, Chaussee MS (2016) The streptococcal proteome. In: Ferretti JJ, Stevens DL, Fischetti VA (eds) Streptococcus pyogenes: basic biology to clinical manifestations. University of Oklahoma Health Sciences Center, Oklahoma City, OK. https://www.ncbi.nlm.nih.gov/books/NBK333416/. Accessed 1 June 2017
7. Rodriguez-Ortega MJ, Norais N, Bensi G et al (2006) Characterization and identification of vaccine candidate proteins through analysis of the group A Streptococcus surface proteome. Nat Biotechnol 24(2):191–197
8. Malmström J, Karlsson C, Nordenfelt P et al (2012) Streptococcus pyogenes in human plasma adaptive mechanisms analyzed by mass spectrometry-based proteomics. J Biol Chem 287(2):1415–1425
9. Sharma A, Arya DK, Sagar V et al (2013) Identification of potential universal vaccine candidates against group a streptococcus by using high throughput in silico and proteomics approach. J Proteome Res 12(1):336–346
10. Lei B, Mackie S, Lukomski S, Musser JM (2000) Identification and immunogenicity of group A Streptococcus culture supernatant proteins. Infect Immun 68(12):6807–6818
11. Cole JN, Djordjevic SP, Walker MJ (2008) Isolation and solubilization of Gram-positive bacterial cell wall-associated proteins. In: Posch A (ed) 2D PAGE: sample preparation and fractionation. Methods in molecular biology, vol 425. Humana Press, New York, NY, pp 295–311
12. Wilk L, Happonen L, Malmström J, Herwald H (2018) Comprehensive mass spectrometric survey of Streptococcus pyogenes subcellular proteomes. J Proteome Res 17(1):600–617
13. Karlsson C, Teleman J, Malmström J (2017) Analysis of bacterial surface interactions with mass spectrometry-based proteomics. In: Nordenfelt P, Collin M (eds) Bacterial pathogenesis. Methods in molecular biology, vol 1535. Humana Press, New York, NY, pp 17–24

Chapter 10

Investigation of Group A Streptococcal Interactions with Host Glycan Structures Using High-Throughput Techniques: Glycan Microarray Analysis Using Recombinant Protein and Whole Cells

Anuk Indraratna, David M. P. De Oliveira, Lauren E. Hartley-Tassell, Christopher J. Day, Mark J. Walker, Michael P. Jennings, and Martina L. Sanderson-Smith

Abstract

Glycans, also known as carbohydrates, are abundant upon cell surfaces, where they often mediate host–pathogen interactions. The specific recognition of host glycans by pathogenic lectins is an important process that allows the adherence of bacteria to the host epithelial surface in many species, including Group A Streptococcus (GAS). Glycan microarrays present a sensitive, high-throughput approach for identifying novel lectin–glycan interactions and can be applied in the context of whole bacteria or purified bacterial proteins.

Key words Glycan, Microarray, Lectin, Adherence, *Streptococcus pyogenes*, M protein

1 Introduction

Glycans, commonly referred to as sugars or carbohydrates, are the most abundant biopolymers in nature and have distinct roles in many biological contexts. Glycans are typically complex oligosaccharide structures, assembled from 10 distinct monosaccharide units that polymerize in many different orientations, generating an enormity of possible conformations. Conjugation of glycans to proteins, termed glycosylation, is thus one of the most diverse forms of post-translational modification and is critical in enabling complex processes in cellular machinery, such as metabolism, signal transduction, and molecular transport [1]. Glycans are present

Thomas Proft and Jacelyn M. S. Loh (eds.), *Group A Streptococcus: Methods and Protocols*, Methods in Molecular Biology, vol. 2136, https://doi.org/10.1007/978-1-0716-0467-0_10,

mostly in the form of glycoproteins, but also occur as proteoglycans or glycolipid conjugates, all of which are abundant at both prokaryote and eukaryote cell surfaces. In multicellular organisms, glycans at the cell surface mediate a number of important processes, such as cell–cell communication, adherence to extracellular matrices, and immune recognition [1].

Glycan moieties are recognized by a specific subset of carbohydrate-binding proteins, termed lectins. The interaction between bacterial lectins and host glycans is now appreciated to be an important process in the adherence to, and colonization of, host tissue by multiple bacterial species, including *Escherichia coli* [2], *Helicobacter pylori* [3], and *Pseudomonas aeruginosa* [4]. Despite recent advancements, the study and application of glycan–lectin interactions remains a significant challenge. Unlike proteins or nucleic acids, glycan synthesis in organisms is neither linear nor template driven, resulting in remarkable diversity and complexity. Furthermore, it is difficult to obtain substantial quantities of glycans from samples, particularly in a homogenous form. Traditional approaches to screen glycan–lectin interactions (e.g., lectin blots, enzyme-linked lectin assays, surface plasmon resonance (SPR), and isothermal calorimetry) require substantial amounts of precious sample and labor-intensive optimization. Glycan array technology presents a fast, high-throughput, and sample-conserving alternative, and thus constitutes a powerful tool in the elucidation of glycan-mediated host–pathogen interactions. A well-known example is the use of glycan arrays to characterize the binding of viral influenza A protein, hemagglutinin, to several host glycan structures [5]. The use of glycan arrays in the examination of host–pathogen interactions is becoming increasingly widespread, with the method being adapted to incorporate whole-cell analyses rather than proteins alone [6, 7]. De Oliveira et al. [8, 9] identified a wide range of novel glycan interactions by the surface-bound M protein of *Streptococcus pyogenes* (Group A Streptococcus; GAS). Although observations of the lectin properties of the GAS M protein had been reported earlier, this study provided a more comprehensive screening method for M protein-dependent binding to 358 candidate glycan structures, yielding 19 positive results. The binding of some of these glycans, such as the histo-blood group antigen H type 1, was confirmed by SPR, revealing high-affinity interactions. Given the abundance of glycans at the host epithelial surfaces colonized by GAS, the importance of glycan recognition in GAS pathogenesis requires further attention. The following method details a protocol that is designed to assess the lectin profile of GAS and the GAS M protein, but can be adapted to other whole-cell bacteria and/or putative lectins.

2 Materials

Prepare all solutions using deionized water and analytical grade reagents. Prior to use, filter all buffers using 22 μm filters. Prepare and store all reagents at room temperature, unless indicated otherwise.

2.1 Preparation of GAS for Whole-Cell Analysis

1. GAS isolates of interest.
2. THY media: Todd-Hewitt broth + 1% (w/v) yeast.
3. Phosphate-buffered saline (PBS): 10 mM Na_2HPO_4, 1.75 mM KH_2PO_4, 2.7 mM KCl, 137 mM NaCl.
4. Array PBS: 17.4 mM Na_2HPO_4, 3.5 mM NaH_2PO_4, 1.75 mM KH_2PO_4, 2.7 mM KCl, 137 mM NaCl, 3.5 mM KCl, 0.9 mM $MgCl_2$, 0.9 mM $CaCl_2$ (*see* **Note 1**). Prepare 200 ml of 20× stock of PBS. Prepare 200 mL of 20× stock of solution B (18 mM $CaCl_2$, 70 mM KCl, 18 mM $MgCl_2$, 2740 mM NaCl) in distilled water (dH_2O). Autoclave both solutions in separate bottles and allow to cool to room temperature. Using separate syringes and sterile filter units, add 20 mL of PBS and 20 ml of solution B to 360 ml of Milli-Q H_2O to a total volume of 400 mL. Adjust the pH to 7.4.
5. CellTrace™ BODIPY® TR methyl ester (*see* **Note 2**) or other fluorescent stain.
6. 1.5 mL microfuge tubes.
7. Fixed rotor centrifuge.
8. Sterile 15 mL centrifuge tubes for bacterial growth.
9. 1 mL syringe.
10. 5 μm pore syringe filter unit.

2.2 Preparation of Purified Protein

1. Purified GAS proteins of interest.
2. Primary antibody against the protein of interest, such as mouse anti-His antibody.
3. Secondary antibody–fluorophore conjugate, such as rabbit anti-mouse Alexa Fluor® 488 antibody conjugate.
4. Tertiary antibody–fluorophore conjugate, such as goat anti-rabbit Alexa Fluor 488 antibody conjugate.

2.3 Preparation of Glycan Microarray Slides

1. Glycan microarray slide.
2. Powder-free gloves.
3. Forceps for slide handling.
4. Petri dish (or similar).

5. 50 mL centrifuge tubes.
6. 0.1% bovine serum albumin (BSA) in array PBS.
7. 0.001% Tween-20 in array PBS.

3 Methods

3.1 Preparing GAS for Whole-Cell Glycan Array Analysis

1. Prepare an overnight culture of GAS grown in THY at 37 °C (5% CO_2).
2. From the overnight culture, inoculate fresh THY media and grow GAS culture to mid-log phase ($OD_{600} = 0.5$) at 37 °C (5% CO_2) without shaking.
3. Harvest bacteria by centrifuging at 5000 × *g* for 10 min at room temperature. Discard the supernatant.
4. Resuspend in PBS and wash by centrifugation, as above.
5. Resuspend 1×10^8 CFU (colony-forming unit) in 250 μL of PBS containing 1 μM CellTrace™ BODIPY® TR methyl ester in a 1.5 mL microfuge tube.
6. Incubate in the dark for 15 min at room temperature to allow staining.
7. Centrifuge cells at 5000 × *g* for 10 min at room temperature to remove excess dye, wash twice via resuspension in 500 μL in array PBS.
8. After the final wash, resuspend cells in 250 μL array PBS.
9. Transfer the bacterial cell suspension into 1 mL syringe and pass through a 5 μm pore Millex-SV syringe filter unit to remove clumps of aggregated bacteria.
10. Keep on ice in the dark until needed.

3.2 Preparing Purified Protein for Glycan Array Analysis

Although this method describes the screening of recombinant M protein, it may be easily adapted for other GAS proteins, provided the appropriate antibodies are employed.

1. Preincubate 1 μg of purified recombinant M protein with mouse anti-His tag antibody, rabbit anti-mouse Alexa Fluor-488 antibody conjugate, and goat anti-rabbit Alexa Fluor-488 conjugate at a ratio of 4:2:1 in a total volume of 65 μl of array PBS (*see* **Note 3**).
2. Incubate for 10 min at room temperature.

3.3 Preparation of Glycan Slides

This protocol does not detail the design and construction of glycan array slides. Readers are directed elsewhere for instruction, should they wish to prepare the slides in-house [6, 10, 11]. Briefly, minute volumes (in the range of pico- to nanoliters) of glycosylamines are printed onto epoxy-functionalized glass slides via robotic microarray spotting technology. We used Arrayit SuperEpoxy 3 slides (Arrayit, USA), printed using contact printing at the Griffith Institute for Glycomics. Given the specific reagents, equipment, and expertise required for these procedures, it is recommended that preprepared slides are sourced commercially. Dextra (UK) produces glycan array slides that consist of 104 unique glycans, and there are many other companies that offer similar products. If a commercial slide is used, the readers are encouraged to consult the specific product manual, in addition to this protocol. In most cases, slides are to be stored at 4 °C, in a vacuum-sealed container, until the day of the assay.

1. Remove the slide from 4 °C storage and allow it to equilibrate at room temperature (*see* **Note 4**).
2. Using forceps, transfer the slide into a 50 mL tube containing array PBS with 0.1% BSA (w/v). Ensure the volume of array PBS is sufficient to submerge the slide.
3. Allow blocking of the slide by incubating at room temperature for 10 min.
4. Rinse the slide by placing it into a clean 50 mL tube containing array PBS and inverting the tube several times for 2 min. Repeat once.
5. Transfer slide to an empty 50 mL tube. Dry by centrifuging at 200 × *g* for 5 min.

3.4 Affixing Gene Frames to Slide

Gene frames (ABgene) are designed to seal and protect reagents on slides from evaporation. Different sizes are available, and an appropriate size must be selected, depending on the dimensions of the given glycan array slide. Some companies provide frames that are compatible with the glycan array slides. This protocol assumes the dimensions of the frame to be 1.5 × 1.6 cm with a working capacity of 65 μl.

1. Remove the protective sheet from the bottom of the frame. The bottom sheet is usually thicker than the top sheet and often has the center square section excised.
2. Without touching the exposed adhesive, press the frame firmly onto the microscope slide. Ensure that the frame does not come into contact with the printed glycans on the glass slide.
3. Affix additional frames along the glass slide in this manner, if appropriate.
4. Press down firmly around the frame with forceps.

3.5 Incubating Sample with Slide

1. Remove the top protective sheet from the frame (*see* **Note 5**).
2. Pipette 65 μL of stained GAS (or protein: antibody solution) into the frame, without touching the slide. Include controls, if appropriate (*see* **Note 3**).

3. Seal the frame quickly by rolling the coverslip on at an angle, ensuring no air bubbles are formed.
4. Incubate for 10 min at room temperature in the dark.
5. Decant roughly 50 mL of array PBS into a large Petri dish or any other clean container.
6. Transfer the slide into the dish and submerge. Using forceps, gently remove the coverslips and frames.
7. Wash the slide for 5 min with gentle agitation in array PBS within the dish.
8. Transfer to a 50 mL tube for a secondary wash in array PBS with 0.001% Tween-20 for 2 min with gentle agitation.
9. Fix the slides by immersing the slides in a 50 mL tube containing 3.7% formaldehyde for 20 min at room temperature.
10. Transfer the slide to an empty 50 mL tube and dry the slide by centrifugation at 200 × *g* for 5 min.

3.6 Analysis

1. Perform a scan on the laser scanner, according to the manufacturer's instructions. An example is given below, but can be adapted depending on the instrument and the fluorophores employed:

Instrument	Innopsys InnoScan 1100 with the following lasers and filters: 488nm laser (520/5- for FITC/GFP), 532nm laser (582/75 - for bodipy ME595 or TR, and 605/11 for AF555) and 635nm laser (680/42 for AF647)
Blue argon laser	FITC setting; 494 nm excitation, 518 nm emission
Gain	40–60%

2. Fluorescence data are typically analyzed by companion software and present the intensity of each spot as a mean fold increase above background fluorescence. Generally, only spots with a fold increase of >1 plus 3 standard deviations, compared to background, can be deemed as binding (*see* **Note 6**).

4 Notes

1. If all components of the array PBS are combined, calcium chloride ($CaCl_2$) will precipitate out of the solution. Therefore, array PBS needs to be prepared afresh using the two stock solutions, as described above.
2. 5 mM CellTrace™ BODIPY® TR methyl ester stock should be diluted 1:5 (v/v) with sterile Milli-Q H_2O to produce a 1 mM working stock. Following incubation with 1 μM CellTrace™

BODIPY® TR methyl ester, wash and resuspend the final pellet in array PBS. The pellet should have a pink appearance.

3. If a 488 nm laser is not available, other Alexa Fluor® antibodies can be used. We routinely use Alexa Fluor® 555 or Alexa Fluor® 647, compatible with standard microarray scanners. Check labeling either by flow cytometry or by fluorescent microscopy using the appropriate excitation/emission filters. To account for false positives, the inclusion of an antibody-only control is recommended.
4. Always wear powder-free gloves when handling the slide, as the glycans are printed and exposed on the surface. Often, there is a sticker or barcode, which is safe to touch. Ensure the slide is oriented correctly, as per the manufacturer's instructions.
5. It is important that no air bubbles are captured between the coverslip and the slide. The presence of air bubbles will create an uneven low-localized concentration of protein within the gene frame and potentially not allow for visualization of positive binding results.
6. After scanning, if nonspecific binding is apparent, the slides can be rewashed once. It is not recommended to rewash slides more than once, as this may "wash-off" medium- to weak-affinity glycan–lectin binding interactions.

References

1. Varki A, Gagneux P (2017) Biological functions of glycans. In: Essentials of Glycobiology, 3rd edn. Cold Spring Harbor Laboratory Press, New York
2. Schembri MA, Christiansen G, Klemm P (2001) FimH-mediated autoaggregation of *Escherichia coli*. Mol Microbiol 41:1419–1430
3. Magalhães A, Marcos-Pinto R, Nairn AV et al (2015) Helicobacter pylori chronic infection and mucosal inflammation switches the human gastric glycosylation pathways. Biochim Biophys Acta Mol basis Dis 1852:1928–1939
4. Imberty A, Wimmerová M, Mitchell EP, Gilboa-Garber N (2004) Structures of the lectins from *Pseudomonas aeruginosa*: insights into the molecular basis for host glycan recognition. Microbes Infect 6:221–228
5. Stevens J, Blixt O, Paulson JC, Wilson IA (2006) Glycan microarray technologies: tools to survey host specificity of influenza viruses. Nat Rev Microbiol 4:857
6. Day CJ, Tiralongo J, Hartnell RD et al (2009) Differential carbohydrate recognition by *Campylobacter jejuni* strain 11168: influences of temperature and growth conditions. PLoS One 4:e4927
7. Mubaiwa T, Hartley-Tassell LE, Semchenko et al (2019) Investigation of whole cell meningococcal glycan interactions using high throughput glycobiology techniques: glycan array and surface plasmon resonance. In: *Neisseria meningitidis*—methods and protocols. Methods in molecular biology. Springer Protocols, New York
8. De Oliveira DM, Hartley-Tassell L, Everest-Dass A et al (2017) Blood group antigen recognition via the Group A *streptococcal* m protein mediates host colonization. MBio 8: e02237-16
9. De Oliveira DM, Everest-Dass A, Hartley-Tassell L et al (2019) Human glycan expression patterns influence Group A *streptococcal* colonisation of epithelial cells. FASEB J 33:10808–10818
10. Bradner JE, McPherson OM, Koehler AN (2006) A method for the covalent capture and screening of diverse small molecules in a microarray format. Nat Protoc 1:2344
11. Waespy M, Gbem TT, Elenschneider L et al (2015) Carbohydrate recognition specificity of trans-sialidase lectin domain from *Trypanosoma congolense*. PLoS Negl Trop Dis 9: e0004120

Part III

Assays to Investigate Group A Streptococcus Virulence Factors and Virulence Mechanisms

Chapter 11

Using *Lactococcus lactis* as Surrogate Organism to Study Group A Streptococcus Surface Proteins

Catherine (Jia-Yun) Tsai, Jacelyn M. S. Loh, Sam Blanchett, and Thomas Proft

Abstract

The isolation of a single Group A Streptococcus (GAS) virulence determinant in functional investigations is challenging, as GAS employs a multitude of virulence factors. The redundancy between many surface proteins such as adhesins also adds complexity and difficulty. *Lactococcus lactis* is a non-pathogenic Gram-positive species related to GAS that can be an ideal surrogate organism to circumvent this problem. Genetic manipulation in *L. lactis* is easy, and the mechanisms for processing and cell wall-anchoring of surface proteins are similar to GAS. Lactococci have been extensively used to express heterologous surface proteins from other bacterial species, and modern molecular cloning tools and protocols have been developed. This chapter describes the workflow of generating recombinant *L. lactis* strains expressing GAS surface proteins and the validation and quantification of their surface expression.

Key words *Lactococcus lactis*, Virulence factor, Surface proteins, Heterologous gene expression, Cell wall extracts, Flow cytometry

1 Introduction

Elucidating the mechanisms of GAS-mediated host-interactions may aid in defining novel targets to impede GAS infections. However, identifying the key components that mediate specific pathogenic function has been hampered by the multiplicity of virulence factors expressed by GAS. For example, GAS carries a number of functionally redundant adhesins to various host components, which is the very first step of host colonization. In an inactivated mutant, redundant adhesins targeting similar host receptors could complement the function, thus masking its function. Furthermore, the expression of important virulence genes is usually tightly regulated by global regulators, and is often associated with a certain growth phase or infection stage [1]. Therefore, classical gene inactivation for the studies of pathogenic role of individual GAS factors could yield ambiguous results.

Thomas Proft and Jacelyn M. S. Loh (eds.), *Group A Streptococcus: Methods and Protocols*, Methods in Molecular Biology, vol. 2136, https://doi.org/10.1007/978-1-0716-0467-0_11,

Lactococcus lactis subsp. *cremoris* 1363 (also known as MG1363) is a strain devoid of virulence factors, including adhesins which facilitate adhesion to mammalian matrix proteins, and its genetic background is well characterized [2]. This bacterium is also plasmid-free, thereby facilitating gene transformation. An additional advantage is that streptococci and lactococci use similar mechanisms to process their cell wall proteins, namely a C-terminal cell wall-anchor domain with the conserved LPXTG motif that is recognized by the house-keeping sorting enzyme Sortase A [3]. Furthermore, the non-pathogenic status of *L. lactis* provides a clean background to isolate the function of the heterologous virulence gene in in vivo experiments, as the gain of function in transformed *L. lactis* is likely to result from the introduction of the specific determinant.

L. lactis has been used as a surrogate host to express an array of virulence factors from Gram-positive organisms [4–7]. In the context of GAS, *L. lactis* has proven to be a useful tool to dissect the functions of a range of virulence determinants in various assays. For example, Chang et al. used a combination of a GAS deletion mutant and an *L. lactis* gain-of-function mutant to characterize a novel GAS virulence factor, *Streptococcus pyogenes* nuclease A (SpnA), and showed that SpnA-expressing *L. lactis* demonstrated better survival in both human whole blood killing assays and neutrophil killing assays [8]. Similarly, Franklin and colleagues showed that expression of AspA, an AgI/II family adhesin of GAS, in the surrogate host *L. lactis* protected bacteria from phagocytosis [9]. Apart from expressing individual virulence genes, *L. lactis* is also capable of accommodating larger heterologous DNA segments such as a pathogenic island with multiple genes. Examples include the GAS pili, which are encoded in the fibronectin-collagen-T-antigen (FCT) genomic island ranging from 5 to 14 kbp. It has been shown that *L. lactis* can be transformed by a plasmid containing the whole GAS pilus operon encoding the structural subunits and assembly enzymes for pilus formation, and express properly assembled GAS pili on the cell surface. These piliated recombinant *L. lactis* strains have been used in adherence and internalization assays [10], as well as live vaccines in immunization studies [11].

Molecular tools for expressing foreign genes in lactococci have been extensively developed. The construction of *E. coli–L. lactis* shuttle vectors has greatly facilitated the cloning of GAS genes via standard molecular techniques available for *E. coli*. The use of the well-characterized lactococcal promoter P23 in these vectors has proven to drive overexpression of streptococcal proteins in *L. lactis* [12]. The constitutive gene expression driven by P23 also provides the additional advantage of studying the function of specific determinants independent of their environmental cues or gene regulation. Here, we describe the pipeline of the experimental procedures for the generation of GAS surface protein expressing *L. lactis* from plasmid transformation to validation of surface protein expression, using the versatile *L. lactis* MG1363 strain as a host. Once the heterologous expression of the GAS protein on the *L. lactis* surface is validated, the gain-of-function mutants can be used in various assays to investigate the function of the GAS protein of interest.

2 Materials

All materials should be prepared under sterile conditions, and be handled with aseptic techniques. Media and buffers should be made up with ultrapure water (such as MilliQ) and sterilized by autoclaving or 0.22-μm membrane filtration.

2.1 Electro-competent *L. lactis*

1. *Lactococcus lactis* MG1363 freezer stock (*see* **Note 1**).
2. GM17 broth: M17 medium supplemented with 0.5% glucose.
3. 50 mM EDTA.
4. 0.3 M sucrose.
5. Ice-cold sterile distilled water.

2.2 Electroporation

1. 2-mm electroporation cuvette.
2. Electroporator.
3. Plasmid DNA with gene of interest constructed with an *E. coli*-*L. lactis* shuttle vector, such as pLZ12km2P23R [11] (*see* **Note 2**). Available tools and considerations during molecular cloning can be found in the review by van de Guchte [13].
4. GM17 agar plates containing appropriate antibiotics.

2.3 Cell Wall Protein Extraction

1. Phosphate-buffered saline (PBS): 2.7 mM KCl, 150 mM NaCl, 10 mM phosphate salts (pH 7.4).
2. Protoplast buffer: 40% sucrose, 10 mM $MgCl_2$, 0.1 M KPO_4 (pH 6.2). Add 2 mg/mL lysozyme, 400 U mutanolysin, and EDTA-free protease inhibitor before use.
3. Rotator or horizontal mixer (that can be placed in the 37 °C incubator).

2.4 Flow Cytometry Analysis of Cell Surface Protein Expression

1. Blocking buffer: PBS with 3% fetal bovine serum (FBS) and 5 mM EDTA.
2. FACS buffer: PBS with 1% FBS and 5 mM EDTA.
3. Primary antibody (against GAS protein of interest) (*see* **Note 3**).
4. Secondary antibody diluted at appropriate concentration (according to manufacturer's instruction) in FACS buffer.
5. 4% Paraformaldehyde in PBS.

6. Round-bottom polystyrene tubes.
7. Water bath sonicator.
8. Flow cytometer.

3 Methods

All procedures should be undertaken according to the containment regulations applicable to the laboratory, and ideally be performed in a sterile hood. Disposal of the used materials and transformed *L. lactis* should follow appropriate decontamination guidelines for biohazard waste.

3.1 Production of Electrocompetent L. lactis

1. Grow 1 mL *L. lactis* in GM17 broth at 28 °C under static conditions overnight.
2. Inoculate the overnight culture in 50 mL of fresh GM17 broth, and grow until the culture reaches mid-log phase (optical density at 600 nm [$OD_{600\ nm}$] of 0.4–0.6).
3. Centrifuge the bacteria at 5000 × *g* for 10 min at 4 °C and remove the culture medium.
4. Wash bacterial pellet twice with 4 mL ice-cold sterile water (*see* **Note 4**) by gently re-suspending the pellet and centrifuging at 5000 × *g* for 5 min at 4 °C.
5. Remove the supernatant, then wash the bacterial pellet once in 2 mL of ice-cold 50 mM EDTA by gently re-suspending the pellet and centrifuging at 5000 × *g* for 5 min at 4 °C.
6. Remove the supernatant, wash bacterial pellet once in 2 mL of ice-cold 0.3 M sucrose by gently re-suspending the pellet and centrifuging at 5000 × *g* for 5 min at 4 °C.
7. Remove the supernatant, and re-suspend the bacterial pellet in 0.4 mL of ice-cold 0.3 M sucrose.
8. Keep the cells on ice and use within 1 h after the preparation.

3.2 Electroporation of L. lactis

1. In a pre-chilled sterile plastic tube (*see* **Note 4**), mix 40 μL of competent *L. lactis* cells with 1–2 μg plasmid DNA (*see* **Note 5**).
2. Leave the mixture on ice for 3–5 min, and then transfer to a pre-chilled electroporation cuvette.
3. Using an electroporation apparatus, apply a single pulse of 2.5 kV, with capacitance at 25 μF and resistance at 200 Ω.
4. After the electric pulse, immediately dilute the cells in 1 mL of GM17 broth and transfer to a sterile plastic tube. Incubate at 28 °C for 2 h under static conditions.
5. Centrifuge the cells at 5000 × *g* for 5 min at RT.

6. Remove the media and re-suspend in 100 μL of GM17 broth, and spread onto a pre-warmed GM17 agar plate containing appropriate antibiotics.
7. Incubate the agar plate at 28 °C overnight.
8. Successful transformants can be selected by screening methods such as PCR using specific primers for the GAS gene.

3.3 *L. Lactis* Cell Wall Protein Extraction

1. Inoculate *L. lactis* clone in 15 mL of GM17 broth containing appropriate antibiotics for plasmid selection. Grow overnight at 28 °C under static conditions.
2. Centrifuge the cells at 5000 × *g* for 15 min at 4 °C.
3. Discard supernatant and wash the cell pellet once in PBS by gently re-suspending the cells and centrifuging at 5000 × *g* for 5 min at 4 °C.
4. Discard the supernatant and re-suspend the cell pellet in 1 mL ice-cold protoplast buffer.
5. Incubate at 37 °C for 3 h on a rotator or horizontal mixer at medium speed (e.g., 300 rpm).
6. Centrifuge at 13,000 × *g* for 15 min at 4 °C.
7. Store the supernatant containing cell surface proteins at −20 °C until use.
8. The expression of surface protein can be analyzed by standard Western Blot analysis using antibodies specific to the GAS protein expressed in *L. lactis*.

3.4 Quantitation of Cell Surface Protein Expression by Flow Cytometry

1. Grow 1.5 mL *L. lactis* culture at 28 °C under static conditions overnight.
2. Measure the OD_{600nm} of the overnight culture.
3. Centrifuge the cells at 5000 × *g* for 5 min at RT.
4. Re-suspend the pellet in blocking buffer at a concentration equivalent to $OD_{600nm} = 0.4$.
5. Put in a water bath sonicator for 2 min, and then incubate on ice for 30 min.
6. Centrifuge at 5000 × *g* for 5 min at RT. Remove the supernatant.
7. Re-suspend the pellet in the same volume (i.e., at $OD_{600nm} = 0.4$).
8. Aliquot 200 μL into a PCR tube (*see* **Note 6**).
9. Centrifuge at 5000 × *g* for 5 min at RT. Remove the supernatant as much as possible without disturbing the pellet.
10. Add 100 μL of primary antibody (diluted in FACS buffer), and incubate on ice for 30 min.

11. Add 100 μL FACS buffer, and then centrifuge at 5000 × *g* for 5 min at RT.
12. Remove 180 μL of the supernatant and wash again with 200 μL FACS buffer. Centrifuge at 5000 × *g* for 5 min at RT, and then remove 200 μL of the supernatant.
13. Re-suspend in 80 μL FACS buffer containing secondary antibody.
14. Incubate on ice for 30 min.
15. Add 100 μL FACS buffer, and then centrifuge at 5000 × *g* for 5 min at RT.
16. Remove 180 μL of the supernatant and wash again with 200 μL FACS buffer. Centrifuge at 5000 × *g* for 5 min at RT, and then remove 200 μL of the supernatant.
17. Re-suspend in 30 μL FACS buffer.
18. Add 50 μL 4% paraformaldehyde to fix the sample. Incubate at 37 °C for 10 min.
19. Add 200 μL FACS buffer and run on a flow cytometer.
20. Include appropriate control samples for gating on the flow cytometer. For single-cell gating, prepare a cell-only sample by re-suspending *L. lactis* cells in 200 μL FACS buffer at $OD_{600nm} = 0.4$. For setting the threshold of the negative cell population, prepare a secondary antibody-only sample by omitting the primary antibody staining procedures (*see* **steps 10–13**).

4 Notes

1. All *L. lactis* subsp. *cremoris* (MG1363) can be obtained from ATCC (ATCC® 19257™). Other plasmid-free *L. lactis* strains like MG1364 can also be used. MoBiTec offers *L. lactis* expression kits with a variety of *L. lactis* MG1363-derived strains.
2. We use the pLZ12km2 P23R plasmid, which derives from the pLZ12km2 plasmid [14] and contains the strong constitutive lactococcal promotor P23 [15] and a kanamycin resistance gene. This plasmid can be requested from our lab. MoBitec offers *L. lactis* expression kits with a variety of plasmids for the nisin-controlled expression (NICE). For expression of cell surface-anchored GAS proteins, the complete open reading frame, including targeting domains such as signal peptides or anchor domains, needs to be cloned. Many expression plasmids lack a ribosome-binding site downstream of the promotor site. Therefore, the ~10–15 nt upstream of the start codon should also be included.

3. A specific antibody (monoclonal or polyclonal) or serum from an immunized animal is required for detection of the surface protein. Some of these are commercially available, including serum against a variety of M proteins and T antigens for serotyping.
4. Ideally, all solutions and container for holding the harvested bacteria should be pre-chilled on ice or to 4 °C, including the centrifuge and the electroporation cuvettes. Solutions can be prepared beforehand and kept in the fridge for at least several hours, and tubes and cuvettes can be placed on ice during the growth time of the *L. lactis* culture. A metal centrifuge rotor is more favorable than a plastic one for maintaining at a cold temperature.
5. The quality and quantity of the plasmid DNA for electroporation is critical for the success of transformation. The DNA should be pure and suspended in sterile water, as impurities such as proteins or salts could lower the electroporation efficiency. High concentration of salts or air bubbles may cause arcing and lower the transformation efficiency too. As the volume ratio between competent cells and the DNA sample is also important (ideally the volume of the DNA solution should not exceed 1/10 of the volume of the competent cells), a low-concentration plasmid DNA sample should be concentrated to reduce the volume required. This can be done by ethanol precipitation or spin column methods.
6. For easy handling, small plastic tubes such as PCR tubes are recommended for the procedures. Depending on the number of the samples, strip PCR tubes or round-bottom 96-well plate can also be used.

References

1. Barrett JF, Hoch JA (1998) Two-component signal transduction as a target for microbial anti-infective therapy. Antimicrob Agents Chemother 42(7):1529–1536
2. Facklam R, Elliott JA (1995) Identification, classification, and clinical relevance of catalase-negative, gram-positive cocci, excluding the streptococci and enterococci. Clin Microbiol Rev 8(4):479–495
3. Schneewind O, Mihaylova-Petkov D, Model P (1993) Cell wall sorting signals in surface proteins of gram-positive bacteria. EMBO J 12 (12):4803–4811
4. Back CR, Douglas SK, Emerson JE et al (2015) Streptococcus gordonii DL1 adhesin SspB V-region mediates coaggregation via receptor polysaccharide of Actinomyces oris T14V. Mol Oral Microbiol 30(5):411–424. https://doi.org/10.1111/omi.12106
5. Sharp JA, Echague CG, Hair PS et al (2012) Staphylococcus aureus surface protein SdrE binds complement regulator factor H as an immune evasion tactic. PLoS One 7(5): e38407. https://doi.org/10.1371/journal.pone.0038407
6. Brennan MP, Loughman A, Devocelle M et al (2009) Elucidating the role of Staphylococcus epidermidis serine-aspartate repeat protein G in platelet activation. J Thromb Haemost 7 (8):1364–1372. https://doi.org/10.1111/j.1538-7836.2009.03495.x
7. Jakubovics NS, Stromberg N, van Dolleweerd CJ et al (2005) Differential binding specificities of oral streptococcal antigen I/II family adhesins for human or bacterial ligands. Mol

Microbiol 55(5):1591–1605. https://doi.org/10.1111/j.1365-2958.2005.04495.x

8. Chang A, Khemlani A, Kang H, Proft T (2011) Functional analysis of Streptococcus pyogenes nuclease A (SpnA), a novel group A streptococcal virulence factor. Mol Microbiol 79 (6):1629–1642. https://doi.org/10.1111/j.1365-2958.2011.07550.x
9. Franklin L, Nobbs AH, Bricio-Moreno L et al (2013) The AgI/II family adhesin AspA is required for respiratory infection by Streptococcus pyogenes. PLoS One 8(4):e62433. https://doi.org/10.1371/journal.pone.0062433
10. Tsai JC, Loh JM, Clow F et al (2017) The group A Streptococcus serotype M2 pilus plays a role in host cell adhesion and immune evasion. Mol Microbiol 103(2):282–298. https://doi.org/10.1111/mmi.13556
11. Loh JMS, Lorenz N, Tsai CJ et al (2017) Mucosal vaccination with pili from group A Streptococcus expressed on Lactococcus lactis generates protective immune responses. Sci Rep 7(1):7174. https://doi.org/10.1038/s41598-017-07602-0
12. Piard JC, Hautefort I, Fischetti VA et al (1997) Cell wall anchoring of the Streptococcus pyogenes M6 protein in various lactic acid bacteria. J Bacteriol 179(9):3068–3072. https://doi.org/10.1128/jb.179.9.3068-3072.1997
13. Vandeguchte M, Kok J, Venema G (1992) Gene-expression in Lactococcus-Lactis. FEMS Microbiol Lett 88(2):73–92. https://doi.org/10.1016/0378-1097(92)90685-H
14. Perez-Casal J, Caparon MG, Scott JR (1991) Mry, a trans-acting positive regulator of the M protein gene of Streptococcus pyogenes with similarity to the receptor proteins of two-component regulatory systems. J Bacteriol 173(8):2617–2624. https://doi.org/10.1128/jb.173.8.2617-2624.1991
15. Biswas I, Jha JK, Fromm N (2008) Shuttle expression plasmids for genetic studies in Streptococcus mutans. Microbiology 154 (8):2275–2282. https://doi.org/10.1099/mic.0.2008/019265-0

Chapter 12

Expression and Purification of Collagen-Like Proteins of Group A Streptococcus

Slawomir Lukomski and Dudley H. McNitt

Abstract

Prokaryotic proteins with extended collagen domain are found in many bacterial species that are pathogenic to humans and animals. The collagen domain is often fused to additional ligand-binding domains and plays both structural and functional roles in modular "bacterial collagens." Here, we describe the step-by-step expression and purification of the recombinant streptococcal collagen-like proteins, rScl, using the *Strep*-tag II system. The integrity and structural characterization of recombinant collagen-like proteins is very important for defining their function.

Key words Streptococcal collagen-like protein, Prokaryotic collagens, Recombinant protein production, Affinity purification, Protein characterization

1 Introduction

The collagen structure has evolved in metazoans as an essential component of extracellular matrix [1], but collagen-like proteins have also emerged in prokaryotes, likely by convergent evolution, and are associated with pathogenesis in numerous bacterial species [2] including pathogenic streptococci [3]. The collagen-like proteins 1 and 2, Scl1 and Scl2 [4–8], have been studied extensively, and Scl1 is involved in Group A Streptococcus pathogenesis, while the role of Scl2 is less clear. Both Scl proteins are homotrimeric and contain two functional extracellular domains: the cell-surface adjacent collagen (CL) domain and attached outermost globular sequence-variable V domain. The Scl-CL domain forms a triple helix [9, 10], which can bind human collagen receptors, integrins α2β1 and α11β1 [11, 12]. The Scl1-V domain binds selected extracellular matrix proteins [13], as well as human plasma components [14–17], thus performing important adhesive and immune defense functions [3]. The crystal structure of the V domain has been solved as a six-helix bundle, comprised of two conserved antiparallel α-helices, in each monomer, connected by surface-

Thomas Proft and Jacelyn M. S. Loh (eds.), *Group A Streptococcus: Methods and Protocols*, Methods in Molecular Biology, vol. 2136, https://doi.org/10.1007/978-1-0716-0467-0_12,

exposed loops [18]. The Scl1-V loops and characteristically distributed surface potential are critical determinants in ligand binding, and the V domain has to be attached to the triple-helical CL domain for ligand-binding function [19, 20].

Collagens are also desirable biomaterials, used in medicinal, cosmetic, and food industries that are extracted from animal sources by harsh procedures (recently reviewed in [21]). In contrast, designed recombinant Scl proteins, rScls, have been expressed in large quantities in *Escherichia coli* by several laboratories [22–26]. The collagen domain in rScl constructs was modified to support new biological functions by inserting binding motifs for heparin, integrins, fibronectin, or matrix metalloproteinases cleavage sites [27–30]. Those collagen mimetics were prepared in forms of surface coatings, hydrogels, or silk-like chimeras [31–33] to support human cell growth and wound healing [34–36].

Here, we describe the expression and purification of the recombinant rScl proteins, using the *Strep*-tag II system. rScl constructs are expressed in the *E. coli* periplasm and are extracted and purified by affinity chromatography under mild near physiological conditions. We also include two "quality control" techniques to detect the presence of collagen triple helix and characteristic domain organization, as a preliminary assessment of rScl functionality.

2 Materials

Prepare all solutions using ultrapure water. Unless otherwise noted, all reagents can be stored at room temperature. For experiments that come into direct contact with bacteria, proper sterile technique should be followed. We utilize the original *Strep*-tag II system (IBA, Lifesciences) [37] to produce rScl proteins. Currently, additional strategies of cloning, expression, and purification using this system are offered, including 6xHis-tag, Flag-tag, GST-tag, etc.; those will require alternative inducing agents and purification resins, but several steps and notes in the materials and methods section will still apply in the production of rScl proteins described here.

2.1 Protein Expression

2.1.1 General Equipment

1. Erlenmeyer flasks (50 mL for *small-scale inductions*).
2. Sterile inoculation loops.
3. Bunsen burner or open flame source.
4. Vortex mixer.
5. Spectrophotometer.
6. Shaking bacterial culture incubator: 30 °C or 37 °C.
7. Microcentrifuge and standard tabletop centrifuge.
8. Centrifuge tubes/bottles (1.5 mL, 35 mL, and 500 mL).

2.1.2 Bacterial Growth Reagents

1. LB-Miller broth (LB-Miller): 10 g/L tryptone, 5 g/L yeast extract, 10 g/L NaCl. A final concentration of 100 μg/mL ampicillin is added to the broth during protein expression for the selection of the pASK-IBA2 plasmid (IBA Lifesciences). Ampicillin stock (100 mg/mL) is made in sterile water and filtered.
2. LB-Miller agar: 10 g/L tryptone, 5 g/L yeast extract, 10 g/L NaCl, 15 g/L agar. A final concentration of 100 μg/mL ampicillin is used for the selection of the pASK-IBA2 vector (IBA Lifesciences).
3. 2 mg/mL anhydrotetracycline (AHTC) stock solution in dimethylformamide (DMF). Store at −20 °C.
4. Frozen bacterial culture: frozen glycerol stocks of *E. coli* colony (ies) harboring clones of *scl* sequences in pASK-IBA2 vector.

2.2 Protein Purification

2.2.1 General Equipment

1. Microcentrifuge or centrifuge.
2. Chromatography column.
3. Support stand with clamps.
4. 15-mL or 50-mL conical tubes.
5. Vortex mixer.
6. 1.5-mL microcentrifuge tube or 35-mL centrifuge tubes.
7. Mini-gel SDS–PAGE running tank.
8. SDS-acrylamide mini-gel casting apparatus.

2.2.2 Affinity Chromatography

1. Buffer P: 100-mM Tris/HCl pH 8.0, 500-mM sucrose, 1-mM ethylenediaminetetraacetic acid (EDTA). High sucrose buffer P is used to release the periplasmic content, including recombinant proteins that are expressed in the periplasm. This approach is referred to as the "buffer P method" throughout the text. The use of buffer P or alternative protein extraction buffer, for example, cell lytic buffer, is determined experimentally.
2. Cell Lytic Buffer: CelLytic B® (Sigma) with 1-mM phenylmethylsulfonyl fluoride (PMSF) and 50 U/mL DNase I. If CelLytic B is not available for purchase, a lab made reagent may be used: 50-mM Tris/HCl pH 8.0, 50-mM NaCl, 2-mM magnesium chloride, 2% Triton X-100, 10-mM β-mercaptoethanol, 0.2 mg/mL lysozyme, 1-mM phenylmethylsulfonyl fluoride (PMSF), 50 U/mL DNase I. This protocol is used if recombinant protein cannot be recovered efficiently with buffer P. This approach is referred to as the "cell lytic method" throughout the text. The use of cell lytic buffer or alternative protein extraction buffer, for example, buffer P, is determined experimentally.

3. 50 U/mL DNase IP, for use in the cell lytic buffer method (see above). Store at 4 °C.
4. 1-mM PMSF, for use in the cell lytic buffer method (see above). Store at 4 °C.
5. Tris–EDTA (TE): 10-mM Tris–HCl, 1-mM EDTA.
6. 2 mg/mL egg white avidin solution, made in sterile water. Store at 4 °C.
7. Strep-Tactin® Sepharose suspension (IBA Lifesciences). Store at 4 °C.
8. Buffer W: 100-mM Tris/HCl pH 8.0, 150-mM NaCl, 1-mM EDTA (*see* **Note 1**).
9. Buffer E: 100-mM Tris/HCl pH 8.0, 150-mM NaCl, 1-mM EDTA, 2.5-mM desthiobiotin. Store at 4 °C.
10. Buffer R: 100-mM Tris/HCl, 150-mM NaCl, 1-mM EDTA, 1-mM hydroxy-azophenyl-benzoic acid (HABA) pH 8.0. Store at 4 °C.
11. SDS–Polyacrylamide 4% stacking gel: 3.04-mL deionized water, 0.48-mL 1.5-M Tris–HCl pH 8.8, 0.4-mL 40% acrylamide, 0.04-mL 10% SDS, 0.04-mL 10% APS, 2-μL TEMED.
12. SDS–Polyacrylamide 10% resolving gel: 8-mL deionized water, 2.5-mL 1.5-M Tris–HCl pH 8.8, 2.5-mL 40% acrylamide (37.5:1 acrylamide/bis-acrylamide ratio), 0.1-mL 10% sodium dodecyl sulfate (SDS), 0.1-mL 10% ammonium persulfate (APS), 5-μL TEMED.
13. SDS–Polyacrylamide 12% resolving gel: 4.3-mL deionized water, 2.5-mL 1.5-M Tris–HCl pH 8.8, 3.0-mL 40% acrylamide, 0.1-mL 10% SDS, 0.1-mL 10% APS, 5-μL TEMED.
14. SDS Polyacrylamide 15% resolving gel: 3.55-mL deionized water, 2.5-mL 1.5-M Tris–HCl pH 8.8, 3.75-mL 40% acrylamide, 0.1-mL 10% SDS, 0.1-mL 10% APS, 5-μL TEMED.
15. SDS–PAGE gel running buffer (1×): 25-mM Tris, 192-mM Glycine, 0.1% SDS.
16. Laemmli sample loading buffer (4×): 0.25-M Tris–HCl pH 6.8, 8% SDS, 40% glycerol, 20% 2-mercaptoethanol, 0.004% bromophenol blue.
17. 25-mM HEPES pH 7.4.
18. Tris–buffer saline (TBS; 1×): 25-mM Tris–HCl, 0.15-M NaCl.
19. Phosphate-buffered saline (PBS): 9.33-mM potassium phosphate, 136-mM NaCl.

2.3 Protein Characterization

1. UV quartz cuvette, 1-mm path length.
2. Circular dichroism (CD) spectrophotometer (e.g., JASCO 810 CD Spectropolarimeter).

3. Nitrogen tank with connected hosing.
4. 1× PBS (*see* Subheading 2.2, **item 19**).
5. Cuvette cleaning solution: 30% concentrated HCl and 70% ethanol.

3 Methods

3.1 Recombinant Protein Expression and Purification

The following steps outline the expression and purification of rScls proteins using the *Strep*-tag II system, specifically, the protein products of the *scl* genes cloned in pASK-IBA2 vector (IBA Lifesciences) [37]. In this system, the rScl coding sequence is fused to the OmpA signal peptide, which results in the accumulation of recombinant protein in the periplasm of *E. coli* (strain BL21 deficient in Lon protease is recommended); in addition to periplasmic fraction, substantial amounts of rScl proteins could be recovered from the culture supernatant (*see* **Note 2**) [22]. rScl constructs have also been produced intracellularly, if the *scl* gene was cloned in pASK-IBA3 vector (*see* **Note 3**).

The protein expression section is broken into three different subsections: 1-mL small-scale induction, 10-mL medium-scale induction, and 1- to 2-L large-scale induction. *Small-scale inductions* serve as a quick screening method to identify the *E. coli* colonies, which produce the recombinant protein of interest. The *10-mL medium-scale inductions* are performed with selected colonies and serve as a method to determine the best conditions for protein expression and purification, such as incubation temperature and protein extraction conditions. *Large-scale inductions* are used to produce large quantities of recombinant protein once the optimal expression and extraction conditions are determined.

3.1.1 Protein Expression—Small-Scale Induction

1. Streak frozen glycerol stock of *E. coli* colony(ies) harboring clones of *scl* sequences in pASK-IBA2 vector onto LB-Miller agar with 100 μg/mL ampicillin. Grow overnight at 37 °C.
2. The next day, inoculate 3-mL LB-Miller broth with 100 μg/mL ampicillin, with an overnight culture. Grow overnight at 37 °C with shaking set at 225 rpm (*see* **Note 4**).

   (*Optional)*: Frozen stocks can be used to directly inoculate 3 mL of LB-Miller broth with 100 μg/mL ampicillin.
3. The next day, for each bacterial culture being tested, setup two 1.5-mL Eppendorf tubes with 1-mL LB-Miller broth/100 μg/mL ampicillin. Inoculate both tubes with 100 μL of overnight culture. Grow for 2 h at 37 °C, shaking at 225 rpm. One tube will serve as a noninduced control.
4. After 2 h, induce protein expression in one tube of each pair by adding to the culture 1 μL of a 1:10 dilution of AHTC stock in

sterile water. Continue to grow both tubes for an additional 2 h at 37 °C, at 225 rpm.

5. After 2 h, pellet the bacteria in the microcentrifuge at >14,000 × *g* for 3 min. Discard the supernatant.
6. Suspend cell pellets in 500-μL water.
7. Remove 10–20 μL of bacterial cell suspension, add 4× gel loading buffer (to final 1×), and boil samples for 5 min.
8. Resolve proteins by SDS–PAGE to determine if expression of the rScl protein is induced; verify expected rScl molecular weight (the molecular weight varies across rScl constructs originating from different GAS strains) and select a colony which is the best rScl producer (Fig. 1). Depending on experiment, standard or gradient (4–20%) gels were used to assess protein mass. Three commonly used concentrations of 10%, 12%, and 15% acrylamide resolving gels are used, combined with the 4% acrylamide stacking gel (10-mL volume is enough to pour 2 mini-gels).

3.1.2 Protein Expression—10-mL Medium-Scale Induction

1. Streak frozen glycerol stock of *E. coli* colony(ies) harboring clones of *scl* sequences in pASK-IBA2 vector onto LB-Miller agar with 100 μg/mL ampicillin. Grow overnight at 37 °C.
2. The next day, inoculate 3-mL LB-Miller broth with 100 μg/mL ampicillin, with an overnight colony. Grow overnight at 30 °C or 37 °C (*see* **Note 5**) with shaking set at 225 rpm.

   (*Optional*): Frozen stocks can be used to directly inoculate 3 mL of LB-Miller broth with 100 μg/mL ampicillin.

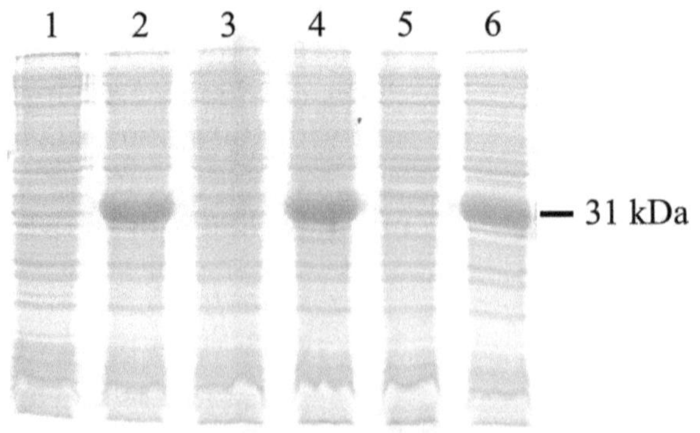

Fig. 1 Screening of *E. coli* colonies for the production of rScl proteins. Example of 12% SDS–PAGE representing the small-scale induction of rScl expression. Three different *E. coli* colonies (lanes 1–2, 3–4, and 5–6), expressing a potential rScl construct, were growth at 37 °C and protein production was compared. Both noninduced (lanes 1, 3, 5) and induced (lanes 2, 4, 6) samples were analyzed on a gel. The induced protein band of expected 31-kDa molecular weight is marked

3. The next day, for each bacterial culture being tested, setup an Erlenmeyer flask with 50 mL of LB-Miller broth with 100 μg/mL ampicillin. Inoculate the broth with 1-mL overnight culture. Grow at 30 °C or 37 °C, shaking at 225 rpm, until the OD_{600} reaches 0.3–0.5.
4. Once culture reaches desired OD_{600}, add AHTC to culture (1 μL of AHTC stock per 10-mL culture). Grow an additional 2–3 h at 30 °C or 37 °C.

   (*Optional*): 10 mL of culture can be transferred to a 35-mL sterile centrifuge tube, prior to AHTC addition, to serve as a noninduced control.
5. After 2- to 3-h incubation, transfer 10 mL of culture to a 35-mL centrifuge tube. Pellet bacterial cells by centrifugation for 10 min at 6500 × *g*, discard supernatant.
6. Two different rScl extraction methods can be utilized at this point: **Steps 7–10** use the buffer P method and **steps 11–15** use the cell lytic method. Both protocols should be initially tested to establish, which method results in greater yield of soluble protein (Figs. 2 and 3).
7. Suspend cell pellet in 100-μL buffer P.
8. Incubate on ice for 15 min.
9. Microcentrifuge for 5 min at >14,000 × *g*.
10. Remove the supernatant and suspend the cellular debris in 500-μL TE buffer; save samples for analysis.
11. Freeze cell pellet for at least 30 min at −20 °C. Pellets can be frozen overnight as well.

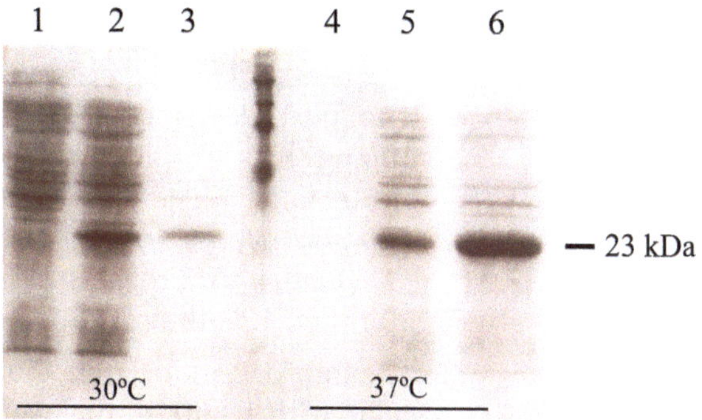

Fig. 2 Impact of incubation temperature on rScl expression. In a 10-mL induction protocol, the same *E. coli* colony was grown either at 30 °C (lanes 1–3) or 37 °C (lanes 4–6). Bacterial cells were treated with buffer P and samples were analyzed by 12% SDS–PAGE. Lanes 1 and 4 are noninduced samples; lanes 2 and 5 were loaded with supernatant fractions and lanes 3 and 6 with cell debris fractions. The induced protein band of 23-kDa molecular weight is shown

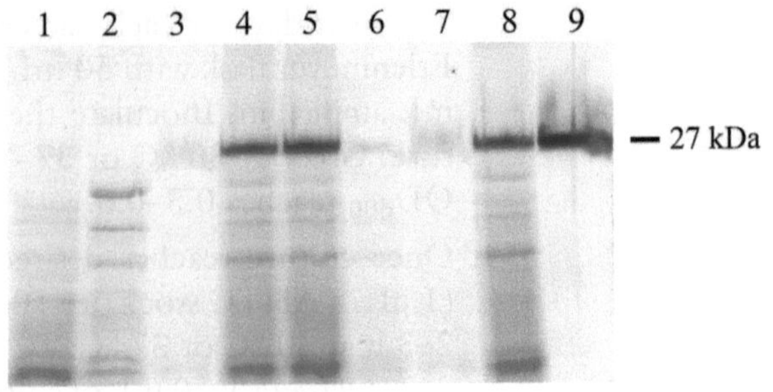

Fig. 3 Comparison of protein extraction methods using buffer P and cell lytic buffer. 10-mL cultures of two different *E. coli* colonies (colony 1, lanes 1–6; colony 2, lanes 7–8) were grown at 37 °C, and cells, noninduced or induced with AHTC, were processed using the buffer P (lanes 3–4, 7–8) or cell lytic B (lanes 1–2, 5–6) methods (*see* **Note 6**). Samples were analyzed by 12% SDS–PAGE. Samples in lanes 1 and 2 show noninduced controls; lanes 3–8 show samples induced for rScl expression. Lanes 1, 3, 5, and 7 show supernatant fraction, and lanes 2, 4, 6 and 8 show cell debris fractions. Lane 9 shows a previously purified rScl preparation

12. Thaw pellet and suspend bacterial cells in 200 μL of CelLytic B®.
13. Vortex seven times (7 cycles of 1 min on and 1 min off).
14. Centrifuge 10 min at 6500 × *g*.
15. Remove the supernatant and suspend the cellular debris in 500-μL TE buffer; save samples for analysis.
16. Analyze 10–15 μL of the supernatant and 2–5 μL of the cellular debris from both methods by SDS–PAGE.

3.1.3 Protein Expression—Large-Scale Induction

1. Streak frozen glycerol stock of *E. coli* colony(ies) harboring clones of *scl* sequences in pASK-IBA2 vector that were selected in protocols onto LB-Miller agar with 100 μg/mL ampicillin. Grow overnight at 37 °C.
2. The next day, inoculate 50 mL of LB-Miller broth with 100 μg/mL ampicillin with overnight culture. Grow overnight at 30 °C or 37 °C (conditions established with *10-mL induction protocol*, *see* **Note 5**) with shaking set at 225 rpm.

   (*Optional*): Frozen stocks can be used to directly inoculate 50 mL of LB-Miller broth with 100 μg/mL ampicillin.
3. The next day, inoculate 1–2 L of LB-Miller/100 μg/mL ampicillin, with 10 mL of the overnight culture per 1 L of medium. Grow at 30 °C or 37 °C, shaking at 225 rpm, until the optical density (OD_{600}) reaches 0.5–0.8.
4. Once culture reaches desired OD_{600}, add AHTC to culture (100 μL of AHTC stock per 1-L culture). Grow for additional 3 h at 30 °C or 37 °C.

5. After 3 h of induction, transfer culture to 500-mL centrifuge bottles. Centrifuge culture sample at 6500 × *g*, for 15 min, at 4 °C, to pellet bacterial cells, discard culture supernatant.
6. Pending the preferred protein extraction method, two different approaches can be used to extract protein: **steps 7–16** describe the buffer P method and **steps 17–25** describe the cell lytic method.
7. Buffer P method: Prewarm the PMSF stock (small crystals typically form at −20 °C), shaking gently for 5 min at RT.
8. Suspend cell pellets in 5 mL of buffer P per each 500 mL of original culture.
9. Combine two samples of cell suspensions (10 mL per 1 L of original culture) into a 35-mL centrifuge tube; add 100-μL prewarmed PMSF stock and 100-μL DNase I per each 10 mL of cell suspension.
10. Incubate on ice for 10 min.
11. Remove from ice and gently rock at RT for 15 min.
12. Centrifuge for 10 min at 17,500 × *g*.
13. Save the supernatant for purification. Suspend the cellular debris in 3–5 mL TE buffer. Store cell debris at 4 °C or −20 °C for analysis by SDS–PAGE (*see* **Note 7**).
14. Add 100 μL of avidin stock, per 10-mL sample, to the supernatant. Incubate for 1 h at 4 °C (can be overnight) (*see* **Note 8**).
15. Centrifuge sample at 17,500 × *g* for 15 min at 4 °C. A small white pellet may be present after centrifugation. Transfer supernatant to a 15- or 50-mL conical tube. If avidin pellet is present, dissolve it in 500 μL of 10 mM Tris–HCl, pH 8.0.
16. Use supernatant during *Affinity Purification* (Subheading 3.2).
17. Cell lytic method: Freeze bacterial cell pellet for at least 1 h at −20 °C, or overnight.
18. Thaw cells and suspend each pellet in 10-mL Cell Lytic B®.
19. Add 100-μL PMSF and 100 μL-DNase I, per 500-mL bottle.
20. Combine all pellets into a 35-mL centrifuge tube. Shake gently at RT for 15 min.
21. Centrifuge at 4 °C, 17,500 × *g*, for 10 min.
22. Save the supernatant for the next step (**step 23**). Suspend the cellular debris in 3- to 5-mL TE buffer. Store cell debris at 4 °C or −20 °C.
23. Add 100 μL of avidin, per 500-mL bottle, to the supernatant. Incubate with avidin for 1 h at 4 °C (can be overnight) (*see* **Note 8**).

24. Centrifuge sample at 17,500 × *g* for 15 min at 4 °C. A small white pellet may be present after centrifugation. Transfer supernatant to a 15- or 50-mL conical tube. If avidin pellet is present, dissolve in 500 μL 10 mM Tris–HCl, pH 8.0.
25. Use supernatant in **step** 7, *Affinity Purification* (Subheading 3.2).

3.2 Protein Purification–Affinity Purification

After isolation of the supernatant fraction, prepare the Strep-Tactin® sepharose column for affinity purification.

1. Add 1-mL Strep-Tactin® sepharose slurry to the column (*see* **Note 9**).
2. Equilibrate resins with 2 column volume Buffer W (*see* **Note 1**).
3. Apply 10–15 mL of the final processed supernatant, after either buffer P or cell lytic method, to the column, collecting the flow through (*see* **Note 10**).
4. Wash the column 5× with 1 column volume Buffer W (*see* **Note 1**). Collect, at least, the first and last washes in Eppendorf tubes for SDS–PAGE analysis to assess if rScl protein is being washed off of the column.
5. Elute the protein with Buffer E, collecting each elution fraction. Elution volumes need to be optimized to maximize protein concentrations and volumes vary between rScl constructs (Fig. 4) (*see* **Note 11**). Standard starting elution pattern is 6 elution fractions, 500 μL each.
6. Save the collected elutions, washes, and flow through for analysis by SDS–PAGE.
7. Regenerate the Strep-Tactin® resin by washing the column 3× with 5 column volumes of Buffer R. Column will turn from white to red after the washes.
8. Remove Buffer R via washing 3× with 5 column volumes of Buffer W. Column should turn from red to white.
9. Store column in 2 column volumes of Buffer W, at 4 °C, until future use.
10. Analyze equal volumes of each collected samples by SDS–PAGE to assess yield of rScl protein in elutions, amount left in flow through, washes, and protein presence in avidin pellet (Fig. 4).
11. Repeat purification **steps 1–9**, using the remaining supernatant sample(s). Purify also samples obtained by combining those flow through washes. Dissolved avidin fractions, which can contain rScl (determined via SDS–PAGE analysis), can be run after remaining sample and flow through are processed.

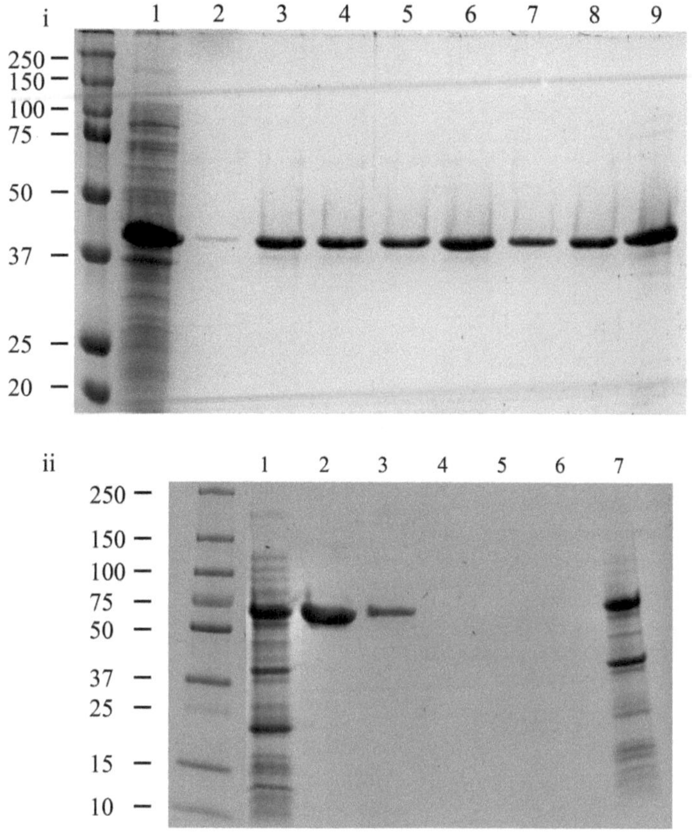

Fig. 4 4–20% SDS–PAGE of two distinct elution patterns represented by rScl constructs 1 and 2, following affinity chromatography. Supernatants from a 1-L induction protocol were loaded onto Strep-Tactin sepharose, washed, and eluted in 500-μL fractions. (**i**) rScl construct 1: (lane 1) flow through, (lane 2) wash, (lanes 3–8) elutions 1–6, (lane 9) purified rScl marker. (**ii**) rScl construct 2: (lane 1) flow through, (lanes 2–6) elutions 1–5, (lane 7) example of avidin pellet fraction showing the presence of rScl protein. Left lane, protein size marker [kDa]

12. After all fractions have been run on affinity column and analyzed by SDS–PAGE, combine fractions with similar band intensities (protein concentrations). Typically, there should be 2–3 different final samples (a high, medium, and low concentration).
13. Once all elutions are combined, dialyze the proteins for either immediate usage or storage. Use a 3000–7000 low-molecular-weight cutoff membrane because rScl proteins are not globular. Dialyze proteins for 36–48 h at 4 °C.

    Immediate use: proteins are typically dialyzed against either 1× TBS pH 7.4 or 1× PBS pH 7.5, when used in either ELISA-binding assays, circular dichroism, or other assays. *See* Subheadings 3.3 (*see* **Note 12**).

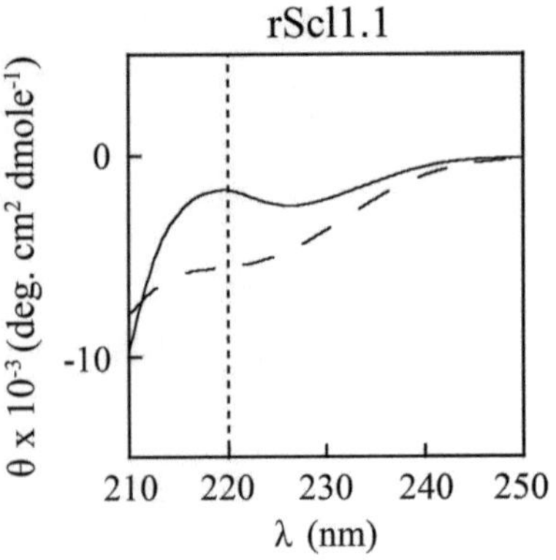

Fig. 5 Circular dichroism (CD) spectra of an rScl1 protein derived from an M1 strain. Wavelength scan showing the collagen triple helix present at 25 °C (solid line—folded conformation) and lacking at 50 °C (dashed line—unfolded conformation); dotted vertical line shows 220_{nm}-maximum molar ellipticity, a characteristics of collagen triple helix. (This research was originally published in the *Journal of Biological Chemistry*. Xu Y, Keene DR, Bujnicki JM, Höök M, Lukomski S. Streptococcal Scl1 and Scl2 proteins form collagen-like triple helices. *J Biol Chem*. 2002; 277:27312-8

Storage: proteins are stored in 25-mM HEPES pH 7.4. After dialysis, proteins are aliquoted into working volumes (50–500 μL) and then frozen in liquid nitrogen (*see* **Note 13**). Frozen samples are stored at −20 °C.

3.3 Protein Characterization (Quality Control)

Circular dichroism (CD) spectroscopy (reviewed in [38]) was used to detect the collagen triple helix in rScl proteins [9, 22]. This is indicated by the presence of molar ellipticity maximum at 220 nm [39] at 25 °C, which is lost at 50 °C when the collagen triple-helix unfolds (Fig. 5). In addition to CD spectroscopy, rScl proteins are characterized by electron microscopy of rotary shadowed protein samples. Rotary shadowing is a technique which shows the domain organization of proteins (reviewed in [40]). rScl proteins have a characteristic "lollipop-like" two-domain organization, with the globular V-domain head and CL-domain stalk [9, 22] (Fig. 6). Outlined below is a general protocol for the preparation and conditions used in CD spectroscopy and rotary shadowing of rScl proteins.

3.3.1 Circular Dichroism Spectroscopy

1. Recombinant streptococcal collagen-like proteins should be dialyzed against 1× PBS for 24–48 h at 4 °C (*see* **Note 14**).
2. Set up the CD Spectrophotometer to perform a wavelength scan, from 240 nm to 200 nm, using the following conditions: pitch: 0.2 nm, mode: continuous, speed: 10 nm/min, response: 1 s, bandwidth: 1 nm, accumulation: 5, nitrogen flow rate: 3–5 λ/min, temperature: 25 °C.

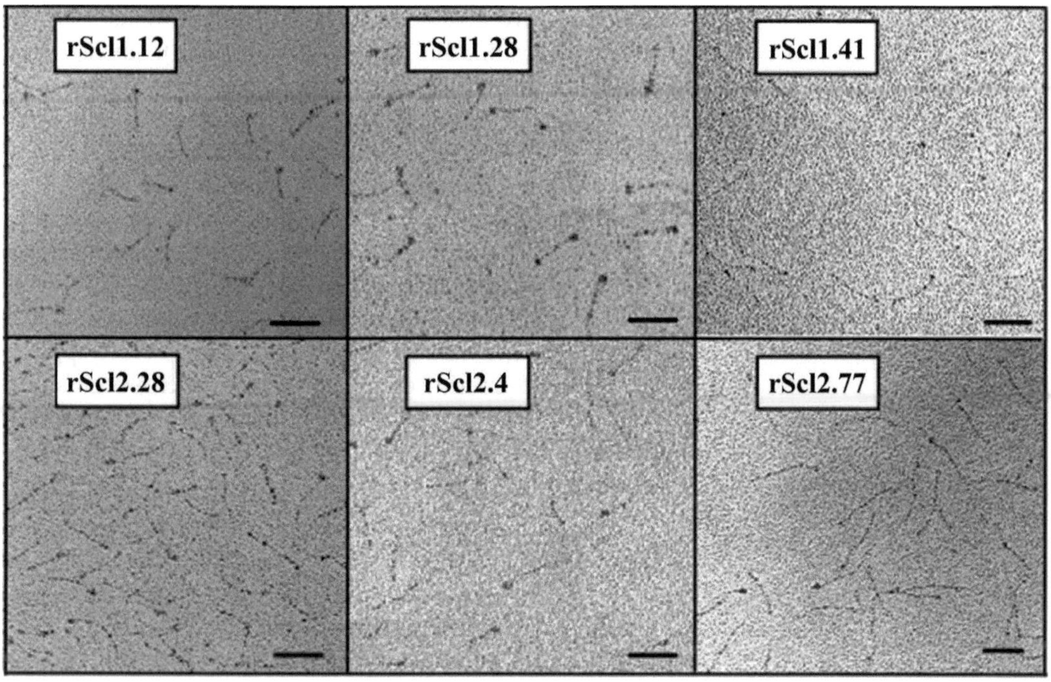

Fig. 6 Electron microscopy images of the rotary shadowed samples of rScl1 and rScl2 constructs. Purified rScl constructs maintain shared "lollipop-like" two-domain organization. Bars: 50 nm

3. Fill UV quartz cuvette with sample and place into the cell of CD spectrophotometer. Start measurement readings.
4. After performing a wavelength scan at 25 °C, use the same settings on the instrument, except change the temperature to 50 °C to assess loss of the collagen triple-helix structure (Fig. 5). This is important because, depending on the contribution of noncollagen domains within the collagen-like protein, the ellipticity maximum at 220 nm may not be well defined.
5. Remove cuvette from the instrument, discard sample, and clean cuvette by repeating rinses with the cleaning solution.
6. Dry the cuvette using a low flow of 100% nitrogen gas until all liquid is removed from the cuvette.

4 Notes

1. If contaminating bands appear with eluted protein of interest, the higher NaCl concentration Buffer W, up to 700 mM, should be tested to increase the stringency washes. Increasing the NaCl concentration may decrease the target protein yield in the process.

2. Recombinant streptococcal collagen-like proteins can be expressed intracellularly (pASK-IBA3 vector); intracellular expression was found to result in a higher yield of the same protein (~20 mg/L) compared to periplasmic expression (~2 mg/L). When expressed in the periplasm (pASK-IBA2 vector), additional amounts of rScl proteins (~5 mg/L) are secreted into the culture medium and can be recovered by precipitation with ammonium sulfate [22]. Extracellular protein fraction precipitated with ammonium sulfate (saturation needs to be established experimentally) and is resuspended in 25 mM Tris–HCl, 1 mM EDTA pH 8.0 buffer before application on the Strep-Tactin sepharose column. rScl proteins produced intracellularly or recovered by precipitation from culture medium maintained triple-helical conformation and lollipop-like organization [22].
3. IBA Lifesciences currently offers numerous additional expression plasmids with alternative affinity tags (e.g., 6xHis-tag and Flag-tag) that will require different inducing agents, extraction methods, and affinity-purification resins. However, in any alternative purification method for recombinant expression, several steps and notes outlined here will still be beneficial.
4. While selecting several of *E. coli* colony(ies) harboring clones of *scl* sequences for sequencing, the same colonies can be screened to test for protein expression, using a 1.5-*mL small-scale induction* protocol. This will enable an initial assessment of protein induction and yields, as well as the size of induced bands.
5. 10-mL medium-scale induction protocols are used to test for the optimal conditions of protein expression. Variables for subsequent production typically include the optimal expression temperature (30 °C or 37 °C) and extraction method (buffer P vs. cell lytic B). Both temperatures and extraction buffers should be tested to establish the optimal condition (Figs. 2 and 3). Lower induction temperatures are recommended when recombinant proteins tend to aggregate.
6. We recommend that three 10-mL samples are tested at the same time: (1) noninduced, (2) induced and processed with buffer P, and (3) induced and processed with cell lytic B.
7. Substantial amounts of rScl protein may still be in the cellular debris, based on SDS–PAGE analysis. In such case, centrifuge the sample at 15,000 × *g* for 10 min and discard supernatant. Treat pellet using the cell lytic method to extract proteins from cellular debris.
8. Egg white avidin is added to bind and mask biotinylated *E. coli* proteins, as well as biotin, that would interfere with purification on Strep-Tactin® sepharose resin (which also binds biotin and biotinylated proteins). The avidin-precipitate fraction may

contain rScl protein, which can be recovered on the Strep-Tactin® column. Addition of the avidin fraction can be done after all elutions, washes, and flow through fractions have been processed and analyzed by SDS–PAGE.

9. Depending on the level of rScl expression, a volume different than standard 1 mL of Strep-Tactin sepharose per 1 L of culture may be used for packing the column.
10. Similarly to **Note 9**, the volume of the supernatant added to the column may be adjusted to minimize protein in the flow through.
11. Different rScl constructs will elute differently. Some rScl variants will still remain on the column following standard elution protocol, thus, requiring more than standard volume of elution buffer E (Fig. 4(i)), whereas other rScls elute completely within elutions 1–2 (Fig. 4(ii)). The elution volumes are typically altered after the initial standard purification and adjusted until a set condition is established.
12. When not in immediate use, proteins should be kept at 4 °C for no longer than 1 week.
13. If an apparatus for freezing tubes with liquid nitrogen is not readily available, an alternative is to freeze samples using a Dewar. Load samples into a storage box and then lower box into liquid nitrogen until it stops bubbling. Bring samples back out of nitrogen and then immediately store at −20 °C.
14. The correct buffer is key, as certain commonly used buffers have lower wavelength limits than others, and some buffers will result in background signals which mask the CD spectrum of the rScl protein. The 1× PBS outlined has a wavelength limit of 200 nm, enabling the evaluation of the collagen triple helix at 220 nm.

Acknowledgments

We are grateful to all previous and current members of Lukomski's laboratory for their contributions to the development of the methods and protocols presented here. We also thank our numerous collaborators, especially Doug Keene for performing rotary shadowing and EM imaging over the years. This work was supported by NIH grants AI50666 and AI083683, as well as West Virginia University HSC Bridge Grant Funding to S.L.

References

1. Exposito JY, Valcourt U, Cluzel C et al (2010) The fibrillar collagen family. Int J Mol Sci 11 (2):407–426. https://doi.org/10.3390/ijms11020407
2. Doxey AC, McConkey BJ (2013) Prediction of molecular mimicry candidates in human pathogenic bacteria. Virulence 4(6):453–466. https://doi.org/10.4161/viru.25180
3. Lukomski S, Bachert BA, Squeglia F et al (2017) Collagen-like proteins of pathogenic streptococci. Mol Microbiol 103(6):919–930. https://doi.org/10.1111/mmi.13604
4. Lukomski S, Nakashima K, Abdi I et al (2000) Identification and characterization of the *scl* gene encoding a group a *Streptococcus* extracellular protein virulence factor with similarity to human collagen. Infect Immun 68 (12):6542–6553. https://doi.org/10.1128/iai.68.12.6542-6553.2000
5. Lukomski S, Nakashima K, Abdi I et al (2001) Identification and characterization of a second extracellular collagen-like protein made by group A *Streptococcus*: control of production at the level of translation. Infect Immun 69 (3):1729–1738. https://doi.org/10.1128/iai.69.3.1729-1738.2001
6. Rasmussen M, Björck L (2001) Unique regulation of SclB - a novel collagen-like surface protein of *Streptococcus pyogenes*. Mol Microbiol 40(6):1427–1438
7. Rasmussen M, Edén A, Björck L (2000) SclA, a novel collagen-like surface protein of *Streptococcus pyogenes*. Infect Immun 68 (11):6370–6377. https://doi.org/10.1128/iai.68.11.6370-6377.2000
8. Whatmore AM (2001) *Streptococcus pyogenes sclB* encodes a putative hypervariable surface protein with a collagen-like repetitive structure. Microbiology 147(Pt 2):419–429. https://doi.org/10.1099/00221287-147-2-419
9. Xu Y, Keene DR, Bujnicki JM et al (2002) Streptococcal Scl1 and Scl2 proteins form collagen-like triple helices. J Biol Chem 277 (30):27312–27318. https://doi.org/10.1074/jbc.M201163200
10. Xu C, Yu Z, Inouye M et al (2010) Expanding the family of collagen proteins: recombinant bacterial collagens of varying composition form triple-helices of similar stability. Biomacromolecules 11(2):348–356. https://doi.org/10.1021/bm900894b
11. Caswell CC, Barczyk M, Keene DR et al (2008) Identification of the first prokaryotic collagen sequence motif that mediates binding to human collagen receptors, integrins α2β1 and α11β1. J Biol Chem 283 (52):36168–36175
12. Humtsoe JO, Kim JK, Xu Y et al (2005) A streptococcal collagen-like protein interacts with the α2β1 integrin and induces intracellular signaling. J Biol Chem 280(14):13848–13857
13. Caswell CC, Oliver-Kozup H, Han R et al (2010) Scl1, the multifunctional adhesin of group a *Streptococcus*, selectively binds cellular fibronectin and laminin, and mediates pathogen internalization by human cells. FEMS Microbiol Lett 303(1):61–68. https://doi.org/10.1111/j.1574-6968.2009.01864.x
14. Caswell CC, Han R, Hovis KM et al (2008) The Scl1 protein of M6-type group A *Streptococcus* binds the human complement regulatory protein, factor H, and inhibits the alternative pathway of complement. Mol Microbiol 67 (3):584–596. https://doi.org/10.1111/j.1365-2958.2007.06067.x
15. Han R, Caswell CC, Lukomska E et al (2006) Binding of the low-density lipoprotein by streptococcal collagen-like protein Scl1 of *Streptococcus pyogenes*. Mol Microbiol 61 (2):351–367. https://doi.org/10.1111/j.1365-2958.2006.05237.x
16. Påhlman LI, Marx PF, Mörgelin M et al (2007) Thrombin-activatable fibrinolysis inhibitor binds to *Streptococcus pyogenes* by interacting with collagen-like proteins A and B. J Biol Chem 282(34):24873–24881. https://doi.org/10.1074/jbc.M610015200
17. Reuter M, Caswell CC, Lukomski S et al (2010) Binding of the human complement regulators CFHR1 and factor H by streptococcal collagen-like protein 1 (Scl1) via their conserved C termini allows control of the complement cascade at multiple levels. J Biol Chem 285(49):38473–38485. https://doi.org/10.1074/jbc.M110.143727
18. Squeglia F, Bachert B, De Simone A et al (2014) The crystal structure of the streptococcal collagen-like protein 2 globular domain from invasive M3-type group A *Streptococcus* shows significant similarity to immunomodulatory HIV protein gp41. J Biol Chem 289 (8):5122–5133. https://doi.org/10.1074/jbc.M113.523597
19. McNitt DH, Choi SJ, Allen JL et al (2019) Adaptation of the group A *Streptococcus* adhesin Scl1 to bind fibronectin type III repeats within wound-associated extracellular matrix: implications for cancer therapy. Mol Microbiol 112(3):800–819. https://doi.org/10.1111/mmi.14317

20. McNitt DH, Choi SJ, Keene DR et al (2018) Surface-exposed loops and an acidic patch in the Scl1 protein of group A *Streptococcus* enable Scl1 binding to wound-associated fibronectin. J Biol Chem 293(20):7796–7810. https://doi.org/10.1074/jbc.RA118.002250
21. Golser AV, Scheibel T (2018) Routes towards novel collagen-like biomaterials. Fibers 6 (2):21. https://doi.org/10.3390/fib6020021
22. Han R, Zwiefka A, Caswell CC et al (2006) Assessment of prokaryotic collagen-like sequences derived from streptococcal Scl1 and Scl2 proteins as a source of recombinant GXY polymers. Appl Microbiol Biotechnol 72 (1):109–115. https://doi.org/10.1007/s00253-006-0387-5
23. Peng YY, Howell L, Stoichevska V et al (2012) Towards scalable production of a collagen-like protein from *Streptococcus pyogenes* for biomedical applications. Microb Cell Factories 11:146–146. https://doi.org/10.1186/1475-2859-11-146
24. Peng YY, Stoichevska V, Madsen S et al (2014) A simple cost-effective methodology for large-scale purification of recombinant non-animal collagens. Appl Microbiol Biotechnol 98 (4):1807–1815. https://doi.org/10.1007/s00253-013-5475-8
25. Cosgriff-Hernandez E, Hahn MS, Russell B et al (2010) Bioactive hydrogels based on designer collagens. Acta Biomater 6 (10):3969–3977. https://doi.org/10.1016/j.actbio.2010.05.002
26. Yoshizumi A, Fletcher JM, Yu Z et al (2011) Designed coiled coils promote folding of a recombinant bacterial collagen. J Biol Chem 286(20):17512–17520. https://doi.org/10.1074/jbc.M110.217364
27. Peng YY, Stoichevska V, Schacht K et al (2014) Engineering multiple biological functional motifs into a blank collagen-like protein template from *Streptococcus pyogenes*. J Biomed Mater Res A 102(7):2189–2196. https://doi.org/10.1002/jbm.a.34898
28. Yu Z, An B, Ramshaw JA et al (2014) Bacterial collagen-like proteins that form triple-helical structures. J Struct Biol 186(3):451–461. https://doi.org/10.1016/j.jsb.2014.01.003
29. Yu Z, Visse R, Inouye M et al (2012) Defining requirements for collagenase cleavage in collagen type III using a bacterial collagen system. J Biol Chem 287(27):22988–22997. https://doi.org/10.1074/jbc.M112.348979
30. Seo N, Russell BH, Rivera JJ et al (2010) An engineered alpha1 integrin-binding collagenous sequence. J Biol Chem 285 (40):31046–31054. https://doi.org/10.1074/jbc.M110.151357
31. Bronk JK, Russell BH, Rivera JJ et al (2014) A multifunctional streptococcal collagen-mimetic protein coating prevents bacterial adhesion and promotes osteoid formation on titanium. Acta Biomater 10(7):3354–3362. https://doi.org/10.1016/j.actbio.2014.04.005
32. An B, DesRochers TM, Qin G et al (2013) The influence of specific binding of collagen-silk chimeras to silk biomaterials on hMSC behavior. Biomaterials 34(2):402–412. https://doi.org/10.1016/j.biomaterials.2012.09.085
33. Parmar PA, St-Pierre JP, Chow LW et al (2016) Harnessing the versatility of bacterial collagen to improve the chondrogenic potential of porous collagen scaffolds. Adv Healthc Mater 5(13):1656–1666. https://doi.org/10.1002/adhm.201600136
34. Peng YY, Yoshizumi A, Danon SJ et al (2010) A *Streptococcus pyogenes* derived collagen-like protein as a non-cytotoxic and non-immunogenic cross-linkable biomaterial. Biomaterials 31(10):2755–2761. https://doi.org/10.1016/j.biomaterials.2009.12.040
35. Browning MB, Dempsey D, Guiza V et al (2012) Multilayer vascular grafts based on collagen-mimetic proteins. Acta Biomater 8 (3):1010–1021. https://doi.org/10.1016/j.actbio.2011.11.015
36. Cereceres S, Touchet T, Browning MB et al (2015) Chronic wound dressings based on collagen-mimetic proteins. Adv Wound Care (New Rochelle) 4(8):444–456. https://doi.org/10.1089/wound.2014.0614
37. Maier T, Drapal N, Thanbichler M et al (1998) Strep-tag II affinity purification: An approach to study intermediates of Metalloenzyme biosynthesis. Anal Biochem 259(1):68–73. https://doi.org/10.1006/abio.1998.2649
38. Greenfield NJ (2007) Using circular dichroism spectra to estimate protein secondary structure. Nat Protoc 1:2876. https://doi.org/10.1038/nprot.2006.202
39. Brodsky-Doyle B, Leonard KR, Reid KB (1976) Circular-dichroism and electron-microscopy studies of human subcomponent C1q before and after limited proteolysis by pepsin. Biochem J 159(2):279–286. https://doi.org/10.1042/bj1590279
40. Sherratt MJ, Graham HK, Kielty CM et al (2000) ECM macromolecules: rotary shadowing and scanning transmission electron microscopy. In: Streuli CH, Grant ME (eds) Extracellular Matrix Protocols. Humana Press, Totowa, NJ, pp 119–132. https://doi.org/10.1385/1-59259-063-2:119

Chapter 13

Detection of Fibronectin-Binding Proteins of *Streptococcus pyogenes* Using Ligand Blot Analysis

Masanobu Nakata and Shigetada Kawabata

Abstract

Streptococcus pyogenes utilizes extracellular cellular matrix (ECM) proteins to adhere to human tissues and internalize into host cells. Fibronectin (Fn) is one of the most abundant ECM proteins and targeted by a wide variety of secreted Fn-binding proteins (Fbps) of *S. pyogenes*. However, prior to detailed kinetic analysis of that binding process, evaluations of the ability of *S. pyogenes* strains to bind to Fn as well as interactions of target molecules with Fn are required. In this chapter, we present routine procedures for ligand blot analysis with labeled human Fn, using bacterial cell wall extracts prepared by either enzymatic digestion of cells or extraction with a denaturing agent.

Key words *Streptococcus pyogenes*, Fibronectin, Fibronectin-binding protein, Biotinylation, Ligand blot analysis

1 Introduction

Fibronectin (Fn), a multidomain glycoprotein consisting of a dimer of approximately 250 kDa monomers, is present in humans in body fluids, on the surface of cells, and in the extracellular cellular matrix (ECM) of various tissues [1]. Secreted Fn binds to cell surfaces by attachment to a large family of integrin heterodimers, mainly α5β1 integrin, via the RGD (Arg-Gly-Asp) loop present in Fn [2, 3], while integrin-bound Fn is capable of binding to ECM components, such as collagen and laminin. Human Fn participates in regulation of cell adhesion, cell migration, and tissue repair [4, 5]. Since discovery of an Fn-binding protein (Fbp) of *Staphylococcus aureus*, it has been revealed that Fn is a common binding partner for bacterial adhesins of both Gram-positive and Gram-negative bacterial pathogens, as well as commensals [6].

Binding of *S. pyogenes* to Fn is mediated by not only Fbps containing Fn-binding repeats, but also a variety of other surface proteins, including M protein, GAPDH/Plr, Shr, and Scl1 [7–11], with the former group including protein F1 (PrtF1)/SfbI, Protein

Thomas Proft and Jacelyn M. S. Loh (eds.), *Group A Streptococcus: Methods and Protocols*, Methods in Molecular Biology, vol. 2136, https://doi.org/10.1007/978-1-0716-0467-0_13, © Springer Science+Business Media, LLC, part of Springer Nature 2020

F2 (PrtF2)/PFBP, FbaA, FbaB, SfbII/serum opacity factor (SOF), SfbX, and Fbp54 [11–20]. Most of those proteins are anchored to the free amino group of the cell wall via recognition of the LPXTG motif by the housekeeping transpeptidase SrtA [21]. Furthermore, it has been reported that Fn-binding repeats interact with the N-terminal Fn domain through a molecular interaction termed "tandem β-zipper" [22]. However, details regarding the molecular interactions between Fn and Fbp that lack such repeats remain elusive. The distribution of each Fbp has been shown to correlate with the M serotype [15, 16], while *S. pyogenes* is known to engage Fn via Fbp, and exploit the affinity between Fn and host integrins. Thus, Fn is utilized as a bridging molecule for adherence of *S. pyogenes* to host cells. This interaction causes activation of signaling pathways that include kinases, such as phosphoinositide 3-kinase and integrin-linked kinase, thereby leading to rearrangement of cytoskeletal actin, as well as distinct morphological changes and induction of bacterial internalization into host cells via several different mechanisms [23, 24]. Some Fbps also contribute to escape from the host immune system via binding to regulators of the complement system [25, 26].

As previously reported, several types of Fbp are capable of cell wall anchorage, and thus, research studies frequently use cell wall fractions for detection or binding assays. Methods often used for preparation of cell-surface proteins include conventional enzymatic digestion of the cell wall and the urea extraction, which was originally utilized for isolation of the cell-associated glucosyltransferase of *Streptococcus mutans*, though is also applied for the surface protein extraction from various bacteria [27]. Utilizing such fractions, ligand blot analysis with labeled Fn has been widely utilized to detect its presence, though that is a time-consuming and labor-intensive process.

2 Materials

All solutions must be prepared with analytical-grade chemicals and ultrapure water.

2.1 Enzymatic Cell Wall Extraction

1. THY medium: Dissolve 30 g of Todd–Hewitt broth and 2 g BBL yeast extract in 1 L of H_2O, and then autoclave.
2. 50 mL centrifugation tubes.
3. Spectrophotometer.
4. Centrifuge.
5. Phosphate buffered saline (PBS, 10×): 100 mM Na_2HPO_4, 18 mM KH_2PO_4, 1.37 M NaCl, 27 mM KCl. Dissolve 14.4 g Na_2HPO_4, 2.4 g KH_2PO4, 80 g NaCl, and 2 g KCl in 800 mL

of H_2O, then adjust pH to 7.4 with HCl, and adjust volume with H_2O to 1 L.

6. PBS (1×): Mix 100 mL 10× PBS and 900 mL H_2O.
7. 1.5 mL Eppendorf tubes.
8. Protoplasting buffer: 0.1 M KPO_4 (pH 6.2), 40% sucrose, and 10 mM $MgCl_2$. Dissolve 40 g sucrose in H_2O, then add 1.92 mL of K_2HPO4, 8.08 mL of KH_2PO_4, and 1 mL of 1 M $MgCl_2$, and adjust the volume with H_2O to 100 mL (*see* **Note 1**).
9. Protease inhibitor stock solution: Dissolve commercial EDTA-free protease inhibitor cocktail in H_2O according to the manufacturer's instructions to make a concentrated solution.
10. Mutanolysin stock solution: 5000 units/mL. Dissolve 0.1 mg mutanolysin from *Streptomyces globisporus* ATCC 21553 in 500 μL of H_2O, and then store the aliquots at −20 °C.
11. Tube rotator.
12. SDS-PAGE sample buffer (2×): 80 mM Tris pH 6.8, 2% SDS, 10% glycerol, 0.0006% Bromophenol blue, 100 mM DTT. Mix 0.8 mL of 1 M Tris pH 6.8, 1 mL 20% SDS, 1 mL glycerol, 60 μL of 0.1% Bromophenol blue, and 1 mL of 1 M DTT, and then adjust the volume with H_2O to 10 mL.
13. Block incubator.

2.2 Extraction of Surface Proteins Under Denaturing Condition

1. 8 M urea solution: Dissolve 48 g urea in H_2O, then add 10 mL of 10× PBS, and adjust the volume with H_2O to 100 mL.
2. Magnetic stirrer and stirring bars.

2.3 Fn Biotinylation

1. Human fibronectin (Fn), plasma.
2. Biotinamidocaproate *N*-hydroxysuccinamide ester (biotin ester) solution: Dissolve 20 mg of biotin ester in 1 mL dimethylformamide.
3. Columns (bed dimensions: 0.97 × 2.8 cm) containing 2.1 mL Sephadex G-25 (*see* **Note 2**).
4. 10% Sodium azide: Dissolve 10 g sodium azide in 100 mL of H_2O.

2.4 Immunoblotting (Ligand Blot Assay)

1. 4–20% precast polyacrylamide gel.
2. Electrophoresis apparatus.
3. SDS-PAGE running buffer (10×): Dissolve 30.3 g Tris, 144.1 g glycine, and 10 g SDS in 800 mL H_2O, and then adjust the volume to 1 L.
4. SDS-PAGE running buffer (1×): Dilute 10× buffer with H_2O.
5. Pre-stained molecular weight standard of protein.

6. Blunt-ended filter forceps.
7. See-saw shaker.
8. Polyvinylidene difluoride (PVDF) membrane.
9. Methanol.
10. Transfer buffer: 25 mM Tris, 192 mM glycine, 20% methanol. Dissolve 3.03 g Tris and 14.4 g glycine in 500 mL H_2O, add 200 mL methanol, and then add 5 mL 10% SDS. Adjust the volume with H_2O to 1 L.
11. Whatman 3MM filter paper.
12. Semi-dry blotting system.
13. Membrane blocking solution: Dissolve 1 g nonfat milk powder in 20 mL TBST (*see* **Note 3**).
14. Tris-buffered saline (TBS, 10×): 0.5 M Tris pH 7.5, 1.5 M NaCl. Dissolve 60.5 g Tris and 87.6 g NaCl in 800 mL of H_2O. Adjust pH to 7.5 with 1 M HCl and make the volume up to 1 L with H_2O.
15. Tris-buffered saline (TBS, 1×): Mix 100 mL 10× TBS and 900 mL H_2O.
16. TBST: TBS containing 0.05% Tween 20.
17. Horseradish peroxidase-conjugated streptavidin.
18. 3,3′,5,5′-Tetramethylbenzidine (TMB) peroxidase substrate solution (*see* **Note 4**).
19. 30% H_2O_2.
20. Coomassie Blue staining solution: Dissolve 2.5 g Coomassie Brilliant Blue R-250 in a mixture of 450 mL methanol, 100 mL acetic acid, and 400 mL H_2O. Adjust the volume to 1 L with H_2O.
21. Coomassie Blue destaining solution: Mix 450 mL methanol, 100 mL acetic acid, and 400 mL H_2O. Adjust the volume to 1 L with H_2O.

3 Methods

3.1 Preparation of Cell-Surface Protein Fraction

3.1.1 Preparation of Cell Wall Extracts with Mutanolysin

1. Grow single *S. pyogenes* colony overnight at 37 °C in THY medium (*see* **Note 5**).
2. Dilute culture to 1:50 in 30 mL of THY medium, followed by incubation until an optical density at 600 nm reaches 1.0 (*see* **Note 6**).
3. Collect cells by centrifugation at 10,000 × *g* for 10 min (*see* **Note 7**).
4. Wash cells twice with 30 mL PBS.

5. Resuspend cells in 1 mL of PBS and transfer the suspension to 1.5-mL tubes.
6. Centrifuge tubes at 10,000 × *g* for 5 min and discard supernatants.
7. Resuspend cells in 200 μL of protoplasting buffer supplemented with 1× EDTA-free protease inhibitor and 20 units of mutanolysin (*see* **Note 8**).
8. Incubate the suspension at 37 °C for 3 h in tube rotator.
9. Centrifuge the tubes at 20,000 × *g* for 20 min.
10. Transfer supernatants into new 1.5 mL tubes.
11. Add the same volume of 2× SDS-PAGE sample buffer.
12. Heat at 100 °C for 2 min in a block incubator and immediately place it on ice.

3.1.2 Preparation of 8 M Urea Extracts

1. Prepare cells as in Subheading 3.1.1, **steps 1–6**.
2. Resuspend the cells in 300 μL of 8 M urea solution.
3. Place a small stirring bar into tubes, and stir the cells with magnetic stirrer (*see* **Note 9**).
4. Incubate the suspension at 4 °C for 3 h, and then the following procedures are performed as in Subheading 3.1.1, **steps 9–12**.

3.2 Biotinylation of Fn

1. Prepare Fn solution containing 500 μg of Fn in 480 μL of PBS.
2. Mix 480 μL of Fn solution with 20 μL of biotin ester solution, and then incubate for 30 min at RT with constant agitation (*see* **Note 10**).
3. During **step 2**, equilibrate the Sephadex G25 resin with PBS (*see* **Note 11**).
4. Apply 500 μL of Fn sample to Sephadex G25 resin.
5. Elute with 1 mL of PBS and collect the eluate fractions. The fraction size is approximately 200 μL. Determine absorbance at 280 nm to verify which fraction contains the eluted protein (*see* **Note 12**).
6. Add 0.2% volume of sodium azide. The resultant sample can be stored in refrigerator for several weeks.

3.3 Fbp Detection with Biotinylated Fn (Ligand Blot Assay)

1. Place polyacrylamide gel in electrophoresis apparatus with buffer reservoir filled with SDS-PAGE running buffer. Subject pre-stained marker and proteins in prepared samples to electrophoresis using acrylamide gel and SDS-PAGE running buffer. Run at current of 20 mA until Bromophenol Blue has migrated to the bottom of the gel (*see* **Note 13**).
2. During **step 1**, prepare one PVDF membrane sheet and four sheets of Whatman 3MM filter paper sized the same as the gel

(*see* **Note 14**). Soak PVDF membrane in methanol for 30 s to 1 min with constant agitation, then wash with deionized water and soak in transfer buffer. Completely saturate filter papers by soaking in transfer buffer. For handling of the membrane and filter papers, use filter forceps so as to avoid contamination (*see* **Note 15**).

3. Following SDS-PAGE, wash gel in water, then equilibrate in transfer buffer to remove salts and detergents (*see* **Note 16**).
4. Place two sheets of filter paper on anode of semi-dry blotting apparatus, and then roll a pipette or blotting roller over the surface to remove air bubbles. Next, place the PVDF membrane onto the filter paper and put gel on the membrane, and then place two sheets of filter paper on the gel. Remove air bubbles in the same manner.
5. Place cathode onto stack and engage latches with guide posts, and then place the safety cover on the unit (*see* **Note 17**).
6. Turn on power supply and transfer proteins to membrane for 30 min to 1 h at 15 V constant voltage.
7. Remove safety cover and cathode assembly, and then place the membrane in solution containing blocking agent and incubate the membrane overnight at 4 °C on see-saw shaker (*see* **Note 18**).
8. Wash the membrane with TBST, and then incubate in TBST containing biotinylated Fn (10 μg/mL) at RT for 1 h.
9. Wash the membrane with TBST for 10 min and repeat three times.
10. Incubate the membrane with HRP-labeled streptavidin diluted 1:2000 in TBST at RT for 1 h on see-saw shaker.
11. Wash the membrane with TBST for 10 min and repeat three times.
12. Add 10 mL of TMB solution containing 5 μL of 30% H_2O_2 and incubate the membrane until color develops (*see* **Note 19**).
13. Stop enzymatic reaction by transferring the blot from substrate solution into tap water.
14. Dry the membrane on filter paper, and determine the sizes of reactive bands.
15. Excise samples and subject to tandem mass spectrometry analysis utilizing acrylamide gel, in which the same samples were electrophoresed, stained with Coomassie Brilliant Blue solution, destained with Coomassie Blue destaining solution (or silver stained), and washed with H_2O; protein bands with the same size as the detected bands in order to identify Fbp [28].

4 Notes

1. The presence of magnesium ion promotes enzymatic activity of mutanolysin from *Streptomyces globisporus* [29]. The optimum pH for this enzyme ranges from 6.5 to 7.0 [29]. Thus, the protoplasting buffer is acidic and contains $MgCl_2$, and the EDTA-free protease inhibitor is used. Sucrose is added as an osmostabilizer to prevent bursting of the membrane, while 30% raffinose instead of sucrose can also be used [8].
2. Commercial prepack columns are available, such as the PD MiniTrap G-25 (GE Healthcare). Alternatively, a self-made column can be prepared based on the manufacturer's instructions.
3. TBST containing 5% bovine serum albumin may be used.
4. A commercial TMB solution for the enzyme-linked immunosorbent assay cannot be used, since oxidized TMB chromogen is soluble and is not deposited on the membrane. Use of a commercial substrate solution suitable for detection on the membrane is required.
5. Since *S. pyogenes* is classified as a biosafety level 2 pathogen, follow guidelines and regulations for handling.
6. When several strains, such as a combination of the wild type and its mutant strains, are tested, it is important to determine their growth rate and adjust the growth phases.
7. If the tested strain produces a significant amount of hyaluronan capsule, the cells may not be adequately precipitated. In that case, increase the centrifugal force or add hyaluronidase to the culture at a concentration of 5 units/mL [30].
8. Due to a high concentration of cells, it is difficult to resuspend them in this buffer volume. We normally suspend cells by mild pipetting using a P200 pipette.
9. A stirring bar small enough to enter the cap of a 1.5-mL tube is required. The tube is turned over and fixed by a tube stander on a magnetic stirrer. Otherwise, increase the culture volume, and use a small beaker and stirring bar of appropriate size.
10. The biotin ester should be added at a 15- to 20-fold concentration. Other biotinylation methods may also be used. Of note, if Fn is already dissolved in an amine-containing buffer, such as Tris buffer, it must be dialyzed.
11. When a commercial prepacked column is used, its equilibration is critical. According to the manufacturer's instructions, the contained stabilizers absorb UV.

12. Alternatively, a spin method using a centrifuge can be also applied when using a column. For this protocol, spin the tubes at $1000 \times g$ for 2 min.
13. Place the gel in electrophoresis device according to the manufacturer's instructions.
14. An overhang due to size differences may result in inappropriate current passing through the membrane.
15. Always wear gloves when handling the gel and membrane during the process.
16. Without this washing step, remaining salts will increase buffer conductivity and the amount of heat generated during the blot transfer. Moreover, equilibration allows the gel to recover its size prior to the transfer. For transfer of large Fbps sized greater than 100 kDa, such as PrtF2, it is better to use a polyacrylamide gel with a concentration less than 8%. Also, large proteins tend to precipitate in the gel, which hinders the transfer. SDS promotes protein elution from the gel, whereas methanol tends to remove SDS from proteins. Therefore, reducing the methanol to 10% or less may reduce the precipitation. Lowering the methanol percentage in the transfer buffer also promotes gel swelling, allowing large proteins to transfer more easily.
17. Ensure that the cathode is set so as to not shift the positioning of the stack.
18. Overnight blocking normally reduces the detection background. In several presented protocols, blocking is conducted at RT for 1 h. A blocking reagent free of biotin is better. Casein-based blocking buffer is not recommended, since milk contains variable amounts of endogenous biotin, which may inhibit the binding of streptavidin to biotinylated Fn.
19. For colorimetric detection, 4-chloro-1-naphtol can be used. Other detection methods using either chemiluminescence or fluorescence can also be employed.

References

1. Frantz C, Stewart KM, Weaver VM (2010) The extracellular matrix at a glance. J Cell Sci 123:4195–4200. https://doi.org/10.1242/jcs.023820
2. Schmidt S, Friedl P (2010) Interstitial cell migration: integrin-dependent and alternative adhesion mechanisms. Cell Tissue Res 339:83–92. https://doi.org/10.1007/s00441-009-0892-9
3. Engvall E, Ruoslahti E (1977) Binding of soluble form of fibroblast surface protein, fibronectin, to collagen. Int J Cancer 20:1–5. https://doi.org/10.1002/ijc.2910200102
4. Singh P, Carraher C, Schwarzbauer JE (2010) Assembly of fibronectin extracellular matrix. Annu Rev Cell Dev Biol 26:397–419. https://doi.org/10.1146/annurev-cellbio-100109-104020
5. Potts JR, Campbell ID (1994) Fibronectin structure and assembly. Curr Opin Cell Biol 6:648–655. https://doi.org/10.1016/0955-0674(94)90090-6

6. Kuusela P (1978) Fibronectin binds to *Staphylococcus aureus*. Nature 276:718–720. https://doi.org/10.1038/276718a0
7. Cue D, Lam H, Cleary PP (2001) Genetic dissection of the *Streptococcus pyogenes* M1 protein: regions involved in fibronectin binding and intracellular invasion. Microb Pathog 31:231–242. https://doi.org/10.1006/mpat.2001.0467
8. Pancholi V, Fischetti VA (1992) A major surface protein on group A streptococci is a glyceraldehyde-3-phosphate-dehydrogenase with multiple binding activity. J Exp Med 176:415–426. https://doi.org/10.1084/jem.176.2.415
9. Fisher M, Huang YS, Li X et al (2008) Shr is a broad-spectrum surface receptor that contributes to adherence and virulence in group A streptococcus. Infect Immun 76:5006–5015. https://doi.org/10.1128/IAI.00300-08
10. Caswell CC, Oliver-Kozup H, Han R et al (2010) Scl1, the multifunctional adhesin of group A *Streptococcus*, selectively binds cellular fibronectin and laminin, and mediates pathogen internalization by human cells. FEMS Microbiol Lett 303:61–68. https://doi.org/10.1111/j.1574-6968.2009.01864.x
11. Hanski E, Caparon M (1992) Protein F, a fibronectin-binding protein, is an adhesin of the group A streptococcus *Streptococcus pyogenes*. Proc Natl Acad Sci U S A 89:6172–6176. https://doi.org/10.1073/pnas.89.13.6172
12. Talay SR, Valentin-Weigand P, Jerlstrom PG et al (1992) Fibronectin-binding protein of *Streptococcus pyogenes*: sequence of the binding domain involved in adherence of streptococci to epithelial cells. Infect Immun 60:3837–3844
13. Jaffe J, Natanson-Yaron S, Caparon MG et al (1996) Protein F2, a novel fibronectin-binding protein from *Streptococcus pyogenes*, possesses two binding domains. Mol Microbiol 21:373–384. https://doi.org/10.1046/j.1365-2958.1996.6331356.x
14. Rocha CL, Fischetti VA (1999) Identification and characterization of a novel fibronectin-binding protein on the surface of group a streptococci. Infect Immun 67:2720–2728
15. Terao Y, Kawabata S, Kunitomo E et al (2001) Fba, a novel fibronectin-binding protein from *Streptococcus pyogenes*, promotes bacterial entry into epithelial cells, and the *fba* gene is positively transcribed under the Mga regulator. Mol Microbiol 42:75–86. https://doi.org/10.1046/j.1365-2958.2001.02579.x
16. Terao Y, Kawabata S, Nakata M et al (2002) Molecular characterization of a novel fibronectin-binding protein of *Streptococcus pyogenes* strains isolated from toxic shock-like syndrome patients. J Biol Chem 277:47428–47435. https://doi.org/10.1074/jbc.M209133200
17. Kreikemeyer B, Talay SR, Chhatwal GS (1995) Characterization of a novel fibronectin-binding surface protein in group a streptococci. Mol Microbiol 17:137–145. https://doi.org/10.1111/j.1365-2958.1995.mmi_17010137.x
18. Rakonjac JV, Robbins JC, Fischetti VA (1995) DNA sequence of the serum opacity factor of group A streptococci: identification of a fibronectin-binding repeat domain. Infect Immun 63:622–631
19. Jeng A, Sakota V, Li Z et al (2003) Molecular genetic analysis of a group A *Streptococcus* operon encoding serum opacity factor and a novel fibronectin-binding protein, SfbX. J Bacteriol 185:1208–1217. https://doi.org/10.1128/JB.185.4.1208-1217.2003
20. Courtney HS, Li Y, Dale JB et al (1994) Cloning, sequencing, and expression of a fibronectin/fibrinogen-binding protein from group A streptococci. Infect Immun 62:3937–3946
21. Mazmanian SK, Liu G, Ton-That H et al (1999) *Staphylococcus aureus* sortase, an enzyme that anchors surface proteins to the cell wall. Science 285:760–763. https://doi.org/10.1126/science.285.5428.760
22. Schwarz-Linek U, Werner JM, Pickford AR et al (2003) Pathogenic bacteria attach to human fibronectin through a tandem beta-zipper. Nature 423:177–181. https://doi.org/10.1038/nature0189
23. Molinari G, Rohde M, Guzman CA et al (2000) Two distinct pathways for the invasion of *Streptococcus pyogenes* in non-phagocytic cells. Cell Microbiol 2:145–154. https://doi.org/10.1046/j.1462-5822.2000.00040.x
24. Wang B, Yurecko RS, Dedhar S et al (2006) Integrin-linked kinase is an essential link between integrins and uptake of bacterial pathogens by epithelial cells. Cell Microbiol 8:257–266. https://doi.org/10.1111/j.1462-5822.2005.00618.x
25. Pandiripally V, Gregory E, Cue D (2002) Acquisition of regulators of complement activation by *Streptococcus pyogenes* serotype M1. Infect Immun 70:6206–6214. https://doi.org/10.1128/IAI.70.11.6206-6214.2002
26. Reuter M, Caswell CC, Lukomski S et al (2010) Binding of the human complement regulators CFHR1 and factor H by

streptococcal collagen-like protein 1 (Scl1) via their conserved C termini allows control of the complement cascade at multiple levels. J Biol Chem 285:38473–38485. https://doi.org/10.1074/jbc.M110.143727

27. Hamada S, Horikoshi T, Minami T et al (1989) Purification and characterization of cell-associated glucosyltransferase synthesizing water-insoluble glucan from serotype *c Streptococcus mutans.* J Gen Microbiol 135:335–344. https://doi.org/10.1099/00221287-135-2-335
28. Mori Y, Yamaguchi M, Terao Y et al (2012) α-Enolase of *Streptococcus pneumoniae* induces formation of neutrophil extracellular traps. J Biol Chem 287:10472–10481. https://doi.org/10.1074/jbc.M111.280321
29. Yokogawa K, Kawata S, Nishimura S et al (1974) Mutanolysin, bacteriolytic agent for cariogenic streptococci: partial purification and properties. Antimicrob Agents Chemother 6:156–165. https://doi.org/10.1128/aac.6.2.156
30. Courtney HS, Ofek I, Hasty DL (1997) M protein mediated adhesion of M type 24 *Streptococcus pyogenes* stimulates release of interleukin-6 by HEp-2 tissue culture cells. FEMS Microbiol Lett 151:65–70. https://doi.org/10.1111/j.1574-6968.1997.tb10395.x

Chapter 14

Morphology and Ultrastructure of Group A Streptococcus Biofilms

Tatyana A. Danilova, T. A. Smirnova, G. A. Danilina, A. A. Adzhieva, N. V. Shevlyagina, and V. G. Zhukhovitsky

Abstract

Light and electron microscopy enables researchers to study the ultrastructure of GAS biofilms formed on abiotic surfaces. Chains of streptococci surrounded by a bluish film are seen under a light microscope after alcian blue staining of preparations grown on coverslips. The extracellular matrix (indicator of biofilm maturity) becomes visible on ultrathin sections in transmission electron microscopy after additional staining with alcian blue; filamentous structures, characteristic of biofilm, are observed in intercellular spaces. The data obtained by scanning electron microscopy also demonstrate the presence of biofilm.

Key words Biofilm, Matrix, Light microscopy, Electron microscopy

1 Introduction

According to the modern concept, a biofilm is a community of bacterial cells fixed to the substrate surface and coated by a biopolymeric matrix. The matrix is a complex structure consisting of proteins, nucleic acids, and lipids; its main components are exogenous polysaccharides. It is assumed to be a mode of defense allowing the bacteria to survive and multiply. An important characteristic of biofilms is their adhesion to various substrates—biotic (human and animal tissues) and abiotic (glass, plastic, metal). The bacteria in biofilms are protected from various destructive factors, including the host congenital and acquired immunity factors, drugs, etc. Biofilms are characterized by close cell–cell contacts, through which they exchange signals necessary for quorum sensing (a way of intercellular bacterial communication) [1–3].

Until recently, the development of Group A Streptococcus (GAS) infection was not associated with the formation of biofilms. For the first time, microcolonies of GAS enveloped in glycocalyx were detected in the analysis of skin lesions by confocal laser

Thomas Proft and Jacelyn M. S. Loh (eds.), *Group A Streptococcus: Methods and Protocols*, Methods in Molecular Biology, vol. 2136, https://doi.org/10.1007/978-1-0716-0467-0_14,

scanning microscopy [4]. It was shown that various GAS types, though to different degrees, were capable of forming biofilms in vivo and in vitro [5]. Bacteria enveloped in the exomatrix are less sensitive to antibiotics than planktonic bacteria, and this creates great problems in medical practice. The formation of biofilms is one of the causes of chronic types of infection, which necessitates an incessant search for new drugs to destroy the biofilms or inhibit their growth [3, 6, 7].

The study of biofilms is usually carried out using photometry on plates. However, it is impossible to study the structure of biofilm on plates, to examine the matrix, or to consider participation of individual cells and chains in biofilm formation. Using light and electron microscopy, we conducted a study of the ultrastructural features of biofilms of two types of GAS strains (serotypes M1 and M30) grown on coverslips and Petri dishes. In light microscopy, when staining preparations grown on coverslips with alcian blue, chains of streptococci are seen surrounded by a bluish film (Fig. 1a, b). On ultrathin sections in the transmission electron microscope, chains of cocci are seen, and division septa are clearly visible (Fig. 2a, b), sometimes with division defects (Fig. 2c). When staining with alcian blue, one can see electron-dense structures next to the dividing cell and chain (Fig. 2d, e). There are numerous electron-dense filaments in the extracellular space (Fig. 2f). Data obtained in scanning electron microscopy revealed a multilayer structure of streptococcal biofilms. The thickness of the layer was uneven and ranged from one to several cells. Exomatrix is mostly pronounced in areas of dense accumulation of microorganisms (Fig. 3a, b).

2 Materials

Prepare all solutions and media using distilled water. Store at 4 °C.

2.1 Formation of Biofilms

1. Cultures of GAS serotype M1 and M30 strains (*see* **Note 1**).
2. Todd-Hewitt broth (THB).
3. Peptone-based medium (PBM): 0.6 g Bacto-peptone, 0.42 g yeast extract, 600 ml distilled water.
4. Petri dishes (5 cm in diameter).
5. Coverslips (24 × 24 mm).
6. 0.9% NaCl.
7. MacFarland Standard.
8. Desiccator.

2.2 Light Microscopy

1. Polysaccharide stain: 1% alcian blue in 95% ethanol [8].

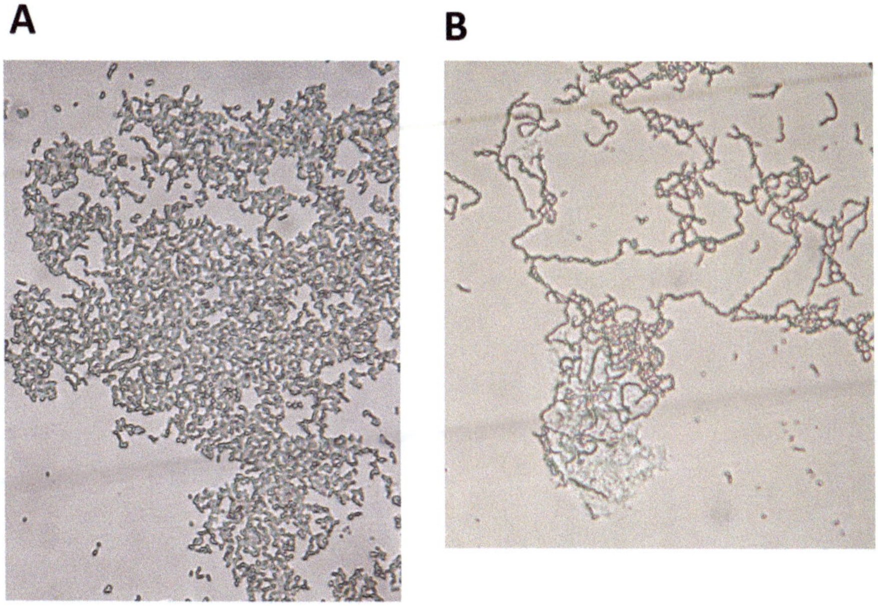

Fig. 1 Biofilms formed by GAS serotypes M1 and M30, alcian blue staining. Magnification ×400. (**a**) GAS serotype M1 strain in THB, chains of streptococci enveloped in biofilm. (**b**) GAS serotype M30 strain in PBM

2.3 Transmission Electron Microscopy

1. Cacodylate buffer: Dissolve 4.28 g of sodium cacodylate in 50 ml of distilled water (0.4 M). Mix 25 ml of 0.4 M solution of sodium cacodylate and 2.1 ml of 0.4 N HCl. Add distilled water to a volume of 100 ml, mix, and adjust pH with HCl to pH 7.2–7.4.
2. Base solution: Weigh 1.5 g of paraformaldehyde, add 50 ml of distilled water, and then heat without boiling. Add 1 N NaOH solution drop by drop and shake up well after each drop. The solution should become translucent after 3–4 drops as a rule. Cool the solution at 4 °C and add an equal volume of 0.2 M cacodylate buffer. Add 0.02 g of picric acid to the prepared solution and mix thoroughly.
3. Fixative: Add 1 ml of 25% glutaraldehyde and 5 ml of distilled water to 4 ml of base solution and mix.
4. Maleate buffer (0.2 M): Two variants—pH 5.2 and pH 6.0. First, prepare 0.8 M maleate buffer. Dissolve 2.32 g of maleic acid and 0.8 g NaOH in 100 ml of distilled water. To prepare 0.2 M buffer, mix 25 ml of 0.8 M solution with 75 ml of distilled water and adjust pH with 1 N NaOH.
5. OsO_4 solution: 1% OsO_4 (osmium tetroxide) solution in 0.2 M cacodylate buffer (*see* **Note 2**).
6. 1% uranyl acetate in 0.2 M maleate buffer, pH 6.0. (When adding uranyl acetate, the pH decreases to 5.2.) Adjust pH.
7. Resin LR White:100% ethanol (2:1).

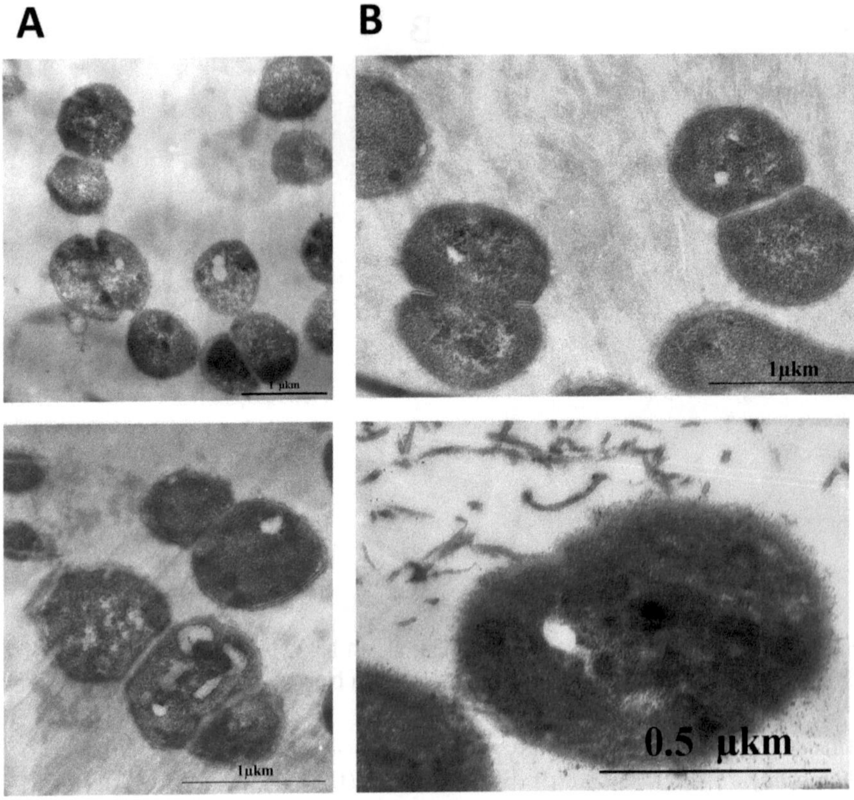

Fig. 2 Ultrathin sections of biofilms formed by GAS serotypes M1 and M30. (**a**) GAS serotype M1 strain, chains of cells with division septa. (**b**) GAS serotype M30, fringy layer of cell wall. (**c**) Cell with division defects. (**d**) Electron-dense structures, stained with alcian blue, outside the cells. (**e**) GAS serotype M1, chain of cocci surrounded by fragments of electron-dense cords (alcian blue staining). (**f**) GAS serotype M30, network of interwoven filaments (alcian blue staining)

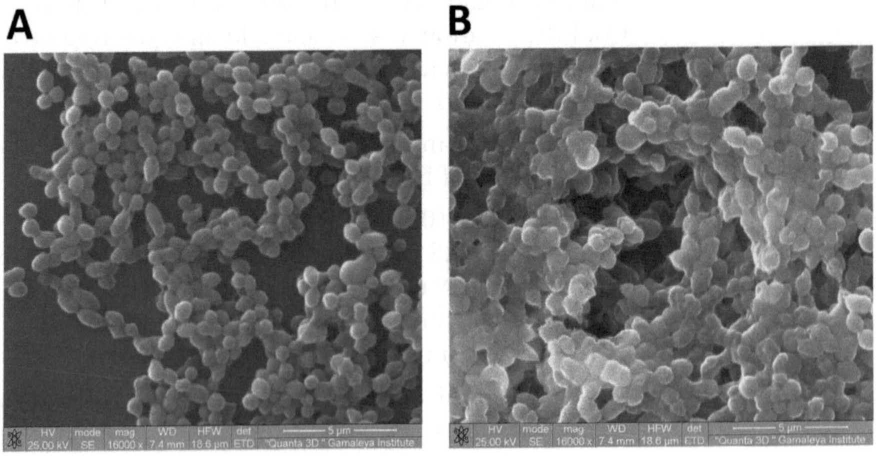

Fig. 3 Biofilm formed by GAS serotype M1. Scanning electron microscopy. (**a**) Accumulation of bacterial cells united by a thin exomatrix layer. (**b**) Multilayer structure formed by cells tightly adhering to each other and partially united by exomatrix

8. 0.3% lead citrate: Dissolve 30 mg of lead citrate in 10 ml of distilled water, add 0.1 ml of 10 N NaOH. Shake until completely dissolved.
9. Sterile scalpel.
10. Ultratome.
11. Glass or diamond knife.
12. Copper grids.
13. Formvar.
14. Dichlorethane.

2.4 Scanning Electron Microscopy

1. Ten percent formalin, pH 7.2–7.4: Mix 10 ml of 37% formaldehyde (purified from acids and stabilized with methanol and calcium carbonate) with 90 ml of distilled water. Mix thoroughly. Adjust pH.
2. SPI-MODULE sputter coater.
3. Aluminum stubs.

3 Methods

3.1 Formation of Biofilms on Coverslips (See Note 3)

1. Prepare 24-h broth culture of the test strain. Dilute in sterile 0.9% NaCl solution to an opacity equivalent to 1.0, according to MacFarland standard. Apply 100 μl of suspension containing 5×10^4 microorganisms (0.005 of MacFarland standard) in PBM onto a sterile coverslip lying horizontally in a Petri dish. Repeat each test twice. Close the Petri dish and incubate for 48 h at 37 °C and 5% CO_2 in a desiccator with 80% humidity.
2. Wash coverslips with three changes of distilled water and air-dry for 15 min. Fix the preparation, holding it at a height of 10 cm above the flame for 20 s.
3. Before staining, dilute the 1% solution of alcian blue ninefold in distilled water. Stain preparations with 0.1% alcian blue for 2–3 min. Wash coverslips with three changes of distilled water and air-dry for 15 min at 37 °C.
4. Examine preparations using a light microscope at 400-times magnification.

3.2 Formation of Biofilms in Petri Dishes

1. Add 2 ml of THB to 1 ml of 24-h culture containing 10^7 bacterial cells. Close the dish and incubate for 48 h at 37 °C and 5% CO_2 in a desiccator with 80% humidity.
2. Scrape off biofilms from the surface of the dish with a sterile scalpel [9] and put into 1 ml of 2.5% solution of glutaraldehyde in cacodylate buffer, pH 7.2–7.4 for 24 h. Centrifuge the material for 10 min at $4000 \times g$ and fix it [10].

3.3 Transmission Electron Microscopy

The ultrastructure of biofilm grown in Petri dishes was studied on ultrathin sections in a transmission electron microscope.

1. Fixation procedure. Fix the material by immersing it in fixative for 1 h at room temperature (RT).
2. Rinse the material 2–3 times with cacodylate buffer.
3. Fix the material in 1% OsO_4 solution for 1 h at RT.
4. Rinse the material three times with cacodylate buffer, then twice with maleate buffer, pH 5.2.
5. Fix the material in 1% uranyl acetate solution for 1–2 h at RT.
6. Rinse the material three times with maleate buffer, pH 5.2.
7. Quickly dehydrate the material in cold 50% ethanol for 10 min, then put it in 70% ethanol for 18 h at 4 °C.
8. Continue dehydration in 96% ethanol for 15 min, then in 100% ethanol three times for 30 min at 4 °C.
9. Put the material in a mixture of 100% ethanol and resin LR White for 1 h at 4 °C, then in pure resin LR White twice for 30 min at 4 °C.
10. Place the material into gelatin capsules, pour it with resin, and put in a thermostat at 56 °C for 24 h.

3.4 Detection of Exopolysaccharides on Ultrathin Sections by Alcian Blue Staining [11–13]

1. Staining of biofilm material with alcian blue. Mix the biofilm with 1 ml of 2.5% glutaraldehyde and 1 ml of 0.15% solution of alcian blue. Incubate for 3 h at RT, then rinse twice with 0.2 M cacodylate buffer.
2. Fix the material in 1% OsO_4 solution for 18 h at RT.
3. Continue the procedure as described in **steps 4–10** in Subheading 3.3.

3.5 Preparation of Ultrathin Sections

1. Prepare sections of the fixed material on ultratome with a glass or diamond knife. A small tray with distilled water is attached to the knife where the sections fall (*see* **Note 4**).
2. Preparation of formvar film. Put a defatted slide in formvar solution in dichlorethane, remove the glass, and air-dry.
3. Cut the film from one end of the glass and put it in distilled water. After floating the film on the water surface, place the grids on the top. Take the grids with formvar film with filter paper and dry.
4. Put sections of GAS biofilms on prepared grids.
5. Contrast the sections with 1% uranyl acetate for 30 min. Rinse three times with distilled water.

6. Contrast the sections with lead citrate solution for 5–10 min at 4 °C without air access.
7. Examine preparations in electron microscope at an ascending voltage of 80 kV and 15,000-times magnification.

3.6 Scanning Electron Microscopy

1. Put coverslips with 48-h GAS biofilms in 10% formalin solution for 24 h at RT. After fixation, dry samples for 5 min at RT.
2. To form an electron-conducting layer on the surface of biofilms, place the dried samples in a SPI-MODULE Sputter Coater with ion plasma sputtering a gold layer on their surface. Gold deposition occurs within 40 s, which corresponds to 5 nm of the layer thickness, according to a quartz sensor measurement.
3. Mount sputtered samples of the GAS biofilm on aluminum stubs with a double-sided carbon adhesive tape.
4. Place stubs with samples in a chamber of the Quanta 200 3D dual-beam scanning electron microscope. Analyze the samples in high vacuum mode and at an accelerating voltage of 5–10 kV (*see* **Note 5**).

4 Notes

1. Other serotypes forming biofilms can be used for this method.
2. Preparation of 1% OsO_4 solution. Considering the high toxicity of osmium tetroxide, the ampoule containing the preparation should be opened as follows: slightly sawed, placed in a dark vessel with a certain amount of liquid, and broken under it. Osmium tetroxide solution should be used after completely dissolving the drug. Osmium tetroxide solution and uranyl acetate should be handled in a fume hood.
3. When growing a biofilm on coverslips, it is necessary to wash and degrease cover glasses very carefully.
4. Only "grey" (fairly thin) sections should be used in the work. Before being applied to copper grids covered with formvar film, the sections should be spread on the water surface with xylene vapor. Then, the sections are applied onto copper grids coated with formvar film.
5. After formalin fixation of samples and before their analysis in a scanning electron microscope, the subsequent chemical (alcohol) and physical (drying at a critical point) dehydration of streptococcal biofilm preparations should be avoided to prevent artifacts during sample preparation.

References

1. Romanova YM, Gintsburg AL (2011) Bacterial biofilms as a natural form of existence of bacteria in the environment and host organism. Zh Mikrobiol Epidemiol Immunobiol 3:99–109
2. Fiedler T, Koller T, Kreikemeyer B (2015) *Streptococcus pyogenes* biofilms—formation, biology and clinical relevance. Front Cell Infect Microbiol 5:15. https://doi.org/10.3389/fcimb.2015.00015
3. Hentzer M, Givskov M (2003) Pharmacological inhibition of quorum sensing for the treatment of chronic bacterial infections. J Clin Invest 112(9):1300–1307
4. Akiama H, Morisane S, Yamasake O et al (2003) Assessment of Streptococcus pyogenes microcolony formation in infected skin by confocal laser scanning microscopy. J Dermatol Sci 32(3):193–199
5. Lembke C, Podbielsky A, Hidalgo-Grass C et al (2006) Characterization of biofilm formation by clinically relevant serotypes of group A streptococci. Appl Environ Microbiol 72 (4):2864–2875
6. Danilova TA, Danilina GA, Adzhieva AA et al (2017) Effects of miramistin and phosprenil on microbial biofilms. Bull Exp Biol Med 163 (4):439–442
7. Marks LR, Mashburn-Warren L, Federle MJ, Hakanson AP (2014) Streptococcus pyogenes biofilm growth in vitro and in vivo and its role in colonization, virulence and genetic exchange. J Infect Dis 210:25–34
8. Doetsch RN (1981) Determinative methods of light microscopy. In: Gerhardt P (ed) Manual of methods for general bacteriology. Amer Soc Microbiol, Washington
9. Marrie TJ, Costerton JW (1984) Morphology of bacterial attachment to cardiac pacemaker leads and power packs. J Clin Microbiol 19:911–914
10. Ito S, Karnovsky M (1968) Formaldehyde-glutaraldehyde fixatives containing trinitro compounds. J Cell Biol 30:168a
11. Mater Y, Karakali S (2001) Fine structural demonstration of anion sites on streptococcal and staphylococcal envelopes by cationic dyes. Turk J Med Sci 31:291–295
12. Herald PJ, Zonolla EA (1988) The use of transmission electron microscopy to study the composition of pseudomonas Fragi attachment material. Food Struct 7(1):53–58
13. Danilova TA, Smirnova TA, Danilina GA et al (2018) Optical and electron microscopic study of the morphology and ultrastructure of biofilms formed by Streptococcus pyogenes. Bull Exp Biol Med 165(1):110–114

Chapter 15

Dynamic Interactions of Group A Streptococcus with Host Macrophages

Angelica Montenegro Riestra and J. Andrés Valderrama

Abstract

Macrophages play a critical role in Group A Streptococcus (GAS) recognition and the consequent activation of innate immunity and inflammatory responses against the pathogen. In parallel, GAS deploys several strategies for escaping detection and elimination by these efficient phagocytic cells. The events that take place in this GAS–macrophage battleground, the cellular consequences for the pathogen and for the immune cell, and the balance between the magnitude of infection and the efficiency of the host immune response can be investigated with a variety of assays presented in this chapter.

Key words Group A Streptococcus, Macrophages, Cell death, Bacteria survival, Protein expression, Cytokine release, Colocalization

1 Introduction

GAS encounters with macrophages occur from the time of initial host colonization and continue during its systemic dissemination in the host. Macrophages constitute an important component of the immune system and are strategically distributed throughout host tissues, with key functions in pathogen recognition, antigen presentation, and direct microbial killing. In response, due to the expression of a multitude of surface-bound and secreted virulence factors, GAS can induce and inhibit a myriad of cell signaling events and processes, including several cell death pathways [1]. Despite the prominent role of macrophages for GAS detection and clearance, the precise molecular mechanisms triggered during GAS–macrophage interactions are less understood, compared to the interaction of the pathogen with other cell types.

The body of knowledge regarding GAS encounters with host macrophages was recently reviewed [2], from the mechanisms utilized by macrophages for pathogen detection to the strategies of GAS to evade macrophage responses. Critical and novel findings to highlight include GAS induction of an inflammatory cell death in

Thomas Proft and Jacelyn M. S. Loh (eds.), *Group A Streptococcus: Methods and Protocols*, Methods in Molecular Biology, vol. 2136, https://doi.org/10.1007/978-1-0716-0467-0_15,

macrophages called pyroptosis [3] and the ability of GAS to replicate within macrophages [4].

In this chapter, we describe tools to further probe uninvestigated aspects of GAS–macrophage interactions. Specifically, we provide methodology to expose the bacterium to macrophages, to evaluate the viability of GAS and host macrophages after their encounter, to identify and quantify the expression of proteins involved in GAS–macrophage contact, as well as the use of fluorescence microscopy to visualize the cellular location of these proteins in order to gain mechanistic insight [5]. Lastly, a protocol to profile the macrophage cytokine response to GAS upon differential experimental conditions or with different GAS strains is also included. Assessment of differences in the macrophage immune response and hypothesis building about the potential downstream effects on the host is important to investigate, as the milieu of cytokines can impact the overall efficacy and the type of host response that is subsequently mounted [6, 7]. In summary, in this chapter we provide methods and techniques to investigate both the bacterium and the host responses, which can contribute to addressing the cellular status depending on the site, stage, and magnitude of GAS infection.

2 Materials

2.1 GAS Medium Preparation and Bacteria Culture

1. Todd-Hewitt Broth (THB). Dissolve 30 g of Todd-Hewitt Broth (THB) in 1 L of purified water. Autoclave at 121 °C for 15 min, cool down the media, and store at room temperature (RT).
2. Todd-Hewitt agar (THA) plates: Add 15 g of agar to THB. Autoclave and pour 20 mL of THA into sterile petri dishes.
3. Group A Streptococcus strain of interest (*see* **Note 1**).
4. Spectrophotometer measuring optical density at 600 nm.
5. PathoDx Strep grouping universal kit.

2.2 Growth and Differentiation of THP-1 Cells

1. THP-1 cells (TIB-202™, ATCC) [8, 9].
2. Roswell Park Medium Institute (RPMI)-1640 medium containing 300 mg/L of L-glutamine.
3. Heat-inactivated fetal bovine serum (HI-FBS) (*see* **Note 2**).
4. RPMI-1640 medium containing 10% HI-FBS, 10 mM 4-(2-hydroxyethyl)-1-piperazineethanesulfonic acid (HEPES), 1 mM sodium pyruvate, 4.5 g/L glucose, 0.05 mM β-mercaptoethanol. Mix well by inversion.

5. Phorbol 12-myristate 13-acetate (PMA): Prepare 16 μM PMA stocks in dimethyl sulfoxide (DMSO). We recommend making 100 μL aliquots for single freeze-thaw use.
6. THP-1 differentiation medium: THP-1 medium +25 nM PMA (prepare fresh).

2.3 Isolation of Bone Marrow-Derived Macrophages (BMDMs)

1. Two 6- to 8-week-old C57BL/6 mice allowed to acclimate in the animal facility for at least 2 days before BMDM isolation.
2. Dulbecco's Modified Eagle Medium (DMEM) containing 4.5 g/L D glucose, 584 mg/L L-glutamine, and 110 mg/L sodium pyruvate.
3. BMDM medium: DMEM containing 20% HI-FBS, 10% filtered macrophage colony-stimulating factor (M-CSF) cell supernatant, 50 units/mL Penicillin, 50 μg/mL Streptomycin. The M-CSF supernatant is obtained by collecting and filtering conditioned media from 3T3-MCSF (*see* **Note 3**).
4. Petri dishes: sterile, non-TC (tissue culture) treated (*see* **Note 4**).
5. Cell strainer.
6. Dulbecco's phosphate-buffered saline (DPBS): without calcium and magnesium + 5 mM ethylenediaminetetraacetic acid (EDTA).
7. BMDM seeding medium: RPMI + 10% HI-FBS.

2.4 Isolation and Differentiation of Human Monocyte-Derived Macrophages (HMDMs)

1. Recombinant Human M-CSF (we utilize M-CSF from Peprotech).
2. Blood collection tubes with sodium heparin.
3. Human whole blood from healthy volunteers.
4. Polymorphprep™.
5. HMDM differentiation medium: RPMI media + 10% HI-FBS + 25 ng/mL human M-CSF. Freshly prepare the media before use, as the recombinant M-CSF can degrade over time.

2.5 Infection of Macrophages with GAS and Assessing GAS Survival

1. Multiwell tissue culture plates.
2. Sterile PBS.
3. Assay medium: RPMI supplemented with only 2% HI-FBS.
4. 0.25% Trypsin, 2.21 mM EDTA solution (without sodium bicarbonate). We recommend making 5 mL aliquots from the stock for single freeze-thaw use.
5. 0.025% Triton X-100/PBS solution. The solution should be filter-sterilized and can be stored at room temperature.
6. THA plates (*see* Subheading 2.1, **item 2**).

7. Antibiotics: 10× solution of penicillin (50 μg/mL) and gentamicin (1000 μg/mL) in PBS. Solution should be filter-sterilized.

2.6 Assessing GAS Differential Protein Expression

1. Mutanolysin stock solution: 3000 U/mL mutanolysin in Tris-EDTA (TE) buffer. Stock solutions can be stored at −20 °C as frozen aliquots.
2. Lysis buffer: 5 U/mL mutanolysin, 5 mg/mL lysozyme, 12.5% sucrose, 10 mM Tris, pH 7.6, and 1 mM EDTA. We recommend preparing the solution right before GAS lysis. Immediately prior to cell lysis, add protease inhibitor to the cell lysis buffer. We recommend the 100× Halt™ Protease Inhibitor Cocktail from Thermo Fisher Scientific™, which is diluted directly into the lysis buffer. Keep this solution on ice or at 4 °C, so that it is cold when ready to be used.
3. Tris-buffered saline (TBS) buffer stock (10×): Dissolve 12.1 g of Tris Base, 90 g of NaCl, and adjust to pH 7.4.
4. TBST: mix 100 mL of 10× TBS stock with 1 mL of Tween-20 and bring up the final volume to 1 L.
5. Five percent milk/TBST or any other desired blocking agent. Dissolve milk powder in half of the volume of TBST. After stirring the solution for about 30 min, bring up to the final volume with TBST.
6. Primary and secondary antibodies: in blocking solution (*see* **Note 5**).

2.7 Assessing Cytokine Release

1. Flat, high protein binding, 96-well plates.
2. Reagent Concentrate 2: 10% BSA/PBS solution from R&D Systems. Dilute the 10× Reagent Concentrate 2 in molecular grade water to obtain a final 1× PBS + 1% BSA concentration (*see* **Note 6**).
3. TMB Substrate Reagent Set.
4. Wash buffer: PBS + 0.05% Tween-20.
5. Stop solution: 2 N H_2SO_4.
6. Microplate reader to record the absorbance at 450 nm and 570 nm.

2.8 Assessing Macrophage Host Cell Death

1. CytoTox-ONE™ Homogeneous Membrane Integrity Assay or a CytoTox 96® Non-Radioactive Cytotoxicity Assay. The first kit is fluorescence based and the second kit is colorimetric (both kits are from Promega) (*see* **Note 7**).
2. Propidium iodide staining solution: 1% BSA + 3 μM propidium iodide (PI).
3. Fluorescence microscope with filter for the detection of red fluorescence from PI. PI has an excitation/emission maxima of 493/636 nm.

2.9 Assessing Differential Macrophage Protein Expression

1. Radioimmuoprecipitation assay (RIPA) buffer: 150 mM NaCl, 25 mM Tris–HCl, pH 7.6, 1% sodium deoxycholate, 1% NP-40, 0.1% sodium dodecyl sulfate (SDS).
2. Lysis buffer for mass spectrometry: 1% *n*-dodecyl-β-D-maltoside, 0.05% Tween-20, 20 mM Tris–HCl, pH 7.5, 150 mM NaCl, 20 mM $MgCl_2$ (*see* **Note 8**).
3. SDS-PAGE (polyacrylamide gel electrophoresis) and Western blot equipment plus reagents: *see* Subheading 2.6, **items 3–6**.

2.10 Fluorescence Microscopy

1. 15 mm diameter coverslips sterilized by autoclaving.
2. 24-well plates.
3. Ten percent sodium azide solution in PBS. Use caution when making the solution and prepare the solution inside a chemical hood.
4. Five percent bovine serum albumin (BSA) in PBS + 0.02% sodium azide solution. This solution will be used to block the coverslips and for antibody dilutions. Prepare the solution by dissolving the BSA in half of the final PBS volume. Add the required amount of sodium azide. Mix for at least 15 min with vigorous vortexing or continuous shaking. Bring up to the final volume with PBS and mix for an additional 15 min. It is critical that the BSA solution is dissolved well in order to ensure optimal blocking and to avoid artifacts due to BSA clumps.
5. Permeabilization solution: PBS + 0.2% Triton X-100. Mix well by vigorous vortexing or continuous shaking.
6. Fixation solution: 4% paraformaldehyde in PBS. Use caution when making the solution and prepare the solution inside a chemical hood.
7. Primary antibody: in 5% BSA in PBS blocking solution (*see* **Note 9**).
8. 1 μg/mL Hoechst solution in PBS. Cover with aluminum foil.
9. Prealiquot 0.5 mL of the fixation, permeabilization, blocking, and Hoechst staining solutions for each sample, in 24-well plates. Use one plate for each solution. Cover the Hoechst aliquot plate with aluminum foil to avoid light exposure.
10. Moisture chamber: cover a Nunc™ Square BioAssay Dish with aluminum foil. Make sure to crease the edges of the aluminum, so that the covered dish can close completely. Add a piece of Whatman® filter paper and wet it using a squirt bottle. Lay pieces of parafilm on top of the paper to cover it completely. On the parafilm, draw lines to outline where coverslips will be placed.
11. Curved end forceps for coverslip handling.

12. Syringe needle. Bend ~5 mm of the bottom end of the needle to make an L-shape. Bent needle will be utilized to lift up coverslips from the plate.

3 Methods

3.1 GAS Culture

1. Take a loop from the cryotube of the selected GAS strain and seed it over the surface of a THA plate.
2. Refreeze the remaining content of the cryotube.
3. Incubate the inoculated plate at 37 °C for at least 24 h.
4. Observe the morphology of GAS colonies at the end of the incubation time. GAS colonies are small, white, and symmetrically round. A mucoid phenotype is characteristic of GAS encapsulated serotypes (Fig. 1, top panel). Simultaneously, representative colonies can be tested by the agglutination method to identify beta-hemolytic Lancefield group A, corroborating the presence of GAS. Follow the PathoDx Strep grouping universal kit manufacturer recommendations (Fig. 1, bottom panel).
5. Take a loop full of GAS colonies from the THA plate and inoculate 3 mL of THB media.
6. Culture at 37 °C for 16–20 h without shaking.

3.2 GAS Inoculum Preparation for Macrophage Infection

1. After incubation time, take a 1 mL aliquot of a GAS overnight culture or an exponential-phase culture (*see* **Note 10**) and measure the optical density (OD_{600nm}) to determine the bacterial count and to prepare the inoculum accordingly.
2. Spin down 1.5 mL of the remaining GAS culture in a sterile 2 mL tube at 15,000 × *g* for 10 min.
3. Resuspend the bacterial pellet in 1.5 mL of assay media to maintain the same initial cellular density.
4. Prepare bacteria dilutions in assay media according to the desired multiplicity of infection (MOI) and utilize the proportion that a GAS suspension with $OD_{600} = 0.4$ is equal to 2×10^8 CFU/mL. We recommend a MOI of 10:1 bacteria: macrophage ratio.

3.3 Growth and Differentiation of THP-1 Cells

1. THP-1 cells grow in suspension. Once a high density has been obtained, the media will look turbid. Passage cells before they get so dense that they start pelleting. Cells should not be grown past 2×10^6 cells/mL.
2. Pass cells by transferring an aliquot of the cells into fresh media. If cells will be used in 2 days a twofold split is recommended. For routine passaging of cells, a fourfold to fivefold split is recommended.

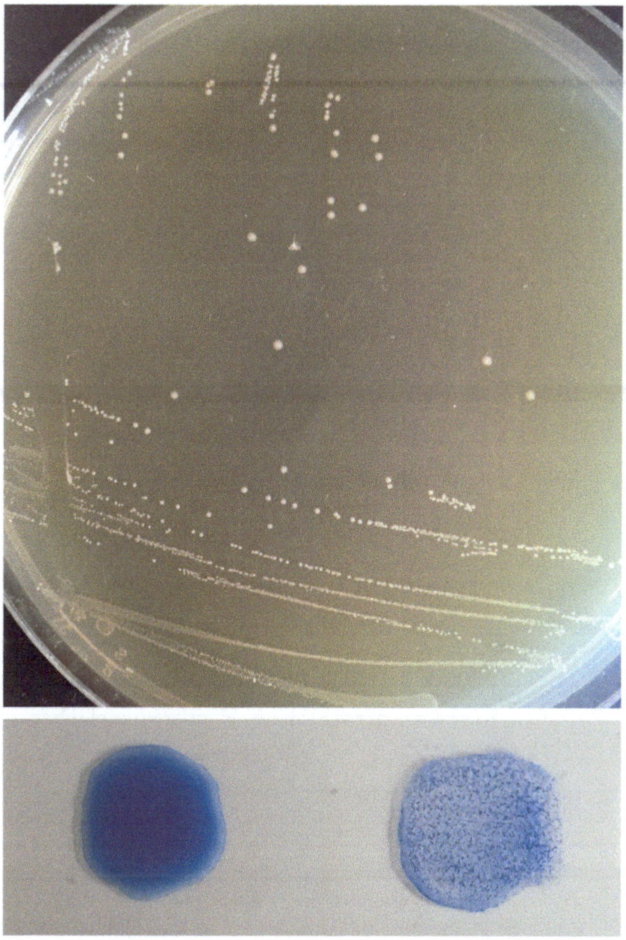

Fig. 1 Isolation of GAS colonies and identification of beta-hemolytic Lancefield group A. Group A Streptococcus colony appearance after cultivation on THA plates at 37 °C overnight (top panel). Beta-hemolytic Lancefield group A assay (bottom panel), showing a negative (left spot, *E. coli*) or positive (right spot, GAS) bacteria for the agglutination test

3. Bring up the final volume of T75 flasks to 20–25 mL and to 35–40 mL for T150 flasks.
4. Pipette the cells up and down about five times in order to mix the cell suspension well.
5. THP-1 cells need to be differentiated into macrophages 1–3 days before the desired experiment (*see* **Note 11**).
6. To differentiate the cells, aliquot the total number of cells needed for the experiment. *See* Table 1 for recommended seeding amounts per well.
7. Spin down the cells at 200 × *g* for 5 min.
8. During the spin, prepare THP-1 media + 25 nM PMA.

Table 1
Recommended volumes and seeding amounts of macrophages to test interactions with GAS

Cell type	Application	Amount of cells to seed per well	Seeding volume/ well	Volume to use when testing interaction with GAS
THP-1	Assessing cytokine production after GAS encounter, LDH release assays	1×10^5 cells/well of a 96-well plate	0.2 mL	0.1 mL
THP-1	GAS viability after encounter with THP-1 cells	5×10^5 cells/well of a 12-well plate	2 mL	0.7 mL
THP-1	Differential protein analysis	2×10^6 cells/well of a 6-well plate	2 mL	1 mL
THP-1	Indirect immunofluorescence assay (IFA)	0.5×10^6 cells/ well of a 24-well plate	1 mL	0.250 mL
THP-1	Microscopy imaging in 35 mm MatTek dishes (with No. 1.5 coverslips)	2×10^6 cells/dish	2 mL	2 mL
BMDM	Assessing cytokine production after GAS encounter, LDH release assays	6×10^4 cells/well of a 96-well plate	0.2 mL	0.1 mL
BMDM	GAS viability after encounter with BMDMs	$2–2.5 \times 10^5$ cells/ well of a 24-well plate	2 mL	0.250 mL
BMDM	Differential protein analysis	1×10^6 cells/well of a 6-well plate	2 mL	1 mL
HMDM	GAS viability after encounter with HMDMs	1.5×10^5 cells/ well of a 24-well plate	2 mL	0.250 mL

9. Decant the media and resuspend in the desired final volume of THP-1 media +25 nM PMA. Mix well by pipetting up and down five times.
10. Aliquot the appropriate volume of cells to each well, depending on the desired experiment. *See* Table 1 for recommended final volumes per well. Figure 2 shows an example of the change in cell morphology after THP-1 cell adherence and differentiation.

3.4 Isolation and Differentiation of Bone Marrow-Derived Macrophages (BMDMs)

1. Sacrifice animals in accordance with institutionally approved procedures.
2. Collect the femur and tibia bones from the rear legs of two mice (eight bones total). Use caution to ensure that you do not break the bones. Start by making an incision in the skin above the lower abdomen.

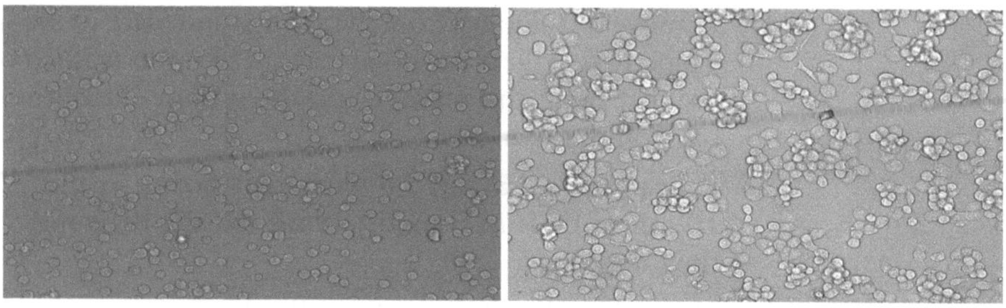

Fig. 2 Differential cell morphology of THP-1 monocyte cell line versus differentiated THP-1 macrophages. THP-1 cells right after PMA addition (left panel) and 18 h after PMA differentiation (right panel)

3. Cut down through the skin as far down the leg as possible, then using your hands peel off the skin surrounding the leg muscles.
4. Feel for the knob located above the top part of the femur and cut above this knob, this will cut the leg out.
5. Remove the mouse paw by flexing it back and forth, the paw will dislodge from the tibia. Be careful not to apply pressure to the tibia, as it can break easily.
6. Gently flex between the femur and the tibia, in order to separate the bones.
7. Gently remove any remaining tissue surrounding the final bones.
8. Place separated bones in a petri dish with sterile PBS. Transport the bones to the tissue culture biosafety cabinet.
9. In an empty petri plate lid, trim the ends of the bones with scissors using a straight cut. Be careful not to accidently shatter the bone while cutting and make sure to maintain the bone above the dish.
10. Place the bones in a new petri dish containing PBS. Incubate for 1 min.
11. Place the bones in a new petri dish containing ethanol. Incubate for 1 min.
12. Place the bones in a new petri dish containing PBS. Incubate for 1 min.
13. Place the washed bones in a new petri dish.
14. Prepare a 10 mL syringe with a 25-gauge needle, fill with 10 mL of DMEM (no additives).
15. Fit the needle of the prepared syringe into the bone and push out the bone marrow (red). Squirt until all the liquid goes through. Bones should appear as lighter white color when the bone marrow has been flushed out.

16. Remove the plunger from the syringe and place to the side. Use a pipette to collect the bone marrow that was washed out from the first bone and place it into the syringe. Place the plunger back on the syringe.
17. Repeat **steps 15** and **16** with each bone, so that the same 10 mL of DMEM is used to flush out the bone marrow from the four bones.
18. Strain the cell solution into a 50 mL conical tube. If the plate still contains a visible amount of cells, wash the plate with an additional 5 mL of DMEM and pass it through the cell strainer to pool with the other cells.
19. Spin down cells at 130 × *g* for 10 min.
20. Resuspend cells in 60 mL of BMDM medium and add 10 mL per petri dish. Incubate at 37 °C in 5% CO_2. Alternatively, cells can be frozen at this stage (*see* **Note 12**).
21. After 3 days of incubation, add 5 mL of BMDM medium to each dish.
22. Monocytes from the bone marrow will differentiate and adhere to the petri plate. BMDMs are ready for use 7 days after initial plating.
23. BMDMs need to be detached from the plate in order to seed cells for an experiment. To start cell detachment, remove the media and wash the cells with 10 mL of PBS. Tilt the plate back and forth a few times and remove the PBS.
24. Add 10 mL of cold Dulbecco's phosphate-buffered saline (without calcium and magnesium) + 5 mM EDTA to each plate.
25. Place the plates in a refrigerator and incubate for 10–15 min.
26. Scrape the cells gently with a rubber policeman five times.
27. Pool cells into a conical tube.
28. Spin down the cells at 200 × *g* for 8 min.
29. Remove the media and resuspend in the appropriate amount of BMDM seeding medium (RPMI + 10% HI-FBS). Take cell counts and adjust to the desired final cell density (*see* Table 1). Cells are ready to be seeded and used for an experiment the next day.

3.5 Isolation and Differentiation of Human Monocyte-Derived Macrophages (HMDMs)

1. Draw 50 mL of blood from healthy volunteers in accordance with institutionally approved procedures and dispense in blood collection tubes.
2. Add 20 mL of Polymorphprep™ into two 50 mL conical tubes.
3. Layer the blood onto the polymorph prep, by slowly dispensing 25 mL of blood to the side of each tube.

4. Spin down at 620 × g for 30 min, without brake.
5. Collect the top white buffy coat layer and transfer it into a new tube.
6. Bring the volume up to 50 mL with PBS. This step serves as a first wash.
7. Spin down at 350 × g for 15 min and remove the supernatant.
8. Resuspend the pellet by finger vortexing. Add 50 mL of PBS to wash cells, this is a second wash step.
9. Spin down at 350 × g for 15 min and remove the supernatant.
10. Resuspend the pellet by finger vortexing and then resuspend pellet in 30 mL of HMDM differentiation medium (RPMI + 10% HI-FBS + 25 ng/mL M-CSF).
11. To three T150 flasks, add 10 mL of cells each.
12. Add 10 mL of HMDM differentiation medium to each flask in order to bring up the final volume to 20 mL. Incubate at 37 °C in 5% CO_2.
13. Three days later, add 10 mL of freshly prepared HMDM differentiation medium to each flask.
14. Cells are ready to be seeded for an experiment 6–9 days after initial plating (*see* **Note 13**).
15. To detach cells from flasks, first remove the supernatant.
16. Perform a quick PBS wash by adding 10 mL of PBS to the flask and rocking plate back and forth five times. Gently remove the PBS by aspiration and repeat the PBS wash.
17. Add 10 mL of Accutase® and incubate the flasks in the refrigerator for 10–15 min (*see* **Note 14**).
18. Gently use a rubber policeman to scrape the cells five times.
19. Spin down the cells at 200 × g for 10 min.
20. Remove the supernatant and gently resuspend the pellet by finger vortexing. Resuspend the cells in 5 mL of RPMI + 10% HI-FBS. Take cell counts and adjust according to desired final volumes and cell density (*see* Table 1).

3.6 Infection of Macrophages with GAS

1. THP-1, BMDMs, or HMDMs cells are seeded in plates accordingly with the specifications listed in Table 1 and Subheadings 3.3, 3.4, and 3.5, respectively. Seed the cells, taking into account that at least three wells (technical triplicates) must be used for each condition tested. It is important to leave triplicate wells with uninfected cells that serve as a negative control for assessing macrophage outputs.
2. Prewarm the assay medium (RPMI + 2% FBS).
3. Remove cell culture media from all wells by gentle aspiration.

4. Add the desired volume of bacteria resuspended in prewarmed assay media and diluted accordingly with the MOI to be tested (*see* Table 1 for recommended assay volumes). Add the same volume of media to the uninfected control wells.
5. Centrifuge plates at 600 × *g* for 5 min at RT to facilitate bacterial contact with macrophages.
6. Infected cells are incubated at 37 °C in 5% CO_2 for the desired time, accordingly with the experimental condition required.

3.7 Assessing GAS Outputs

3.7.1 GAS Total Bacteria Survival

1. GAS and macrophages are coincubated at 37 °C in 5% CO_2 for up to 2 h (*see* **Note 15**).
2. Gently aspirate supernatants and wash the cells with 1 mL PBS to eliminate cell debris and extracellular bacteria.
3. Detach the cells by adding 100 μL of Trypsin/EDTA to each well and mixing. Incubate for 5 min at 37 °C in 5% CO_2.
4. Lyse the cells by adding 900 μL of 0.025% Triton X-100 solution.
5. Mix vigorously by pipetting up and down several times to ensure total macrophage lysis.
6. Serially dilute samples in PBS (10^{-1} to 10^{-8}). Plate 20 μL spots of each dilution in triplicate for each condition on THA plates. We recommend using a multichannel pipette to perform dilutions and spotting the samples to minimize error due to excessive pipetting.
7. Incubate plates at 37 °C overnight for colony-forming units' (CFUs') enumeration.

3.7.2 GAS Intracellular Bacteria Survival

1. GAS and macrophages are coincubated at 37 °C in 5% CO_2 for 30 min.
2. Add the 1× solution of penicillin and gentamicin to each well (for example, 70 μL of 10× stock solution for 700 μL total volume per each well) to kill extracellular bacteria. Incubate plates for an additional 1 h or desired intracellular replication time.
3. At various time points (*see* **Note 16**), gently aspirate the media and rinse once with PBS to remove extracellular dead bacteria. Afterwards, cells are detached and lysed following the same Trypsin/EDTA and Triton X-100 treatments (*see* Subheading 3.7.1, **steps 3–5**).
4. Serially dilute samples in PBS (10^{-1} to 10^{-8}) and plate 20 μL spots of each dilution and each condition on THA plates. We recommend using a multichannel pipette to perform dilutions and spotting of the samples to minimize error due to excessive pipetting.

5. Incubate plates at 37 °C overnight for CFUs' enumeration.
6. Calculate the recovered CFU/mL by multiplying the CFU counts by the dilution factor plated and the volume factor.

3.7.3 GAS Differential Protein Expression

GAS protein expression can be analyzed from total bacteria or from the fraction of intracellular bacteria after infection (*see* Subheadings 3.7.1 and 3.7.2, respectively). The most critical part is ensuring efficient lysis of the GAS peptidoglycan layer on the bacterial cell surface following the directions of the steps below.

1. To assess differential protein expression from total bacteria, immediately after the infection time point spin down the plate 500 × *g* for 5 min at RT. Supernatants can be collected for assessing macrophage outputs (*see* Subheading 3.8).
2. The total amount of bacteria will be together with the macrophages at the bottom of the plate after the above spin. To dislodge the macrophages and collect all bacteria (*see* Subheading 3.7.1, **steps 3–5**). For analysis of protein expression of intracellular GAS, collect only bacteria within the macrophage (*see* Subheading 3.7.2, **steps 2** and **3**).
3. Transfer the entire cell/bacteria mix (1 mL approx.) to a 1.5 mL tube and spin down at 15,000 × *g* for 10 min.
4. Wash with PBS and spin down again as above.
5. Discard supernatant by gentle aspiration and keep the pellet-containing GAS on ice.
6. Resuspend pellets in 200 μL lysis buffer-containing mutanolysin.
7. Incubate at 37 °C for 2 h with constant rotation.
8. Spin down for 5 min at maximum speed. This step will pellet any insoluble material.
9. Transfer the supernatant to a new tube. GAS lysates can be frozen at −80 °C at this step.
10. To compare the differential presence of proteins in whole cell lysates, first quantify the amount of protein in each sample using a protein quantification assay of choice. Calculate the amount of protein in the whole cell lysates and in dilutions of the sample (for example, prepare a 10× dilution of whole cell lysate using the lysis buffer as diluent).
11. Separate and blot proteins by standard SDS gel electrophoresis and Western blot methods.
12. Block the blot using 5% milk/TBST or BSA solution for 1 h at RT or overnight at 4 °C (*see* **Note 5**).
13. Dilute the primary antibody in the appropriate diluent and at the final desired concentration (*see* **Note 5**).

14. Allow antibody to bind for 1 h at RT or overnight at 4 °C on a shaking platform.
15. Remove the antibody solution and add a sufficient amount of TBST to cover the membrane. Incubate for 10 min on a shaking platform. Repeat two additional times, for three washes.
16. Dilute the secondary antibody in the appropriate diluent and at the final desired concentration (*see* **Note 5**).
17. Allow antibody to bind for 1 h at RT.
18. Remove the antibody solution and add a sufficient amount of TBST to cover the membrane. Incubate for 10 min on a shaking platform. Repeat two additional times, for a total of three washes.
19. Develop the blot using the commercial kit of your choice and acquire multiple exposures.
20. Annotate the molecular weights from the protein ladder on the film if utilizing film to collect the exposures.
21. Compare the absence/presence and intensity of the bands corresponding to the protein that was probed for among the different GAS experimental conditions, compared to untreated GAS control (background levels of protein).

3.8 Assessing Macrophage Outputs

3.8.1 Cytokine Release

After exposing GAS to macrophages, the cell supernatant can be transferred to a new plate/tubes and frozen down in order to analyze cytokine production at a later time point. The differential presence of cytokines in experimental cell supernatants can be tested using commercially available enzyme-linked immunosorbent assay (ELISA) kits. The following protocol is an example with our commonly used Human IL-1β/IL-1F2 DuoSet® ELISA kit (R&D Systems, DY401).

1. Using PBS, dilute the capture antibody to the final concentration indicated by the kit.
2. Using a multichannel pipette, add 100 μL of anti-mouse interleukin-1β (IL-1β) capture antibody dilution to the appropriate number of wells that will be used for the assay. Remember to also account for wells that will be utilized to assay the standard curve samples in triplicate and negative control wells (reagent diluent only).
3. Incubate the plate overnight at RT.
4. Wash the plates three times with wash buffer. For each wash, decant the entire contents of the plate, fill up the wells with wash buffer using a squirt bottle, multichannel pipette (400 μL), or an automatic plate washer.

5. Remove the remaining liquid, by inverting the plate and forcefully blotting it against a stack of paper towels. Flip the paper towel after each blot and keep blotting until no liquid is observed on the paper towels.
6. Block the plate by adding 300 μL of reagent diluent to each well and incubate at RT for at least 1 h.
7. Spin down samples at the appropriate speed for the cell type (*see* Subheadings 3.3–3.5). Transfer the cell-free supernatant to a new plate/tube.
8. During the blocking step, prepare the appropriate dilutions of the samples and the standards using reagent diluent (*see* **Note 17**). Prepare the standard curve by performing serial dilutions according to the manufacturer's specifications. We recommend preparing the dilutions in a 96-well round-bottom well plate with the desired sample and standard curve positioning that will be used in the ELISA. Prepare a sufficient amount of the dilutions to be able to assay each supernatant in triplicate plus extra volume for pipetting.
9. After the 1 h blocking step is finished, wash the plates three times, as indicated in Subheading 3.8.1, **steps 4** and **5**.
10. Add a total of 100 μL of sample/sample dilutions, and standard curve dilutions. Add 100 μL of reagent diluent to the negative control wells and incubate for 2 h at RT.
11. Near the end of the 2 h incubation, dilute the detection antibody in reagent diluent, according to the manufacturer's instructions.
12. After 2 h incubation, wash the plate three times using wash buffer.
13. Add 100 μL of detection antibody/reagent diluent to each well and incubate for 2 h at RT.
14. Near the end of the 2 h incubation, dilute the Streptavidin–horseradish peroxidase (HRP) solution in reagent diluent, according to the manufacturer's instructions. The solution is light sensitive, therefore cover the solution with aluminum foil.
15. After 2 h incubation, wash the plate three times using wash buffer.
16. Add 100 μL of the working dilution of Streptavidin–HRP to each well. Cover the plate with aluminum foil and incubate for 20 min at RT.
17. Near the end of the incubation, prepare the TMB substrate solution according to the manufacturer's instructions. Cover the solution with aluminum foil. The substrate solution should be colorless or will have a very faint blue color.

18. After 20 min incubation, wash the plate three times using wash buffer.
19. Add 100 μL of the prepared TMB substrate solution and incubate for 20 min. Cover the plate with aluminum foil. The standard curve wells will show a gradient of blue color intensity. Wells with a strong positive signal will also have a strong blue color.
20. Add 50 μL of 2 N H_2SO_4 stop solution to each well. Gently tap the plate to ensure thorough mixing. The wells will now turn yellow. Cover the plate with aluminum foil and transport to the plate reader.
21. Using a microplate reader, record the optical density (OD) at 450 nm. Also take a reading at 540 nm or 570 nm to subtract and correct for optical imperfections in the plate.
22. Calculate the amounts of antigen in each sample. Start by subtracting the background values (negative control values) from all the corrected OD_{450} values (OD_{450}–$OD_{540/570}$). Graph the background-subtracted standard curve OD_{450} values versus the standard protein concentration. Add a trendline and record the regression line. Calculate the amount of protein in each sample using the regression line equation.

3.8.2 Assessing Macrophage Host Cell Death Using the LDH Release Assay

1. When setting up the macrophage infections (*see* Subheading 3.6), make sure to include six extra wells where cells are not infected or treated with any reagent. Three wells will be utilized to determine the "average background" values that will be subtracted out from all the experimental samples. The other three wells will be treated with lysis buffer at the end of the experiment and thus serve as the "100% lysis" wells.
2. Add the appropriate amount of substrate reagent to a 96-well plate and cover the plate with aluminum foil. We recommend assaying 50–75 μL of supernatant. Add the corresponding amount of substrate in order to have a 1:1 ratio of supernatant and substrate reagent. For example, if 75 μL of supernatant will be assayed add 75 μL of substrate reagent. Cover the plate with aluminum foil to avoid direct light exposure.
3. Thirty minutes before the end of the desired coincubation time, add the appropriate amount of lysis solution (provided by kit) to the "100% lysis" wells. At the end of the incubation, check that the cells have physically lysed by comparing the morphology of the "100% lysis" wells to untreated cells.
4. Transfer the appropriate amount of all the samples to the plate containing an equal volume of the prealiquoted substrate reagent. Cover the plate with aluminum foil.

5. Incubate the plate until the "100% lysis" control wells develop a hot pink color (if using the fluorescence-based kit) or a red color (if using the colorimetric-based kit). A 30 min incubation is typically recommended.
6. Add the appropriate amount of stop solution. For example, add 38 μL of stop solution if performing the assay in 96-well plates and the supernatant + substrate reagent volume is 150 μL.
7. Tap the sides of the plate to mix and pop any existing bubbles with a syringe needle.
8. Read the plate at the appropriate wavelength. If using the fluorescence-based assay, utilize an excitation wavelength of 560 nm and an emission wavelength of 590 nm. If using the colorimetric-based assay, record the absorbance values at 490 or 492 nm.
9. To calculate the % lysis of macrophages. First subtract the average of the background noninfected and untreated control cells from all experimental values. Afterwards, calculate the average value of the "100% lysis" wells. Divide all the background-subtracted values by the average "100% lysis" values and multiply by 100 to calculate the % cytolysis.

3.8.3 Assessing Macrophage Host Cell Death Using Propidium Iodide Uptake Assay

1. Seed macrophages in 24-well plates according to Table 1 and infect with GAS according to Subheading 3.6.
2. After the coincubation time, remove the supernatant and wash the cells three times with PBS.
3. Add 0.5 mL of PI staining solution to the cells and incubate for 10 min. Cover the plate with aluminum foil to avoid direct light exposure. Immediately image the samples.
4. Acquire at least 10 independent fields of each sample throughout the well for quantitative microscopy analysis. In each image, count the number of cells that display red fluorescence, these are the PI^+ population. In order to calculate the total number of cells per field, also count the number of unstained cells. Calculate the average number of cells with PI uptake as the PI^+ cells/total number of cells.

3.8.4 Macrophage Differential Protein Expression

1. Seed macrophages in 6-well plates according to Table 1 and infect with GAS according to Subheading 3.6.
2. After the coincubation time, collect the cell supernatant if you desire to analyze differences in released proteins after GAS exposure. Place the supernatants on ice.
3. Wash the cells by adding 1 mL of PBS and quickly rocking the plate back and forth five times.
4. Add 0.5 mL of lysis buffer + protease inhibitors to lyse the cells.

5. Place the plate on a nutator at 4 °C for 20 min (*see* **Note 18**).
6. Place the plate on top of ice, then using a rubber policeman scrape the cells by moving the rubber policeman back and forth around the plate five times.
7. Spin down the cells at 15,000 × g for 15 min at 4 °C. This step will pellet any insoluble material.
8. Transfer the cell supernatant into a fresh microfuge tube. We recommend that the bottom 50 μL of cell supernatant is not transferred, in order to avoid dislodging or accidently touching the insoluble pellet. Whole cell lysates can be frozen at −80 °C at this step.
9. To compare differences in released proteins, prepare samples for Western blot analysis by comparing equal volumes of cell supernatant. Mix the same amount of cell supernatant with the appropriate amount of Laemmli loading buffer.
10. To compare differences in protein from whole cell lysates (*see* Subheading 3.7.3, **steps 10–20**).
11. Compare the absence/presence and intensity of the bands corresponding to the protein that was probed for in GAS-exposed samples, compared to the uninfected control (background levels of protein).

3.8.5 Fluorescence Microscopy to Observe GAS–Macrophage Encounters and to Perform Colocalization Analysis

1. Place autoclaved coverslips in 24-well plates and seed the cells, according to Table 1.
2. Infect macrophages with GAS (*see* Subheading 3.6).
3. After the desired coincubation time, pick up each coverslip by sliding a bent syringe needle under the edge of the coverslip and lifting it up. Keeping track of the side that is facing up, use the curved end forceps to pick up the coverslip and transfer it into the 24-well plate containing the prealiquoted paraformaldehyde solution.
4. Fix the coverslips for 20 min.
5. Wash the coverslips three times with PBS. We recommend that washes are performed in a 24-well plate. Using a pipette, fill the wells all the way to the top with PBS in order to have a convex meniscus extending from the top of each well. Dip the coverslip into a clean well for each wash and swing the coverslip back and forth 10 times during each wash. If the immunofluorescence assay (IFA) will not be performed on the same day, after the washes place the coverslips in 0.5 mL of PBS in 24 well plates, cover the plate with aluminum, and store at 4 °C.
6. If samples will be permeabilized, place the coverslips into the PBS + 0.2% Triton X-100 permeabilization solution prealiquoted in a 24-well plate. Incubate for 30 min.

7. Block all samples for 1 h in the PBS + 5% BSA solution.
8. Aliquot 50 μL drops of your primary antibody dilution onto the parafilm inside of your moisture chamber (*see* Subheading 2.10, **item 10**). You should include double and single secondary antibody controls that are only stained with both of the secondary antibodies and singly with one of the secondary antibodies. For these samples, use 50 μL drops of the PBS + 5% BSA blocking solution for the first antibody incubation step. If you are performing colocalization of two proteins, make sure to also include single stain controls in which samples will receive only one of the primary antibodies. All other samples receive both of the primary antibodies (*see* Table 2).
9. Invert the coverslips onto the primary antibody or blocking solution drops. Incubate for 1 h.

Table 2
Suggested controls for colocalization analysis of proteins of interest (Example proteins A and B)

Primary antibody/antibodies	Secondary antibody/antibodies	Purpose
Mouse anti-A and rabbit anti-B	Anti-mouse-fluorophore#1 and anti-rabbit-fluorophore#2	Allows you to observe the colocalization of proteins A and B
Anti-A	Anti-mouse-fluorophore#1	1. Positive control for protein A staining. Allows you to observe that the staining for protein A detection worked if the sample expresses protein A. 2. Observation of this sample with the acquisition condition to detect the B protein will allow you to visualize potential bleed-through from the A signal into the B signal detection-channel
Anti-B	Anti-rabbit-fluorophore#2	2. Positive control for protein B staining. Allows you to observe that the staining for protein B detection worked if the sample expresses protein B. 2. Observation of this sample with the acquisition condition to detect the A protein will allow you to visualize potential bleed-through from the B signal into the A signal detection-channel.
None	Anti-mouse-fluorophore#1 and anti-rabbit-fluorophore#2	Allows visualization of the background signal due to both secondary antibodies
None	Anti-mouse-fluorophore#1	Allows visualization of the background signal due to the anti-mouse-fluorphore#1 secondary antibody
None	Anti-rabbit-fluorophore#2	Allows visualization of the background signal due to the anti-rabbit-fluorphore#2 secondary antibody

This table describes the positive and negative controls whose inclusion in an indirect immunofluorescence assay (IFA) will allow the researcher to conclude that the observed signal in colocalization analysis is the true signal above background

10. At the end of the incubation, transfer the whole primary staining setup (Whatman® paper + parafilm + coverslips) onto an empty square plate. Be careful not to cause the coverslips to slide around.
11. Add a new piece of Whatman® paper to your moisture chamber, wet it, and cover with parafilm.
12. Aliquot 50 μL drops of the appropriate secondary antibody/antibodies' dilution and close the moisture chamber to protect the antibody from light (*see* **Note 9**). For colocalization studies, "both secondary antibody controls" get stained with a mixture of the two secondary antibodies. Secondary only controls get stained with a single secondary antibody. The single stained controls receive the secondary antibody appropriate for the primary antibody isotype. All other samples receive both of the secondary antibodies (*see* Table 2).
13. Wash the coverslips three times with PBS. At the end of the third wash, tap the edge of the coverslip on a Kimwipe and roll the coverslip, in order to allow the excess liquid remaining on the coverslip to drip down onto the Kimwipe.
14. Place the coverslips onto the secondary antibody dilution drops. Make sure that the samples are protected from light. Incubate for 1 h.
15. Wash the coverslips three times with PBS and tap off the excess PBS, as described in Subheading 3.8.5, **step 13**.
16. Place the coverslips into the 24-well plate with prealiquoted Hoechst dye. Incubate for 15 min.
17. Wash the coverslips three times with PBS and tap off the excess PBS, as described in Subheading 3.8.5, **step 13**.
18. Mount the coverslip onto a glass slide utilizing a mounting reagent.
19. Dry off the bottom part of the moisture chamber, invert it, and place it over the samples to cover them from light. Allow slides to mount well overnight.
20. Seal the edges of the coverslip with clear nail polish.
21. Image the slides using the microscope of choice. We recommend scanning the experimental samples and comparing the signal to the secondary only control and experimental controls to set the proper exposure times for image acquisition.
22. Acquire multiple images from all the samples for downstream quantitative analysis using the method or program of choice. We utilize Fiji and ImageJ, which are freely downloadable software. Figure 3 shows an example of the IFA results for a GAS/macrophage colocalization experiment.

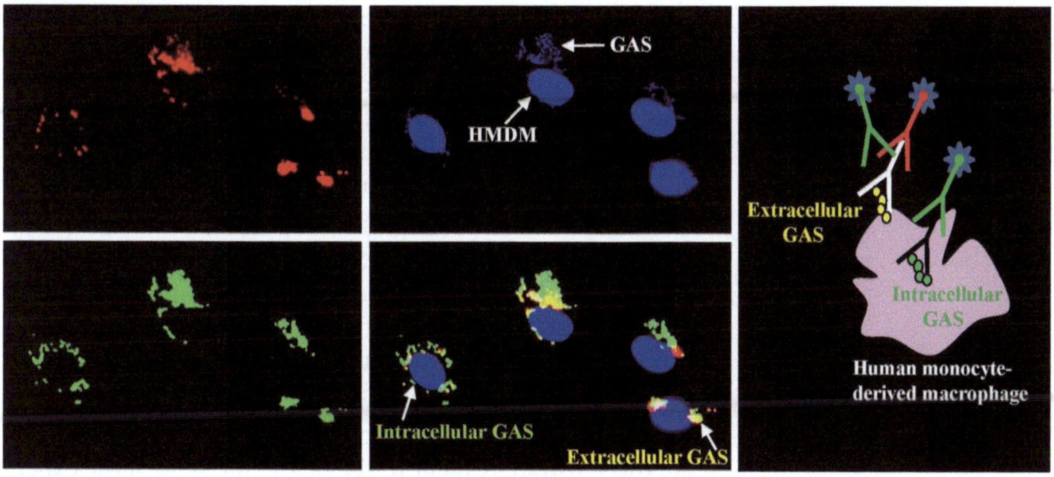

Fig. 3 Example of an indirect immunofluorescence assay (IFA) to observe GAS–macrophage colocalization. HMDMs were exposed to the GAS ΔSpeB mutant for 2 h, extracellular bacteria were killed with gentamicin, and an additional 4 h incubation was performed. Extracellular bacteria can be distinguished from intracellular bacteria by performing an IFA with differential permeabilization. An anti-Group A Streptococcus carbohydrate antibody (Abnova) and an anti-rabbit Alexa Fluor® 595 conjugated secondary antibody are used to stain all the bacteria. Subsequently, cells are permeabilized and stained again with the anti-Group A antibody and an Alexa Fluor® 488 conjugated secondary antibody. Intracellular bacteria are protected from the initial stain and thus only display single staining with the green secondary antibody. Extracellular bacteria display colocalization of both secondary antibodies, areas of strong colocalization are observed as yellow in the merged image. Both the HMDMs and GAS are detected with Hoechst dye nuclear staining (blue). A schematic of the colocalization experiment is shown on the right panel

4 Notes

1. We commonly utilize the GAS M1T1 5448 strain [10].
2. Thaw serum at 37 °C, then heat-inactivate the serum at 56 °C for 30 min. Make sure to mix the serum by inverting the bottle several times at least once halfway through the incubation.
3. If M-CSF producing cells are not available for the preparation of BMDM medium, purified recombinant murine M-CSF can be commercially purchased and used at a final concentration of 20 ng/mL. Another alternative is to use RPMI medium supplemented with 33% filtered L929 conditioned media supernatant +55 U/mL Penicillin +55 μg/mL Streptomycin +20% HI-FBS.
4. It is very difficult to dislodge BMDMs from traditional plates used for tissue culture work, thus ensure that the proper plates are utilized for BMDMs' differentiation.
5. Blocking and dilution of antibodies should be performed using the diluent recommended by the manufacturer. Certain antibodies bind optimally in BSA-based solutions versus milk

solutions, and vice versa, therefore this factor may need to be optimized for the primary antibody being utilized. If high background antibody binding is observed, we recommend the use of ECL™ Blocking Agent (GE Healthcare) for both blocking and antibody dilutions. For commercially available antibodies, a 1:1000 dilution for primary antibodies and a 1:5000 dilution for HRP-conjugated secondary antibodies are a good range of dilutions to utilize.

6. An alternative to using and diluting the commercially purchased reagent diluent 2 is to prepare a 1% BSA/1× PBS solution. Ensure that the solution dissolves well by mixing for at least 1 h and then filter-sterilize the solution.

7. Depending on the coincubation times, prepare and preequilibrate the substrate reagent to RT before the start of the experiment, or during the macrophage coincubation with the bacteria. Resuspend the vial of substrate mix in the amount of assay buffer indicated by the manufacturer. The remaining substrate reagent that is not utilized in the lactate dehydrogenase (LDH) assay can be frozen down at −20 °C and used in future experiments to ensure that the substrate reagent is equilibrated to RT prior to use. Make sure to cover the reagents and experimental plate with aluminum foil, as reagents are light sensitive.

8. For optimal protein identification by mass spectrometry, the use of a nonionic detergent, such as *n*-dodecyl maltoside, is recommended above other commonly used detergents like Triton X-100 and NP-40 [11].

9. The optimal antibody dilution needs to be experimentally determined. However, for commercially purchased monoclonal antibodies, we typically dilute the antibody 1:1000. Polyclonal sera are typically diluted 1:100–1:500. We utilize Alexa Fluor® 488 and 594 conjugated secondary antibodies at a 1:5000 dilution.

10. When GAS exponential-phase growth culture is required, dilute the overnight inoculum (1/10) in 3 mL fresh THB media and culture again at 37 °C for 2–3 h. Monitor cellular density by measuring the optical density (OD_{600nm}). GAS cultures between OD_{600} 0.4 and 0.6 are considered to be in exponential phase.

11. It is recommended that the researcher determines the length of macrophage differentiation that is required to obtain the optimal THP-1 response being investigated. We have found that differentiating cells for 1 day is optimal.

12. To freeze BMDMs, we recommend freezing the cells isolated from two mice in three vials (freezing media contains DMEM + 10% DMSO + 20% HI-FBS). When cells are thawed from these

vials, resuspend each vial in 20 mL of BMDM medium and seed two petri plates with 10 mL of cells/dish. Three days later, add 5 mL of BMDM medium. Cells are ready for use 7 days after initial thaw.

13. It is recommended that the researcher determines the length of HMDM differentiation time that is required for the optimal response being investigated. We have had success seeding the cells 6 days after differentiation and experimental use of the HMDMs the day after.
14. HMDMs can also be dislodged using cold PBS + 5 mM EDTA, however, we found detachment with Accutase® to be more efficient. If PBS + 5 mM EDTA is utilized, we recommend the following procedure. Add 15 mL of ice-cold PBS + 5 mM EDTA to each flask and then incubate the flask for 20 min in the refrigerator. Afterwards, scrape the cells five times with a rubber policeman, transfer the cells to a conical tube, and spin down the cells.
15. Times of infection to assess total GAS survival can vary accordingly with the GAS strain to be tested. For wild-type strains, prolonged infection times (>2 h) can result in 100% macrophage cell death.
16. Use an independent multiwell plate for each time point tested to avoid exposure of the cells to several temperature changes. Time points of 1 h, 2 h, and 4 h are the most appropriate to assess GAS intracellular survival.
17. Depending on the cytokine being probed and the strength of the host cell response, cell supernatants may need to be diluted at varying dilutions. The goal is to find the right dilution that allows detection of the cytokine if it is present and that absorbance values from the assay fall within the values of the standard curve.
18. Cells can also be lysed by placing the cells on ice and vortexing the cells every 5–10 min for a total of 20–30 min. Alternatively, a sonicator can also be used to lyse the macrophages.

References

1. Pancholi V (2019) Group A Streptococcus-mediated host cell Signaling. Microbiol Spectr 7(1)
2. Valderrama JA, Nizet V (2018) Group a streptococcus encounters with host macrophages. Future Microbiol 13:119–134
3. Valderrama JA, Riestra AM, Gao NJ et al (2017) Group a streptococcal M protein activates the NLRP3 inflammasome. Nat Microbiol 2(10):1425–1434
4. O'Neill AM, Thurston TL, Holden DW (2016) Cytosolic replication of group a streptococcus in human macrophages. MBio 7(2): e00020–e00016
5. Dunn KW, Kamocka MM, McDonald JH (2011) A practical guide to evaluating colocalization in biological microscopy. Am J Physiol Cell Physiol 300(4):C723–C742
6. Murray PJ, Wynn TA (2011) Protective and pathogenic functions of macrophage subsets. Nat Rev Immunol 11(11):723–737

7. Mosser DM, Edwards JP (2008) Exploring the full spectrum of macrophage activation. Nat Rev Immunol 8(12):958–969
8. Tsuchiya S, Yamabe M, Yamaguchi Y et al (1980) Establishment and characterization of a human acute monocytic leukemia cell line (THP-1). Int J Cancer 26(2):171–176
9. ATCC THP-1 (ATCC® TIB-202™) https://www.atcc.org/products/all/TIB-202.aspx
10. Chatellier S, Ihendyane N, Kansal RG et al (2000) Genetic relatedness and superantigen expression in group a streptococcus serotype M1 isolates from patients with severe and non-severe invasive diseases. Infect Immun 68(6):3523–3534
11. Hess S (2013) Sample preparation guide for mass spectrometry-based proteomics. Spectroscopy 12:12–17

Chapter 16

Identification of Group A Streptococcus-Containing Autophagosome-Like Vacuoles

Takashi Nozawa and Ichiro Nakagawa

Abstract

Group A Streptococcus (GAS) is one of the major human pathogens that can invade nonphagocytic cells. GAS internalized through endocytosis secretes the pore-forming toxin Streptolysin O (SLO) to escape into the cytoplasm. The cytosolic GAS is selectively captured by autophagic membranes (GAS-containing autophagosome-like vacuoles, GcAVs) and delivered to lysosomes for degradation. Macroautophagy (referred to as autophagy hereafter) is a highly conserved lysosome-mediated catabolic process, which is critical for cellular homeostasis. Autophagy also acts as an intracellular immune system. In this section, we describe how to identify GcAVs in infected cells using fluorescent microscopy.

Key words Group A Streptococcus, Autophagy, Group A Streptococcus -containing autophagosome-like vacuole, Fluorescent microscopy

1 Introduction

1.1 Group A Streptococcus Infection of Epithelial Cells

Group A Streptococcus (GAS) is an important Gram-positive bacterium that causes localized respiratory tract and skin infections as well as severe human diseases such as toxic shock-like syndrome [1]. Adherence to and internalization by host epithelial cells is the first critical step of the GAS infection process and a number of GAS strains are known to efficiently invade the epithelial cells [2, 3]. During GAS invasion, the pore-forming toxin Streptolysin O (SLO) enables GAS to disrupt endosomes and escape into the cytoplasm [4]. In response to GAS invasion of the cytosol by SLO, host cells activate the autophagy pathway to protect cells from intracellular GAS [4]. In human keratinocyte cells, Streptolysin S is also reported to damage endosomal vacuoles and trigger the autophagy pathway [5]. A number of in vitro studies established that autophagy in epithelial cells protects them from infection. However, there are no sufficient in vivo data to validate these findings. Given the previous data that the entry of GAS into host cells is relevant to its pathogenesis and globally disseminated serotypes of

Thomas Proft and Jacelyn M. S. Loh (eds.), *Group A Streptococcus: Methods and Protocols*, Methods in Molecular Biology, vol. 2136, https://doi.org/10.1007/978-1-0716-0467-0_16,

GAS circumvent autophagy to efficiently replicate in cells [6], studying autophagy response to GAS infection in epithelial cells might be important to understand the initial stage of the GAS infection process and host response to GAS.

1.2 Autophagy

Macroautophagy (referred to as autophagy hereafter) is an intracellular degradation system, in which a portion of cytoplasm is sequestered in an autophagosome and subsequently degraded upon fusion with a lysosome [7]. Although autophagy is recognized as a nonselective degradation system, recent studies have established that autophagy can selectively target various unwanted intracellular materials, such as excessive or damaged organelles and invading microbes [7, 8]. Autophagy against microbes is also called xenophagy. Autophagy is initiated by the formation of phagophore membrane (isolated membrane) that originates from the endoplasmic reticulum (ER) or other cellular membranes, such as mitochondria and plasma membrane. Phagophore membranes elongate and then enwrap targets. After the scission of the membrane edges, the structure becomes a complete double-membrane vacuole known as autophagosome. Autophagosomes fuse with lysosomes and form autolysosomes [7].

Several markers have been used to detect autophagic vacuole formation process. Microtubule-associated protein 1 light-chain 3 (LC3) is an autophagic membrane marker for detecting phagophore membranes, autophagosomes, and autolysosomes [9]. Since the autophagy-related protein 5/autophagy-related protein 16-1 (ATG5/ATG16L1) complex specifically localizes to the phagophore membrane, ATG5 and ATG16L1 are used as phagophore membrane markers. Lysosome marker, lysosomal-associated membrane protein 1 (LAMP1)-positive LC3 vacuole, is used to indicate autolysosomes. When autophagy is induced, LC3 is lipidated by phosphatidylethanolamine, which is then recruited to the autophagosomal membrane [9]. Contrastingly, it has also been demonstrated that LC3 can be recruited to other membranes, including single-membrane phagosomes, through a process known as LC3-associated phagocytosis (LAP) [10]. Therefore, the observation of LC3-positive structures alone can no longer be considered an evidence of autophagy.

1.3 Autophagy During GAS Infection

Autophagy against invading microbial pathogens was first reported in GAS-infected cells in 2014. A LC3-positive double-membrane structure against GAS was designated as GAS-containing autophagosome-like vacuole (GcAV) [4]. Following this research, various studies have revealed that autophagy can target a variety of invading pathogens, including *Salmonella enterica* serovar Typhimurium, *Shigella flexneri*, and *Mycobacterium tuberculosis* [11]. Given the evidences that LC3 surrounded only SLO-expressing GAS (cytosolic GAS) and GAS was captured by double-membrane structures

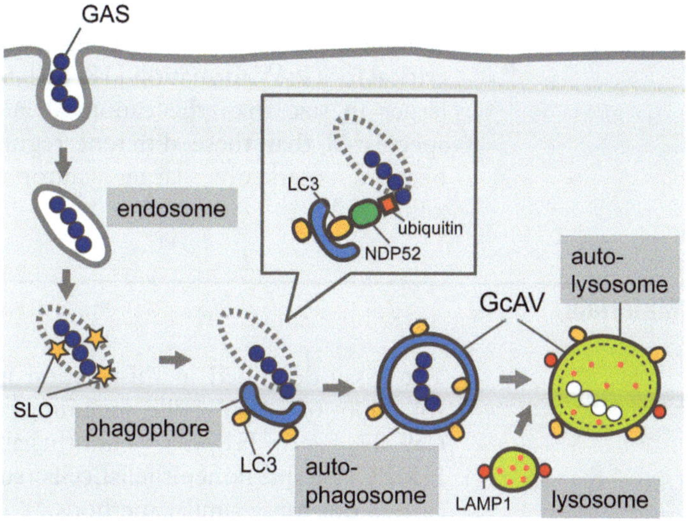

Fig. 1 Autophagy pathway targeting invading GAS. After internalization through endocytosis, GAS escape into the cytosol by using SLO. The cytosolic GAS is targeted for autophagy through the ubiquitin–NDP52–LC3 axis. GAS captured by GcAV is delivered to lysosome and subsequently degraded

in an ATG5-dependent mechanism in several epithelial cells [4, 12], it was demonstrated that GAS is targeted for autophagy in nonphagocytic epithelial cells. During autophagy of GAS, cytosolic GAS is marked with ubiquitin, bound by the nuclear dot protein 52 kDa (NDP52), and engulfed by GcAVs (Fig. 1) [5, 13]. Since the LAP process does not involve ubiquitin or NDP52, GcAVs in epithelial cells are not processed by LAP.

Intracellular M6, M49, and M89 serotype GAS are efficiently degraded in an ATG5-dependent manner [4, 14, 15], indicating that various GAS strains are eliminated by autophagy in epithelial cells. In contrast, the globally disseminated serotype of GAS, that is, M1T1, evades autophagic killing by using streptococcal pyrogenic exotoxin B (SpeB) protease, which degrades the NDP52 [6]. In addition, GAS can inhibit GcAV–lysosome fusion and canonical autophagy pathway by using nicotinamide adenine dinucleotide (NAD)-glycohydrolase [5, 16]. Recently, autophagic response to GAS infection in endothelial cells has been reported [4]. Endothelial cells failed to form GcAVs and allowed GAS to efficiently grow inside cells. This deficiency seemed to arise from reduced recruitment of ubiquitin and core autophagy proteins in endothelial cells during GAS infection.

Although GcAV formation involves canonical ATG proteins such as ATG5, authophagy-related protein 7 (ATG7), and LC3 autophagy-related protein 8 (ATG8), some autophagy regulators, including Beclin 1 authophagy-related protein 6 (ATG6), authophagy-related protein 14 (ATG14), and phosphatidylinositol-3-kinase (PI3K), seem to be unnecessary [5, 16, 17]. In addition, distinct sets of RAB GTPases that are dispensable in canonical starvation-induced autophagy have been reported to be critical for GcAV formation [18–20]. Since GcAVs can be ten times larger in size than the canonical autophagosomes, it has been speculated that these different regulators enable the GcAVs to become markedly larger autophagosomes for efficiently killing GAS.

2 Materials

The assay described in this section is optimized for HeLa cells. Infection of other cells, including epithelial (such as Caco-2, Ca9-22, and HEp-2), endothelial (such as HUVEC and HMEC-1), and nonepithelial cells (such as macrophages), can be carried out using similar methods.

1. HeLa cells (American Type Cell Collection): Kept in frozen stocks in cell preservation medium at −80 °C.
2. Standard cell culture medium: Dulbecco's Modified Eagle Medium (DMEM) supplemented with 10% heat-inactivated fetal calf serum (FCS) and 50 μg/mL gentamicin.
3. Infection medium: DMEM supplemented with 10% heat-inactivated FCS without antibiotics.
4. Extracellular bacteria killing medium: DMEM supplemented with 10% heat-inactivated FCS and 100 μg/mL gentamicin.
5. Phosphate-buffered saline (PBS).
6. GAS strain JRS4 (our own laboratory collection) (*see* **Note 1**): Kept in frozen glycerol stocks at −80 °C.
7. 3% Todd-Hewitt 0.2% yeast extract (THY) medium.
8. 24-well plates.
9. Glass coverslips: We generally used 12-mm diameter glass coverslips from Matsunami Glass Co. We sterilized coverslips in advance by dry-heat sterilization.
10. PBS supplemented with 0.1% gelatin.
11. Fixation solution: 4% paraformaldehyde (PFA) in PBS.
12. Permeabilization buffer: 0.1% Triton X-100 in PBS.
13. Blocking buffer: 2% bovine serum albumin (BSA) in PBS.
14. Anti-LC3 antibody.
15. Fluorescent-labeled secondary antibodies.
16. Green fluorescent protein (GFP)–LC3 or mCherry–LC3 plasmid.
17. Transfection reagent.
18. 1 mg/mL of 4′,6-diamidino-2-phenylindole (DAPI).

19. Mounting medium: We used Vectashield mounting medium for preserving fluorescent labeling.
20. Microscope slide.
21. Clear nail varnish.
22. Confocal laser scanning microscope: Olympus FluoView FV1000.

3 Methods

To visualize the GcAVs under a fluorescence microscope, we observed the localization of LC3 (autophagic membrane marker) using immunostaining or fluorescent protein-tagged LC3.

3.1 Immunostaining for LC3

1. Place glass coverslips in a 24-well plate immersed in gelatin/PBS solution and incubate for 10 min at room temperature (RT).
2. Remove the gelatin/PBS solution and seed epithelial cells onto the coverslips. The number of cells depends on the experiment (*see* **Note 2**).
3. Incubate at 37 °C in a humidified atmosphere containing 5% CO_2.
4. The day before infection, prepare a liquid culture of GAS by inoculating a glycerol stock of GAS in 5 mL of THY liquid medium. Let the culture grow overnight at 37 °C.
5. On the day of infection, transfer 50 μL of overnight grown culture to a fresh tube containing 5 mL of THY liquid medium and incubate at 37 °C until the bacteria grow to log-phase ($OD_{600} = 0.6$–0.8).
6. Measure the bacterial density at 600 nm and estimate the number of bacteria (JRS4 strain, $OD_{600} = 0.8$ is equivalent of 7.2×10^8 CFU/mL) (*see* **Note 3**).
7. Prepare the inoculum by diluting bacteria in an appropriate volume of infection medium (*see* **Note 4**).
8. Aspirate the medium of HeLa cells, wash with 1 mL of PBS three times to completely remove gentamicin in standard medium, and add 450 μL of infection medium.
9. Add 50 μL of the inoculum to each well and incubate at 37 °C in a humidified atmosphere with 5% CO_2 for 1 h to allow GAS to infect HeLa cells (*see* **Notes 5** and **6**).
10. After 1 h of infection, wash the wells once with PBS and then replace the medium with 500 μL of extracellular bacteria killing medium.
11. Incubate cells at 37 °C in a humidified atmosphere containing 5% CO_2 for the required time (*see* **Note 7**).

12. Extract each coverslip from the 24-well plate using tweezers, transfer to new 24-well plate filled with fixation solution, and allow fixation to occur for 15 min at room temperature (RT) (*see* **Note 8**).
13. Wash the coverslips with PBS and incubate with permeabilization buffer for 10 min.
14. Wash the coverslips with PBS and incubate with blocking buffer for at least 1 h.
15. Incubate cells with 50 μL of primary antibody; in this case, the anti-LC3 antibody is diluted in blocking buffer for at least 1 h at RT (if convenient, overnight at 4 °C) (*see* **Note 9**).
16. Wash the coverslips with PBS three times and incubate cells with secondary antibody (we used Alexa488-conjugated anti-mouse antibody from Molecular Probes/Invitrogen) diluted in blocking buffer for 1 h at RT in the dark.
17. Wash the coverslips with PBS twice and incubate cells with DAPI solution for 30 min at RT (*see* **Note 10**).
18. Wash the coverslips with PBS three times, remove coverslips from the wells, and wipe the backside and edges of the coverslips to remove excess liquid.
19. Mount coverslips by placing cells upside down on microscope slides with a small drop of mounting medium. Remove excess mounting medium with a piece of filter paper.
20. Paint around the edge of coverslips with a clear nail varnish.
21. Examine cells using a fluorescence microscope with appropriate filters and lenses (Fig. 2).

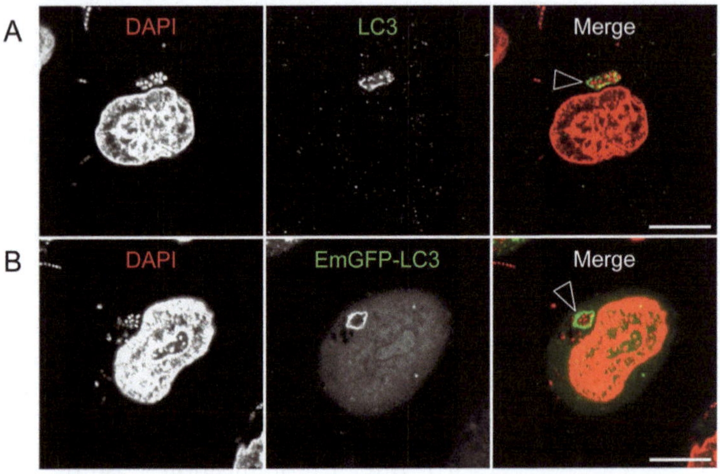

Fig. 2 Confocal microscopy images of GcAVs. (**a**) HeLa cells infected with GAS for 4 h were immunostained for LC3 and DNA was stained with DAPI. (**b**) HeLa cells expressing EmGFP–LC3 were infected with GAS for 4 h and stained with DAPI. Scale bars = 10 μm. Arrowheads indicate GcAVs

3.2 Fluorescent Protein-Tagged LC3

1. Culture HeLa cells on coverslips in a 24-well plate, as described in **steps 1–3** of Subheading 3.1 of the LC3 immunostaining protocol.
2. Mix transfection reagents with LC3 plasmids, according to the instructions of the manufacturer of the transfection reagents (*see* **Note 11**).
3. Infect cells and fix, as described in Subheading 3.1, **steps 4–12**.
4. Remove the fixation buffer and wash with PBS.
5. Incubate with DAPI solution for 30 min at RT in the dark to protect the fluorescent probe.
6. Wash cells with PBS and prepare slides, as described in Subheading 3.1, **steps 18–20**.
7. Acquire images using a fluorescence microscope (Fig. 2) (*see* **Note 12**).

4 Notes

1. We usually use this JRS4 strain (M6) or streptococcal toxic shock syndrome (STSS)-derived NIH35 strain (M28), because these strains efficiently invade into human epithelial cells. These strains can be substituted by other GAS strains.
2. When we perform infection experiment on the next day, we usually seed 6×10^4 HeLa cells per well. When we need to transfect the plasmid and perform an infection experiment the day after transfection, we seed 4×10^4 HeLa cells per well. Since appropriate numbers of cells are different among cell types, we need to examine this in advance.
3. We usually use this method to obtain mid-log phase bacteria, other growth parameters can also be used.
4. In this step, it is recommended to wash bacteria with PBS before infection, since the supernatant contains secreted cytotoxin Streptolysin O and other GAS effectors.
5. We mostly incubate cells with GAS for 1 h to allow GAS to infect the cells, but other protocols can also be used. For example, after inoculation, centrifugation of a 24-well plate at $500 \times g$ for 5 min ensures simultaneous infection of cells.
6. The specific multiplicity of infection (MOI) of the inoculum needs to be adjusted, according to the GAS strain being examined.
7. In the case of JRS4 strain infection, GcAVs start appearing 2 h after infection and increase with time. The frequency of GcAV formation reaches a peak 4 h after infection. Therefore, we usually fix infected cells 2 and 4 h after infection.

8. Overfixation (more than 1 h) by 4% PFA can mask the epitope. Therefore, we fix the cells for 10–15 min during immunostaining.
9. Not all internalized bacteria escape into the cytosol (the frequency of bacteria escaping into the cytosol depends on the bacterial strain and experimental conditions). To distinguish between GAS inside intact endosomes and GAS exposed to the cytoplasm, it is helpful to stain ubiquitin or galectin 3 (marker for damaged endosome) by immunostaining postpermeabilization.
10. DAPI staining is used to visualize bacteria (host cell nucleus is also stained). Other DNA staining reagents (Hoechst, SYTO) can also be used. Immunostaining using anti-group A carbohydrate antibody can also be used to visualize GAS.
11. Lipofection may cause extensive unintended changes in gene expression of the transfected cells.
12. Transfection of fluorescent protein-tagged LC3 is useful to observe GcAVs; however, the overexpression of LC3 could affect the efficiency of GcAV formation (the overexpression of LC3 suppresses GcAV formation) and cause nonspecific recruitment of various proteins to GcAVs.

References

1. Walker MJ, Barnett TC, McArthur JD et al (2014) Disease manifestations and pathogenic mechanisms of group A Streptococcus. Clin Microbiol Rev 27:264–301
2. Greco R, De Martino L, Donnarumma G et al (1995) Invasion of cultured human cells by *Streptococcus pyogenes.* Res Microbiol 146:551–560
3. LaPenta D, Rubens C, Chi E, Cleary PP (1994) Group A streptococci efficiently invade human respiratory epithelial cells. Proc Natl Acad Sci U S A 91:12115–12119
4. Nakagawa I, Amano A, Mizushima N et al (2004) Autophagy defends cells against invading group A Streptococcus. Science 306:1037–1040
5. O'Seaghdha M, Wessels MR (2013) Streptolysin O and its co-toxin NAD-glycohydrolase protect group A Streptococcus from xenophagic killing. PLoS Pathog 9:e1003394
6. Barnett TC, Liebl D, Seymour LM et al (2013) The globally disseminated M1T1 clone of group A Streptococcus evades autophagy for intracellular replication. Cell Host Microbe 14:675–682
7. Mizushima N, Komatsu M (2011) Autophagy: renovation of cells and tissues. Cell 147:728–741
8. Gatica D, Lahiri V, Klionsky DJ (2018) Cargo recognition and degradation by selective autophagy. Nat Cell Biol 20:233–242
9. Mizushima N, Yoshimori T, Levine B (2010) Methods in mammalian autophagy research. Cell 140:313–326
10. Heckmann BL, Boada-Romero E, Cunha LD et al (2017) LC3-associated phagocytosis and inflammation. J Mol Biol 429:3561–3576
11. Sharma V, Verma S, Seranova E et al (2018) Selective autophagy and xenophagy in infection and disease. Front Cell Dev Biol 6:147
12. Lu SL, Kawabata T, Cheng YL et al (2017) Endothelial cells are intrinsically defective in xenophagy of *Streptococcus pyogenes.* PLoS Pathog 13:e1006444
13. Minowa-Nozawa A, Nozawa T, Okamoto-Furuta K et al (2017) Rab35 GTPase recruits NDP52 to autophagy targets. EMBO J 36:2790–2807
14. Joubert PE, Meiffren G, Grégoire IP et al (2009) Autophagy induction by the pathogen receptor CD46. Cell Host Microbe 6:354–366
15. Thurston TL, Ryzhakov G, Bloor S et al (2009) The TBK1 adaptor and autophagy receptor NDP52 restricts the proliferation of ubiquitin-coated bacteria. Nat Immunol 10:1215–1221

16. Toh H, Nozawa T, Minowa-Nozawa A et al (2019) Group A Streptococcus modulates RAB1- and PIK3C3 complex-dependent autophagy. Autophagy. https://doi.org/10.1080/15548627.2019.1628539
17. Nakajima S, Aikawa C, Nozawa T et al (2017) Bcl-xL affects group A Streptococcus-induced autophagy directly, by inhibiting fusion between autophagosomes and lysosomes, and indirectly, by inhibiting bacterial internalization via interaction with Beclin 1-UVRAG. PLoS One 12:e0170138
18. Nozawa T, Aikawa C, Goda A et al (2012) The small GTPases Rab9A and Rab23 function at distinct steps in autophagy during group A Streptococcus infection. Cell Microbiol 14:1149–1165
19. Haobam B, Nozawa T, Minowa-Nozawa A et al (2014) Rab17-mediated recycling endosomes contribute to autophagosome formation in response to group A Streptococcus invasion. Cell Microbiol 16:1806–1821
20. Nakajima K, Nozawa T, Minowa-Nozawa A et al (2019) RAB30 regulates PI4KB (phosphatidylinositol 4-kinase beta)-dependent autophagy against group A Streptococcus. Autophagy 15:466–477

Chapter 17

Flow Cytometry-Based Assays to Quantify Complement Deposition and Neutrophil Uptake of Group A Streptococcus

Nicola N. Lynskey

Abstract

Quantification of complement deposition and subsequent neutrophil-mediated uptake provides an important way to assess the role of different bacterial factors in evasion of the host innate immune response. Here, we describe flow cytometry-based methods to allow quantification of deposition of the complement opsonin C3 on the bacterial surface and subsequent uptake by primary human neutrophils. The assays outlined below provide key methods to determine whether specific bacterial factors are involved in the evasion of complement-mediated immunity, using widely accessible reagents and equipment.

Key words Complement, Neutrophils, Opsonization, Group A Streptococcus, Phagocytosis

1 Introduction

The complement pathway comprises a tightly regulated, self-perpetuating cascade of plasma serine proteases that collectively results in the clearance of pathogens by a combination of opsonization, anaphylatoxin release, and formation of the lytic membrane attack complex (MAC) [1]. Opsonization of Group A Streptococcus (GAS) with the central complement component C3b plays a pivotal role in the innate immune clearance of this pathogen, and is mediated predominantly by neutrophils, which provide the first line of defense against streptococcal pathogenesis [2].

The complement cascade plays a key role in coordinating the immune response to invading GAS, and as such has provided a strong selective pressure resulting in the evolution of numerous mechanisms of complement evasion by this pathogen [3–7]. This chapter will focus on some basic techniques that facilitate the quantification of bacterial opsonization by C3, and thus allow the role of specific bacterial proteins in disrupting this process to be determined. We also describe a classical methodology for the purification of human neutrophils from whole blood, which allows the role of both bacterial and host factors in neutrophil-mediated

Thomas Proft and Jacelyn M. S. Loh (eds.), *Group A Streptococcus: Methods and Protocols*, Methods in Molecular Biology, vol. 2136, https://doi.org/10.1007/978-1-0716-0467-0_17,

uptake of GAS strains to be assessed. These assays provide a starting point from which to study bacterial evasion of complement and neutrophil-mediated host defenses.

2 Materials

2.1 Phlebotomy

1. BD Vacutainer Safety-Lok Blood Collection sets with preattached holder (21-gauge needle).
2. Six milliliter BD Vacutainer blood collection tubes (for serum preparation) and lithium heparin spray coated (for whole blood collection).
3. Alcohol swabs.
4. Tourniquet.

2.2 C3 Deposition Assay

1. Frozen stocks of GAS strains.
2. Columbia blood agar (CBA) plates.
3. Todd-Hewitt broth (THB) (*see* **Note 1**).
4. Sterile phosphate-buffered saline (PBS) without magnesium or calcium (*see* **Note 2**).
5. PBS with 0.1% Tween 20.
6. Fresh human serum from consented adults diluted to 50% in sterile PBS (*see* **Note 3**).
7. Fluorescein isothiocyanate (FITC)-conjugated anti-C3 antibody (MP Biomedicals) (*see* **Note 4**).
8. Spectrophotometer.
9. Waterbath sonicator.
10. Rotator at 37 °C (*see* **Note 5**).
11. Flow cytometer (*see* **Note 6**).
12. Two milliliter lo-bind tubes (Eppendorf).
13. 96-well plates.

2.3 Neutrophil Purification

1. Whole blood from a consented donor.
2. 3% (w/v) dextran (MW > 100,000) in 0.9% (w/v) saline.
3. End-over-end rotator (at room temperature).
4. Ficoll-paque plus.
5. Sterile water.
6. Sterile PBS without magnesium or calcium (*see* **Note 2**).
7. Sterile 10 × PBS without magnesium or calcium.
8. Improved Neubauer Hemocytometer.
9. Trypan blue.

2.4 Neutrophil Uptake Assays

1. Freshly purified neutrophils (5×10^5 cells/reaction).
2. Human serum.
3. FITC-conjugated anti-Group A carbohydrate (GAC) antibody (Abcam) (*see* **Note 7**).
4. Spectrophotometer.
5. Flow cytometer (*see* **Note 6**).
6. Two-milliliter lo-bind tubes.
7. Stop solution: 0.02% ethylenediaminetetraacetic acid (EDTA) in 0.9% saline.

3 Methods

3.1 GAS Culture

1. Streak out bacterial strains of interest onto Columbia blood agar (CBA) plates and culture overnight at 37 °C, 5% CO_2.
2. Use a sterile loop to lift 5–10 bacterial colonies from the CBA plate and inoculate into 5–10 ml sterile THB. Culture overnight static at 37 °C, 5% CO_2.
3. See below for the preparation of optimal GAS cultures for each protocol.

3.2 Preparation of Fresh Human Serum

1. Aseptically collect 10 ml whole blood from a consented donor into plastic vacutainer tubes (*see* **Note 8**).
2. Leave to stand at room temperature for 20 min to allow blood to clot.
3. Centrifuge blood at $1000 \times g$, 10 min, room temperature to remove the clot.
4. Remove the supernatant (serum) with a Pasteur pipette and transfer to a new, sterile tube.
5. Dilute in sterile PBS as necessary and mix by gentle inversion to prevent activation.

3.3 Complement Deposition Assay

1. Centrifuge 10 ml of overnight culture of desired GAS strains ($3000 \times g$, 10 min, 4 °C) and wash in 1 ml sterile PBS in 2 ml tubes.
2. Check the optical density of the culture and resuspend to give 1×10^8 cfu/ml (*see* **Note 9**).
3. Aliquot 100 μl bacterial culture to 2 ml tube (approx. 1×10^7 GAS cfu) and centrifuge (max speed, 1 min) (*see* **Note 10**).
4. Discard supernatant and resuspend pellet in 50% human serum (diluted in sterile PBS) (*see* **Note 11**).
5. Incubate samples at 37 °C with end-over-end rotation for 30 min.

6. Centrifuge samples (13,000 × *g*, 1 min) and wash by resuspending in 250 μl PBS with 0.1% Tween 20.
7. Transfer samples to a lo-bind 96-well plate.
8. Centrifuge 3000 × *g*, 3 min, 4 °C.
9. Repeat wash (**steps 6–8**).
10. Prepare anti-C3 antibody at working concentration, according to manufacturer's guidelines (*see* **Note 12**).
11. Resuspend pellets in 50 μl anti-C3 antibody at working concentration (*see* **Note 13**).
12. Incubate for 30 min on ice, protected from light.
13. Wash samples twice with 250 μl PBS with 0.1% Tween 20.
14. Resuspend samples in 100 μl PBS.
15. Transfer samples to flow cytometry tubes containing 400 μl PBS (stored on ice).
16. Sonicate all samples for 30 s in waterbath sonicator to break up bacterial chains (*see* **Note 14**).
17. Cover samples with aluminum foil to protect from light.
18. Run samples on flow cytometer (*see* **Note 6**).
19. To visualize GAS populations clearly on the flow cytometer, both forward and side scatter axes should be set to logarithmic scale. For analysis, a gate is drawn around the GAS population. The proportion of fluorescent GAS (i.e., those associated with C3) and the geometric mean fluorescent intensity (gMFI) of the population are quantified. Data are reported as the fluorescence index, calculated as the proportion of C3-positive GAS multiplied by the gMFI [8, 9].

3.4 Neutrophil Purification

1. Aseptically collect 20 ml heparinized whole blood from a consented donor (*see* **Notes 8** and **15**).
2. Pool heparinized blood from each individual donor in a 50 ml conical tube and combine 1:1 with 3% dextran in 0.9% saline (to encourage erythrocyte rouleaux and subsequent sedimentation).
3. Incubate on ice for at least 30 min, until a clear interface forms between layers.
4. Remove the upper, yellow layer (containing serum and leukocytes) to a new 50 ml conical tube and pellet cells by centrifugation (250 × *g*, 10 min, room temperature; no brake).
5. Resuspend the pelleted cells in 20 ml of 0.9% saline.
6. Layer cell solution onto 10 ml Ficoll-paque plus in a 50 ml conical tube. To layer cells, use a 25 ml stripette connected to a laboratory pipettor set to gravity flow (*see* **Note 16**).

7. Place the 50 ml conical tube carefully in a centrifuge and pellet cells (700 × *g*, 30 min, 20 °C; no brake). The resulting pellet contains granulocytes and erythrocytes.
8. Erythrocytes can be removed by water lysis (pellet is resuspended in 9 ml sterile distilled water (dH_2O) for 20 s and 1 ml of 10 × PBS added immediately to make the solution isotonic and prevent lysis of granulocytes).
9. The resulting cell suspension is centrifuged (300 × *g*, 10 min, 20 °C) and the granulocyte pellet resuspended in 5 ml PBS.
10. For cell enumeration, 10 μl cell solution is mixed with 10 μl Trypan blue dye. A coverslip is placed on the hemocytometer and 10 μl of this mixture is loaded into the well (*see* **Note 17**).
11. Centrifuge cell suspension (300 × *g*, 10 min, 20 °C) and resuspend in sterile PBS at a concentration of 1×10^6 cells/ml (*see* **Note 18**).
12. Other methods of neutrophil purification can also be used (*see* **Note 19**).

3.5 Neutrophil Uptake Assays

3.5.1 Preparation of GAS

1. Pellet overnight cultures by centrifugation (3000 × *g*, 10 min, 4 °C) and wash in 5 ml sterile PBS. Pellet bacteria again, resuspend in 5 ml sterile PBS, and check optical density of the bacterial suspension. Resuspend cultures to an A_{600} of 0.35 in sterile PBS. 100 μl (approximately 5×10^6 bacterial cells) of this suspension is required for each reaction.
2. Pellet required volume of GAS suspension (100 μl/reaction) and resuspend in 200 μl working solution of FITC-conjugated anti-GAC antibody in a 2 ml tube (for Abcam antibody, dilute 1:100 in sterile PBS [8, 10], for other antibodies check manufacturer's guidelines).
3. Incubate at room temperature for 20 min, protected from light.
4. Wash samples twice in sterile PBS, to remove unbound antibody.
5. Resuspend labeled bacterial pellets in sterile PBS to a concentration of 5×10^7 cells/ml, as calculated in **step 1**.
6. Opsonize bacteria with fresh human serum. For each reaction, to a 2 ml tube add 350 μl sterile PBS, 100 μl labeled GAS (5×10^6 cells), and 50 μl serum (*see* **Note 20**).
7. Incubate at 37 °C with end-over-end rotation for 30 min.

3.5.2 Neutrophil Uptake

1. Prepare neutrophils, as outlined in Subheading 3.4 (*see* **Note 21**).
2. Add 500 μl neutrophils (5×10^5 cells) to each tube prepared in **steps 3** and **4**. The multiplicity of infection (MOI) for this experiment is 10:1 (GAS: neutrophils).

3. Incubate neutrophils with opsonized bacteria at 37 °C with end-over-end rotation for 30 min (*see* **Note 22**).
4. During the incubation, flow cytometry tubes are labeled and incubated on ice with 500 μl stop solution.
5. After the incubation, tubes are placed on ice and 500 μl neutrophil reaction added to a prechilled flow cytometry tube prepared in **step 9**. Stop solution will inhibit any further uptake of bacteria by neutrophils.
6. Samples are then run on a flow cytometer (*see* **Note 6**). For analysis, a gate is drawn around the neutrophil population, and the proportion of fluorescent neutrophils (i.e., those associated with labeled GAS), and the gMFI of the population are quantified. Data are reported as the proportion of neutrophils associated with labeled bacteria (*see* **Note 23**).

4 Notes

1. THB may be supplemented with 0.5% yeast extract to facilitate bacterial growth.
2. Sterile PBS can be purchased or prepared in house (For 1 L: 8 g NaCl, 0.2 g KCl, 1.44 g Na_2HPO_4, 0.24 g KH_2PO_4. Adjust the pH to 7.4. Sterilize by autoclaving or filtering).
3. Other dilutions of serum can be tested, which may be especially useful where samples are limited (for example from patients with a complement deficiency). Freshly prepared serum is recommended for this assay.
4. It is also acceptable to use an unconjugated primary antibody, and following additional wash steps to use a secondary antibody conjugated to the desired fluorophore.
5. If a rotator cannot be plugged into an incubator, a hybridizer (normally used for Southern and Northern blotting) can be set to 37 °C and used to rotate samples.
6. Check with the owner of flow cytometer that appropriate health and safety protocols are in place to permit running of live bacteria through the machine, and that appropriate wash steps are included in machine cleanup, to ensure all residual bacteria are killed. Cells can otherwise be fixed in paraformaldehyde following the final wash step (resuspend bacterial pellets or neutrophils in 100 μl of 2% paraformaldehyde and incubate for 10 min at room temperature. Wash twice with 250 μl PBS and prepare for flow cytometry, as described in Subheadings 3.3 and 3.5.2, respectively). Make sure to confirm that the antibody used will still be functional following fixation.

7. Antibodies can also be made inhouse, or obtained from other manufacturers. Additionally, unlabeled primary antibodies may be used in combination with a secondary antibody conjugated to the desired fluorophore.
8. For work with human blood, it is imperative to obtain appropriate ethics approval from the relevant authority in advance, and signed consent is required from each donor for use of blood in these assays. Drawing of blood must be performed by a trained phlebotomist. All human blood samples should be treated as potentially infectious and handled with appropriate personal protective equipment.
9. For M1T1 strain H598 [8, 11], an A_{600} reading of 8 gives approximately this concentration. For each strain, the necessary optical density must be ascertained by quantitative culture, prior to carrying out this assay. Different serotypes (and strains) can have markedly different A_{600} readings for the same number of bacterial cfu.
10. Multiple technical replicates for each bacterial strain are recommended for each experiment.
11. Use a P200 tip to remove the supernatant. Take care not to displace the pellet. The pellet can be resuspended in untreated serum, or serum which has been incubated with a recombinantly expressed bacterial protein hypothesized to interact with complement deposition. Pretreatment with recombinantly expressed C3 cleaving enzymes produced by both GAS and *Staphylococcus aureus* at a concentration of 0.5 μM has been demonstrated to be sufficient to impair C3 deposition on the bacterial surface [8, 12].
12. For FITC-conjugated antibody purchased from MP Biomedicals, dilute 1:300 in PBS with 0.1% Tween 20 [8, 9]. Check manufacturer's guidelines for working concentration of alternative antibodies.
13. Make sure to prepare negative control samples that are resuspended in PBS without antibody for setup on the flow cytometer.
14. Vortexing is not sufficient to break up chains of GAS. Failure to break up chains can result in artificial differences in fluorescence intensity between GAS strains that vary in chain length and that do not reflect real differences in C3 deposition per individual coccus.
15. The number of neutrophils purified from the same volume of blood can vary considerably between donors.
16. It is imperative that the cells are layered slowly and carefully and the interface is not disturbed. Initially, touch the tip of the

stripette onto the side of the conical tube and allow cell solution to flow gently onto the Ficoll layer.

17. The hemocytometer contains a grid, the central square of which is divided into a 5 × 5 grid of triply bound cells. Using a light microscope, count the total number of cells within this 5 × 5 square (equates to 100 nl cell suspension). Multiply cell number by 2 to correct for trypan dilution and then by 1×10^4 to give cells/ml. NB: Trypan blue diffuses into dead cells, count only cells with a blue membrane, and not blue cytoplasm.
18. Do not store cells in PBS for long periods of time, as without serum neutrophils will die.
19. As listed for the main method, aseptically collect heparinized whole blood from a consented donor. Purification of neutrophils from whole blood can be performed more rapidly using the MACSxpress® Whole Blood Neutrophil Isolation Kit (the protocol requires MACSxpress Separator; Miltenyi Biotec: #130–098-308). Using this method, approximately 2×10^6 neutrophils can be purified per milliliter of whole blood; however, there can be wide variation between donors. The separation beads provided in this kit purify neutrophils by negative selection, thus preventing activation of purified cells. Lysis of erythrocytes and quantification of neutrophils can be carried out, as described in Subheading 3.4, **steps 8–11**.
20. Serum can be treated prior to use in this assay. For example, heat inactivation of serum (56 °C, 30 min) will inactivate complement, and thus can be used to ascertain the relative role of complement in subsequent neutrophil uptake experiments. Other opsonizing agents can also be added, a well-documented example of which is pooled human intravenous immunoglobulin (IVIG) [13].
21. Neutrophils can be pretreated, prior to use in this assay, to further determine the role of specific bacterial factors or complement components in neutrophil uptake of GAS. For example, complement receptors can be blocked by addition of specific antibodies or antagonistic chemicals such as the C5a receptor antagonist PMX205 [8].
22. Negative control reactions, which are incubated at 4 °C, can be set up to give background levels of association of GAS with neutrophils.
23. Trypan blue dye can be added to samples before running through the flow cytometer. This quenches certain wavelengths of fluorescent light and eliminates signal from bacteria not internalized by neutrophils.

References

1. Janssen BJ, Huizinga EG, Raaijmakers HC et al (2005) Structures of complement component C3 provide insights into the function and evolution of immunity. Nature 437 (7058):505–511. https://doi.org/10.1038/nature04005
2. Laabei M, Ermert D (2019) Catch me if you can: Streptococcus pyogenes complement evasion strategies. J Innate Immun 11(1):3–12. https://doi.org/10.1159/000492944
3. Ermert D, Weckel A, Agarwal V et al (2013) Binding of complement inhibitor C4b-binding protein to a highly virulent Streptococcus pyogenes M1 strain is mediated by protein H and enhances adhesion to and invasion of endothelial cells. J Biol Chem 288(45):32172–32183. https://doi.org/10.1074/jbc.M113.502955
4. Honda-Ogawa M, Ogawa T, Terao Y et al (2013) Cysteine proteinase from Streptococcus pyogenes enables evasion of innate immunity via degradation of complement factors. J Biol Chem 288(22):15854–15864. https://doi.org/10.1074/jbc.M113.469106
5. Pandiripally V, Gregory E, Cue D (2002) Acquisition of regulators of complement activation by Streptococcus pyogenes serotype M1. Infect Immun 70(11):6206–6214. https://doi.org/10.1128/iai.70.11.6206-6214.2002
6. Carlsson F, Sandin C, Lindahl G (2005) Human fibrinogen bound to Streptococcus pyogenes M protein inhibits complement deposition via the classical pathway. Mol Microbiol 56(1):28–39. https://doi.org/10.1111/j.1365-2958.2005.04527.x
7. Ji Y, McLandsborough L, Kondagunta A et al (1996) C5a peptidase alters clearance and trafficking of group A streptococci by infected mice. Infect Immun 64(2):503–510
8. Lynskey NN, Reglinski M, Calay D et al (2017) Multi-functional mechanisms of immune evasion by the streptococcal complement inhibitor C5a peptidase. PLoS Pathog 13 (8):e1006493. https://doi.org/10.1371/journal.ppat.1006493
9. Hyams C, Trzcinski K, Camberlein E et al (2013) Streptococcus pneumoniae capsular serotype invasiveness correlates with the degree of factor H binding and opsonization with C3b/iC3b. Infect Immun 81(1):354–363. https://doi.org/10.1128/IAI.00862-12
10. Lynskey NN, Banerji S, Johnson LA et al (2015) Rapid lymphatic dissemination of encapsulated group A Streptococci via lymphatic vessel endothelial Receptor-1 interaction. PLoS Pathog 11(9):e1005137. https://doi.org/10.1371/journal.ppat.1005137
11. Turner CE, Dryden M, Holden MT et al (2013) Molecular analysis of an outbreak of lethal postpartum sepsis caused by Streptococcus pyogenes. J Clin Microbiol 51 (7):2089–2095. https://doi.org/10.1128/JCM.00679-13
12. Laarman AJ, Ruyken M, Malone CL et al (2011) Staphylococcus aureus metalloprotease aureolysin cleaves complement C3 to mediate immune evasion. J Immunol 186 (11):6445–6453. https://doi.org/10.4049/jimmunol.1002948
13. Reglinski M, Gierula M, Lynskey NN et al (2015) Identification of the Streptococcus pyogenes surface antigens recognised by pooled human immunoglobulin. Sci Rep 5:15825. https://doi.org/10.1038/srep15825

Chapter 18

Live Cell Microscopy and Flow Cytometry to Study Streptolysin S-Mediated Erythrocyte Hemolysis

Dustin L. Higashi, Charles R. Tessier, and Shaun W. Lee

Abstract

The ability to induce hemolysis, the rupturing of erythrocytes with the consequent release of their intracellular contents, is a phenotypic hallmark of a number of microbial toxins. *Streptococcus pyogenes* or Group A Streptococcus (GAS) is a human pathogen responsible for a wide range of diseases from mild pharyngitis to severe conditions such as toxic shock syndrome. GAS produces a powerful hemolytic toxin called streptolysin S (SLS). Herein, we describe a procedure for the preparation of SLS toxin and the use of two complementary approaches, live microscopy and flow cytometry, to study the effects of the SLS toxin on erythrocytes. In addition to providing insights into SLS-mediated hemolysis, these assays have the potential to be modified for the study of other hemolytic toxins and compounds.

Key words Streptolysin S, Hemolysis, Live imaging, Flow cytometry

1 Introduction

A number of pathogenic bacteria, viruses, parasites, and fungi produce hemolytic toxins. Many of these toxins have been associated with disease severity. *Streptococcus pyogenes* or Group A Streptococcus (GAS) produces a potent cytolytic toxin called streptolysin S (SLS). The production of SLS has been linked to disease severity in both in vitro and in animal models of infection [1–3].

Recent studies have demonstrated that SLS is a ribosomally produced peptide toxin that undergoes extensive posttranslational modification [2, 4, 5]. SLS belongs to a family of bacteriocins known as the thiazole/oxazole-modified microcins (TOMMs) [6]. While peptide toxins like SLS were believed to work through the disruption of general membrane structure and function [7–9], the specific mechanism of other TOMMs led to the thought that SLS may have a more specific mode of action [10]. Indeed, recent work has demonstrated the ability of SLS to target specific host proteins to exert its toxic effect [11, 12].

Thomas Proft and Jacelyn M. S. Loh (eds.), *Group A Streptococcus: Methods and Protocols*, Methods in Molecular Biology, vol. 2136, https://doi.org/10.1007/978-1-0716-0467-0_18, © Springer Science+Business Media, LLC, part of Springer Nature 2020

SLS activity is typified by hemolysis (Red Blood Cell [RBC] destruction resulting in zones of clearance) around GAS colonies when grown on blood agar or through the release of hemoglobin into the extracellular milieu in liquid media containing RBCs [4, 5]. While these indicators have proved to be reliable methods in demonstrating toxin activity, the need for other tools is essential in providing further insights into toxin function. In this chapter, we present a procedure for the preparation of cell-free, concentrated SLS toxin. We describe two complementary techniques—(1) live cell imaging and (2) flow cytometry, to study the SLS-induced fluid flow into erythrocytes, its effect on intracellular chloride levels, and hemolysis. In addition to providing insights into SLS-mediated hemolysis, these assays have the potential to be generalized for detailed examination of other hemolytic toxins and compounds.

2 Materials

2.1 Bacteria Culture

1. Freezer stocks (−80 °C) of GAS bacteria; wt5448, 5448:*sagAΔcat*, and 5448:*sagAΔcat* + pDCerm*sagA* [2, 11, 13] or other wild-type SLS-producing and isogenic *sagA* deletion strains (non-SLS-producing) (*see* **Note 1**).
2. Antibiotic stock solutions: 10 mg/mL (w/v) chloramphenicol in ethanol;. 5 mg/mL (w/v) erythromycin in ethanol. Store aliquots of antibiotics at −20 °C until use.
3. Todd-Hewitt Broth (THB): Dissolve 30 g of media into 1 L of distilled water with stirring. Autoclave all media for 15 min at 121 °C.
4. Todd-Hewitt agar plates (TH plates): Dissolve 30 g of media into 1 L of distilled water with stirring. Add 15 g of agar with stirring. For the growth of mutant strains, add the appropriate amount of antibiotic stock solutions to achieve a final antibiotic concentration of 5 μg/mL (w/v) erythromycin or 1 μg/mL (w/v) chloramphenicol.
5. Group A Selective Strep Agar with 5% Sheep Blood.
6. Polyester tipped swabs.
7. Humidified 37 °C, 5% CO_2 incubator.

2.2 SLS Preparations

1. Polyester tipped swabs.
2. Phosphate-buffered saline (PBS), pH 7.4 (PBS; 10×): 10.59 mM KH_2PO_4, 1551.72 mM NaCl, 29.66 mM $Na_2HPO_4 \cdot 7H_2O$.
3. PBS, pH 7.4 (1×): add 100 mL of 10× PBS into 900 mL of distilled water and mix well. Filter through a 1 L, 0.22 μm filter unit.

4. Todd-Hewitt Broth (THB).
5. 1.5 mL, 15 mL, and 50 mL polypropylene tubes.
6. 1.5 mL cuvettes, disposable.
7. 50 mL and 1000 mL, 0.22 μm pore size filter unit.
8. 3K Molecular Weight Cutoff (MWCO) Filtration Unit.
9. Vortex mixer.
10. Spectrophotometer.
11. Refrigerated centrifuge equipped with adaptor for 50 mL polypropylene tubes.

2.3 Blood Preparation

1. Defibrinated whole sheep blood.
2. PBS, pH 7.4 (*see* Subheading 2.2, **item 3**).
3. 1.5 mL and 50 mL polypropylene tubes.
4. Vortex mixer.
5. Refrigerated microcentrifuge with rotor equipped for 1.5 mL tubes.

2.4 Chloride Indicator Loading of Red Blood Cells

1. Washed defibrinated whole sheep blood.
2. Loading buffer: 1.05 mM KH_2PO_4, 2.97 mM Na_2HPO_4, 155.17 mM $NaHCO_3$. Per 1 L of distilled water, add 0.144 g KH_2PO_4, 0.795 g $Na_2HPO_4 \cdot 7H_2O$, and 13.035 g $NaHCO_3$. Filter through a 1 L, 0.22 μm filter unit.
3. DiH-MEQ (6-methoxy-*N*-ethyl-1,2-dihydroquinoline): prepare fresh (within 2 weeks when stored at −80 °C) from MEQ, 6-methoxy-*N*-ethylquinolinium iodide (Molecular Probes), according to manufacturer's instructions (*see* **Note 2**).
4. Hypertonic PBS (hPBS), pH 7.4: 1.1 mM KH_2PO_4, 466 mM NaCl, 2.97 mM $Na_2HPO_4 \cdot 7H_2O$. Add 18.25 g NaCl per 100 mL of 10× PBS and mix. Add 900 mL of distilled water and mix until completely dissolved. Filter using a 1 L, 0.22 μm pore filter unit.
5. 1000 mL, 0.22 μm pore size filter unit.
6. 1.5 mL and 50 mL polypropylene tubes.
7. Vortex mixer.
8. 37 °C water bath or incubator.
9. Refrigerated centrifuges with rotor equipped for 1.5 mL and 50 mL tubes.

2.5 Live Imaging

1. A Widefield Inverted Fluorescence Microscope that is suitable for automatic image acquisition with a controlled camera and shutters. Equipped with an automated focus adjustment control system (*see* **Note 3**), a 60× oil immersion objective (1.4 N.A.),

Differential Interference Contrast (DIC) attachments, 4′,6-diamidino-2-phenylindole (DAPI) and Texas Red Filter sets (or similar), and a complementary metal-oxide semiconductor (CMOS) or charge-coupled device (CCD) camera.

2. Environmental incubation system capable of maintaining a humidified, 37 °C temperature.
3. Microscope stage clips for 35 mm dishes.
4. Immersion oil with proper refractive index at 37 °C.
5. Fifty milliliter polypropylene tubes.
6. Glass-bottom dishes, 35 mm No. 1.5 coverslip thickness.
7. PBS, pH 7.4 (*see* Subheading 2.2, **item 3**) and hPBS, pH 7.4 (*see* Subheading 2.4, **item 4**).
8. ImageJ software (NIH).

2.6 Flow Cytometry

1. Flow cytometer.
2. Cytometer-specific sheath fluid.
3. 5 mL polystyrene round-bottom tubes 12 × 75 mm.
4. SLS preparations.
5. Washed RBC preparations.
6. PBS, pH 7.4 (*see* Subheading 2.2, **item 3**) and hPBS, pH 7.4 (*see* Subheading 2.4, **item 4**).
7. Vortex mixer.
8. 37 °C incubator or water bath.
9. FlowJo software (Becton Dickinson).

3 Methods

3.1 Bacteria Culture

The following method describes the culturing of GAS bacteria for the purpose of generating SLS toxin (*see* **Note 4**).

1. From −80 °C freezer glycerol stocks of GAS bacteria (wt5448, 5448:*sagA*Δ*cat*, and 5448:*sagA*Δ*cat* + pDCerm*sagA*), scrape a portion from the frozen surface using a sterile pipette tip and spot on the agar (±appropriate antibiotic) surface. Using a sterile loop, streak onto agar plate for single bacteria. Grow in a humidified, 5% CO_2 incubator for 20–24 h.
2. Using a swab, pick 3–5 single colonies for each strain from the overnight plate and lawn onto a fresh TH plate (no antibiotics).
3. Using the same swab in **step 2**, streak a line onto Group A Selective Strep Agar with 5% Sheep Blood (optional).
4. Grow all plates in a humidified, 5% CO_2 incubator for 16–18 h (*see* **Note 5**).

3.2 SLS Preparation

SLS activity is typically associated with the bacterial surface. However, SLS cytolytic activity can be preserved using carrier molecules such as lipoteichoic acid, RNA core, or serum albumin [14–16]. The following method describes the production of a concentrated, cell-free SLS toxin preparation. This method does not require the addition of carrier molecules outside of those present in the growth media. The toxin preparation is stable for extended periods of time when stored at ≤ -20 °C.

1. Using a swab for each strain, collect the entire lawn of bacteria from the overnight TH plate (*see* Subheading 3.1, **step 4**) and transfer into 2–5 mL of TH broth in a 15 mL tube. Swirl the swab to facilitate the release of bacteria into broth. Vortex the suspension at medium setting for 1 min. Dilute 1/10 by adding 0.1 mL of bacterial suspension into 0.9 mL of TH broth in a polypropylene tube and vortex for 30 s. Transfer the mixture to a cuvette. Blank the spectrophotometer at $\lambda_{600\text{nm}}$ optical density (OD_{600}) using TH broth and determine the OD_{600} for each diluted bacterial suspension. Calculate the concentration of bacteria colony-forming units (CFUs)/mL in the original suspension. The OD_{600} to CFU/mL correlation can be determined empirically. In our hands, we have found an OD_{600} of 1.0 corresponds to ~1.6×10^8 CFUs/mL.
2. Adjust the original bacterial suspensions for equal OD_{600} values by the addition of TH broth. Vortex at medium setting for 30 s. Add the appropriate amount of bacteria inoculum to chilled PBS in a 50 mL tube for a final volume of 21 mL and a final bacteria concentration equal to 1×10^8 CFUs/mL.
3. Vortex for 2 min at medium setting and place on ice for 30 min.
4. Pellet the bacteria by centrifugation at 2000 × *g* for 10 min at 4 °C in a refrigerated centrifuge.
5. Transfer the supernatant into a fresh 50 mL tube and filter-sterilize through a 50 mL, 0.22 μm filter unit.
6. Concentrate 20 mL of the filtrate using a 3K MWCO Filtration Unit. Spin at 4 °C, 2500 × *g* checking intermittently until the filter reservoir volume is 2 mL.
7. Aliquot the 10× SLS preparations into chilled 1.5 mL polypropylene tubes and store ≤20 °C (*see* **Note 6**). Thaw on ice and bring to room temperature (RT) immediately before use.

3.3 Blood Preparation

SLS has been shown to target RBCs over a broad range of animal species including sheep, mouse, avian, and human [17]. Below, we describe a method for the preparation of sheep RBCs for use in SLS-induced hemolysis studies.

1. In duplicate, take a 0.1 mL volume of fresh defibrinated whole sheep blood and suspend in 1 mL of cold PBS in a 1.5 mL polypropylene tube by gentle vortexing.

2. Pellet RBCs by centrifugation at 400 × *g* for 5 min at 4 °C. Discard the supernatant that will appear between clear and cherry red.
3. Resuspend each pellet in 1 mL of cold PBS by gentle vortexing.
4. Repeat **steps 2** and **3** twice (for a total of three washes). The supernatant should appear clear after the final centrifugation step. If the supernatant is not clear, additional wash steps may be added.
5. After the final wash, pellet the RBCs by centrifugation at 400 × *g* for 5 min at 4 °C and discard the supernatant. The RBC pellet is now ready to be suspended into the appropriate buffer solution (*see* **Note** 7).

3.4 Chloride Indicator Loading of Red Blood Cells

The regulation of intracellular chloride levels is central in maintaining proper RBC function. The following method describes the noninvasive loading of DiH-MEQ into RBCs. Upon entry into the cytosol, the DiH-MEQ undergoes rapid oxidation and is converted to the positively charged and membrane-impermeable, chloride-sensitive MEQ [18].

1. Resuspend washed defibrinated blood in 1 mL of room temperature loading buffer.
2. Pellet RBCs by centrifugation at 400 × *g* for 5 min and discard the supernatant.
3. Repeat **steps 1** and **2**.
4. Resuspend cells in 1 mL of room temperature loading buffer and add to 39 mL of 37 °C loading buffer in a 50 mL tube.
5. Prepare a 50 mM solution of chloride indicator concentrate by dissolving the DiH-MEQ oil into an appropriate volume of loading buffer or dimethyl sulfoxide (DMSO). Protect from light by wrapping the tube in aluminum foil.
6. Add 40 μL of 50 mM DiH-MEQ solution to the RBC suspension from **step 4** and mix well by gentle vortexing. Wrap the tube in foil and incubate in a 37 °C water bath for 15–20 min (*see* **Note 8**).
7. Pellet MEQ-loaded RBCs by centrifugation at 400 × *g* for 5–10 min and discard the supernatant.
8. Resuspend the RBC pellet in 1 mL of loading buffer by pipetting and transfer to a 1.5 mL polypropylene tube.
9. Pellet RBCs by centrifugation at 400 × *g* for 5 min and discard the supernatant.
10. Repeat **steps 8** and **9** (for a total of two washes).
11. The RBC pellet is now ready for suspension in assay buffer.

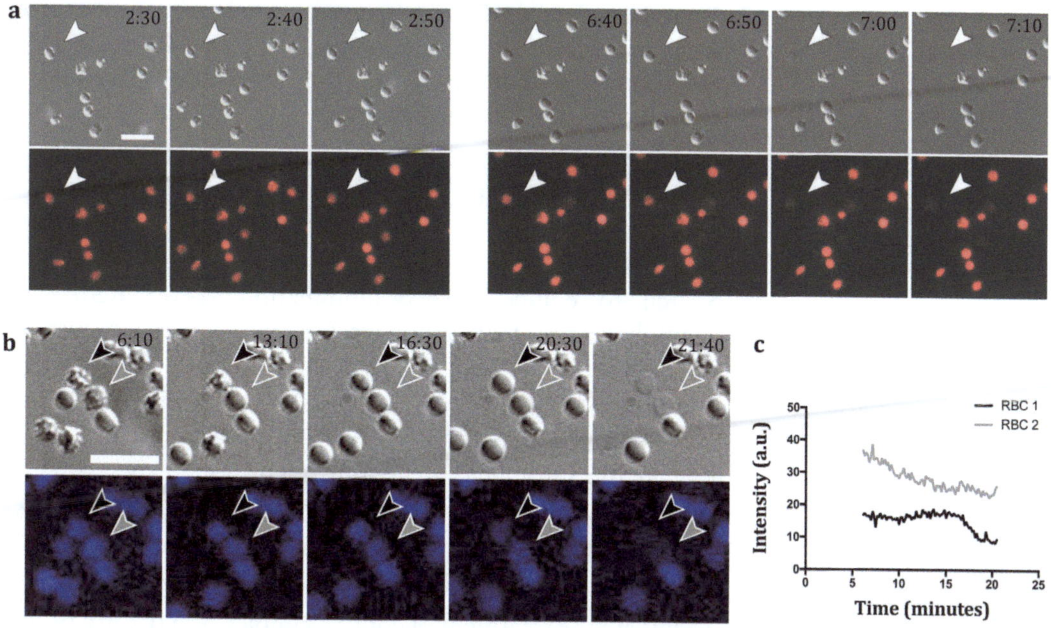

Fig. 1 Live microscopy of RBC hemolysis. (**a**) Using DIC microscopy (top panels) and the fluorescence of hemoglobin (Hb) (bottom panels). Water influx is indicated by the rounding up of the RBCs (left panel set) while hemolysis results in the rapid loss of hemoglobin from the cell along with a change in the RBC membrane structure (right panel set). Examples of each are indicated by arrowheads. Scale bar, 11 μm. (**b**) Using DIC microscopy and the fluorescence of the chloride-sensitive indicator MEQ. Select time points illustrating water influx, increases in intracellular chloride (drop in fluorescence), and hemolysis are indicated by arrowheads. Scale bar, 11 μm. (**c**) Fluorescence intensity (arbitrary units, a.u.) of MEQ over time (6:10 to 20:30) in individual cells prior to hemolysis. RBC 1 (black) and RBC 2 (gray) corresponding to cells indicated by the black and gray-filled arrowheads in (**b**)

3.5 Live Imaging

The following method utilizes live, time-lapse imaging to study the dynamics of SLS-induced hemolysis. Taking advantage of the fluorescence of RBC hemoglobin and the contrast of DIC microscopy, the process of hemolysis can be studied in detail. Cell rupture is associated with the rapid release of hemoglobin and a concomitant change in membrane contrast (*see* Fig. 1a). For the measurement of intracellular chloride levels, chloride influx can be monitored in conjunction with fluid flow and cell rupture (*see* Fig. 1b and c).

1. Power up microscope and equilibrate the humidified environmental chamber to 37 °C.
2. For the observation of SLS-induced hemolysis in an isotonic buffer, suspend the RBC pellet (*see* Subheading 3.3, **step 5**) in 40 mL of PBS and vortex (proceed to **step 3**). For the observation of fluid flow/hemolysis, suspend the RBC pellet (*see* Subheading 3.3, **step 5**) in 40 mL of hPBS and vortex (proceed to **step 3**). For the observation of SLS-induced chloride influx and fluid flow, suspend the RBC pellet (*see* Subheading 3.4, **step 11**) in 40 mL of hPBS and vortex.

3. Place 1.8 mL of RBC cell suspension in a 35 mm glass-bottom dish and allow to equilibrate to 37 °C in the microscope environmental chamber. Note: the immersion oil should be equilibrated to 37 °C before use.
4. Remove the lid of the optics dish. Place the dish on the stage, securing it using stage clips. Using the 60× oil objective and DIC illumination, find the focal plane of the RBCs associated with the glass surface.
5. Switch to epifluorescence illumination, Texas Red Filter (hemolysis) or DAPI Filter set (chloride influx). Adjust the intensity of the excitation light, to maximize the signal-to-noise ratio. Ideally, aim for minimum light exposure to prevent photobleaching and phototoxicity.
6. Engage the automated focus adjustment system.
7. In the experimental parameters, set the image acquisition in DIC and Fluorescence (Texas Red or DAPI) mode to an interval of 10 s for the duration of each experiment (30 min to 1 h).
8. To initiate the treatment of control samples, add 0.2 mL of PBS or hPBS to the dish containing 1.8 mL preequilibrated cells and mix by gently pipetting. For SLS treatments, add 0.2 mL of SLS preparations (*see* Subheading 3.2, **step** 7) from wt5448, 5448:*sagA*Δ*cat*, or 5448:*sagA*Δ*cat* + pDCerm*sagA* to each 1.8 mL dish of RBC and mix by gently pipetting. After the addition of treatments, allow cells to settle for 2 min.
9. Initiate image capture (*see* **Note 9**).
10. Export monochromatic Tagged Image File Format (TIFF) images for movie construction and image analyses using the ImageJ software.
11. For analysis of fluorescence intensity, use ImageJ analyze/measure functions. Selecting a cell of interest can be done using any of the drawing or selection tools. In addition, select a clear region near the cell for determining the background.

3.6 Quantitating Water Influx by Flow Cytometry

Normal RBCs have a specific forward scatter (FSC) versus side scatter (SSC) profile when measured on a flow cytometer. In a hypertonic-buffered environment, however, RBCs will experience a net loss of water and undergo significant changes in these cytometric parameters. These changes allow for the visualization and quantitation of water influx in a cell population (*see* Fig. 2). The following method describes the quantitation of SLS-induced fluid movement in an RBC population.

1. From Subheading 3.3, **step 5**, suspend one washed RBC pellet in 20 mL of room temperature PBS and a second in 20 mL room temperature hPBS, vortex gently to suspend cells.

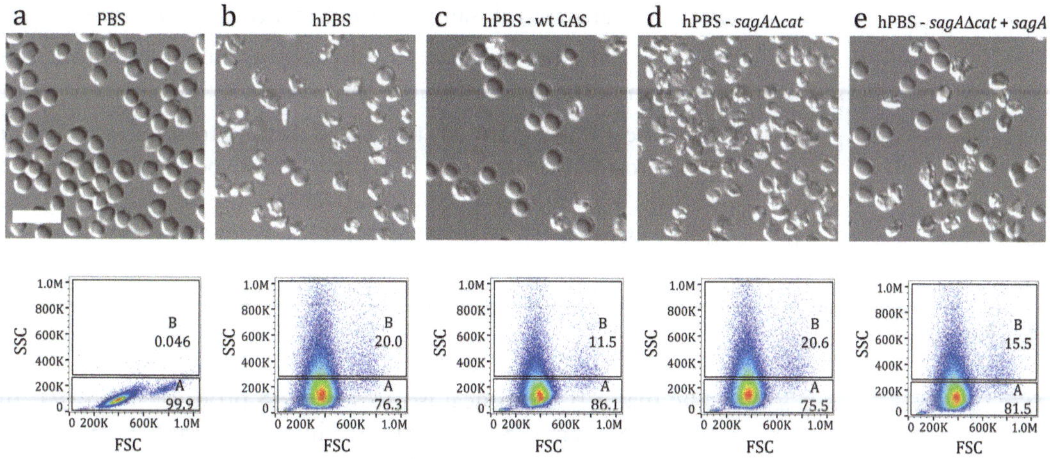

Fig. 2 Microscopy and flow cytometry illustrating water content within RBCs. (**a**) DIC Microscopy images (top panels) and corresponding flow cytometry pseudocolor plots (bottom panels) of RBCs in isotonic PBS, (**b**) in hPBS control treatment, (**c**) or in hPBS with SLS treatments from wt GAS, (**d**) *sagA*Δ*cat*, (**e**) and complemented strain (*sagA*Δ*cat* + *sagA*). Scale bar, 11 μm. Gate A denotes the percentage of agranular RBCs and Gate B denotes the percentage of granular RBCs. Colors indicate population density; Red (very high); Yellow (high), Green (medium), Blue (low)

2. Prepare triplicate samples for each condition to be tested. Aliquot 0.9 mL volumes of PBS suspended cells into three tubes and 0.9 mL hPBS suspended cells into 12 tubes (polystyrene round-bottom 5 mL).
3. For control samples, add 0.1 mL of PBS to each 0.9 mL PBS and hPBS suspended RBC sample. For SLS treatments, add 0.1 mL of 10× SLS preparations (*see* Subheading 3.2, **step 7**) from wt5448, 5448:*sagA*Δ*cat,* or 5448:*sagA*Δ*cat* + pDCerm-*sagA* into each 0.9 mL hPBS cell suspension. Mix with gentle vortexing.
4. Place tubes in a 37 °C incubator or water bath for 10–20 min (*see* **Note 10**).
5. Transfer tubes directly to ice to stop the hemolytic process. Keep tubes on ice until analysis by flow cytometry. Vortex briefly, directly before loading into flow cytometer.
6. Prepare the flow cytometer as per manufacturer's recommendations.
7. Set the instrument sample flow rate to medium. Do not exceed the instrument's detection rate capacity. Set the acquisition limit to 100,000 events.
8. Use PBS suspended, untreated cells to optimize detector voltages. Adjust FSC and SSC voltages, so that the primary cell populations are in the lower half of the SSC plot (*see* Fig. 2). This is necessary in order to accurately visualize the large expansion of cells in hypertonic conditions.

9. Collect data from all samples using identical cytometer settings.
10. Export data files and use FlowJo software for population analyses and plot generation.
11. Use the PBS suspended, untreated sample to set a rectangular gate around normal cells. Use the hPBS suspended cells to set a second rectangular gate around cells with an increased SSC value (*see* Fig. 2). Analyze all samples using these gates.

4 Notes

1. The majority of clinical isolates of GAS produce SLS [19]. In addition to an SLS-producing strain, we recommend the inclusion of an isogenic *sagA* deletion strain (non-SLS-producing) and a *sagA* complemented strain in additional control preparations.
2. DiH-MEQ (6-methoxy-*N*-ethyl-1,2-dihydroquinoline) is a cell-permeable version of the fluorescent indicator MEQ. MEQ is quenched in the presence of chloride [18].
3. An automated focus adjustment control system is necessary for the real-time correction of axial focus fluctuations. These have various trademark names depending on the microscope manufacturer. Some examples include the Nikon (Perfect Focus System), Zeiss (Definite Focus), Leica (Adaptive Focus), and Olympus (Zero Drift).
4. GAS is a Biosafety level 2 (BSL-2) organism and should be handled under BSL-2 guidelines using proper personal protective equipment. Sterile technique should be used in handling bacteria at each step.
5. The use of Group A Selective Strep Agar allows for the confirmation of SLS production. Growth on agar will result in blood clearance under and around bacterial growth of wild-type and complemented strains. No blood clearance should be seen under the SLS mutant strain.
6. SLS activity is temperature sensitive above 4 °C. Avoid prolonged storage or handling of SLS preparations at elevated temperatures prior to use.
7. RBC concentration may vary with the manufacturer and lot of defibrinated whole sheep blood. Resuspension volume may be adjusted to achieve optimal concentrations of cells (~1–2 × 10^6 cells per mL) for imaging and flow cytometry assays.
8. Depending on the quality of the DiH-MEQ used, the loading time and concentrations can be optimized. If using higher concentrations for loading, additional wash steps may be required.

9. We recommend measuring the final volume in the optic dish at the end of each experiment to determine if a loss of volume has occurred (through evaporation) during imaging. This can cause aberrant results. In the event that the environmental chamber humidifier is inadequate, we suggest adding distilled water-saturated Kimwipes to the chamber near the optic dish to provide additional humidity.
10. Time course studies may be done here as storage on ice halts the hemolytic process. Samples should be kept on ice until being run through the flow cytometer.

Acknowledgments

We would like to thank all the members of the Lee Lab for their assistance with lab studies and troubleshooting with regard to the protocols. We would especially like to thank Dr. Henry Vu for assistance with manuscript preparation. Special thanks to Dr. Brandon Ashfeld at University of Notre Dame, Dr. Nicolas Biais at Brooklyn College CUNY, Dr. Victoria Ploplis and Dr. Francis Castellino at the W. M. Keck Center for Transgene Research, Notre Dame, IN for their valuable insight and assistance.

References

1. Betschel SD, Borgia SM, Barg NL et al (1998) Reduced virulence of group A streptococcal Tn916 mutants that do not produce Streptolysin S. Infect Immun 66:1671–1679
2. Datta V, Myskowski SM, Kwinn LA et al (2005) Mutational analysis of the group A streptococcal operon encoding streptolysin S and its virulence role in invasive infection. Mol Microbiol 56:681–695. https://doi.org/10.1111/j.1365-2958.2005.04583.x
3. Flaherty RA, Puricelli JM, Higashi DL et al (2015) Streptolysin S promotes programmed cell death and enhances inflammatory Signaling in epithelial keratinocytes during group A streptococcus infection. Infect Immun 83 (10):4118–4133. https://doi.org/10.1128/IAI.00611-15
4. Lee SW, Mitchell DA, Markley AL et al (2008) Discovery of a widely distributed toxin biosynthetic gene cluster. Proc Natl Acad Sci U S A 105:5879–5884. https://doi.org/10.1073/pnas.0801338105
5. Nizet V, Beall B, Bast DJ et al (2000) Genetic locus for Streptolysin S production by group A streptococcus. Infect Immun 68:4245–4254
6. Molloy EM, Cotter PD, Hill C et al (2011) Streptolysin S-like virulence factors: the continuing *sagA*. Nature Rev Microbiol 9:670–681. https://doi.org/10.1038/nrmicro2624
7. Carr A, Sledjeski DD, Podbielski A et al (2001) Similarities between complement-mediated and Streptolysin S-mediated Hemolysis. J Biol Chem 276:41790–41796. https://doi.org/10.1074/jbc.M107401200
8. Duncan JL, Mason L (1976) Characteristics of streptolysin S hemolysis. Infect Immun 14:77–82
9. Shai Y (1999) Mechanism of the binding, insertion and destabilization of phospholipid bilayer membranes by alpha-helical antimicrobial and cell non-selective membrane-lytic peptides. Biochim Biophys Acta 1462:55–70
10. Heddle JG, Blance SJ, Zamble DB et al (2001) The antibiotic microcin B17 is a DNA gyrase poison: characterisation of the mode of inhibition. J Mol Biol 307:1223–1234. https://doi.org/10.1006/jmbi.2001.4562

11. Higashi DL, Biais N, Donahue DL et al (2016) Activation of band 3 mediates group A streptococcus streptolysin S-based beta-haemolysis. Nat Microbiol 1:15004. https://doi.org/10.1038/nmicrobiol.2015.4
12. Sumitomo T, Nakata M, Higashino M et al (2011) Streptolysin S contributes to group A streptococcal translocation across an epithelial barrier. J Biol Chem 286:2750–2761. https://doi.org/10.1074/jbc.M110.171504
13. Chatellier S, Ihendyane N, Kansal RG et al (2000) Genetic relatedness and superantigen expression in group A streptococcus serotype M1 isolates from patients with severe and non-severe invasive diseases. Infect Immun 68:3523–3534
14. Alouf JE (1980) Streptococcal toxins (streptolysin O, streptolysin S, erythrogenic toxin). Pharmacol Ther 11:661–717
15. Ginsburg I (1999) Is streptolysin S of group a streptococci a virulence factor? APMIS 107:1051–1059
16. Theodore TS, Calandra GB (1981) Streptolysin S activation by lipoteichoic acid. Infect Immun 33:326–328
17. Schwab JH (1956) An intracellular hemolysin of group A streptococci. J Bacteriol 71:100–107
18. Biwersi J, Verkman AS (1991) Cell-permeable fluorescent indicator for cytosolic chloride. Biochemistry 30:7879–7883
19. Yoshino M, Murayama SY, Sunaoshi K et al (2010) Nonhemolytic Streptococcus pyogenes isolates that lack large regions of the sag operon mediating streptolysin S production. J Clin Microbiol 48:635–638. https://doi.org/10.1128/JCM.01362-09

Chapter 19

Isolation of Monoclonal Antibodies to Group A Streptococcus Antigens Using Phage Display

Jeremy M. Raynes, Mei Lin Tay, Sok H. By, John D. Steemson, and Nicole J. Moreland

Abstract

High-affinity monoclonal antibodies are valuable tools for studying the humoral immune response to Group A Streptococcus (GAS) antigens. This protocol describes a method for the selection of monoclonal antibody fragments that bind to GAS antigens using either naïve or immune repertoires displayed on the surface of M13 bacteriophage. Clones that specifically bind to GAS antigens are enriched for during the biopanning process, in which antibody-phage clones bind to an immobilized GAS antigen and are then washed, eluted, and amplified for subsequent rounds of selection. After the final round of biopanning, individual clones are screened by phage enzyme-linked immunosorbent assay (ELISA), and unique clones are identified by DNA fingerprinting and sequencing. The isolated monoclonal antibodies can be used to explore antibody–antigen interactions in molecular detail and provide insight into the protective mechanisms from GAS infection.

Key words Phage display, Biopanning, Group A Streptococcus, Monoclonal antibody, Antigen

1 Introduction

Group A Streptococcus (GAS), also known as *Streptococcus pyogenes* or StrepA, is a nonmotile, Gram-positive coccus that exclusively infects humans causing over 500,000 deaths globally per year [1]. While GAS cause a vast array of clinical diseases, including invasive and noninvasive pathologies of varying severity, the majority of GAS-related mortality is attributed to rheumatic heart disease—a chronic autoimmune condition caused by repeated episodes of acute rheumatic fever [2]. Currently, there is no prophylactic vaccine available for GAS but a large number of vaccine candidates are in preclinical development and early stage clinical trials [3]. GAS produces a vast array of virulence factors, which enable the bacteria to cause such a broad range of diseases. These include surface-bound and secreted antigens that aid in colonization and interfere with host immune responses. Different GAS

Thomas Proft and Jacelyn M. S. Loh (eds.), *Group A Streptococcus: Methods and Protocols*, Methods in Molecular Biology, vol. 2136, https://doi.org/10.1007/978-1-0716-0467-0_19, © Springer Science+Business Media, LLC, part of Springer Nature 2020

strains express varying suites of virulence factors and this affects their ability to cause certain diseases [4]. Monoclonal antibodies are powerful tools for studying the humoral immune response to GAS virulence factors in molecular detail, thereby providing insight into mechanisms of virulence, pathology, and antibody-mediated protection. Ultimately, this molecular knowledge could inform the rational design of improved vaccine candidates.

Antibody-phage display technology uses extremely large antibody libraries to screen high numbers of clones against a target antigen in vitro. These libraries are generated by amplifying the variable regions of antibodies from a large number of antibody-producing cells from naïve or immunized animals. Naïve libraries are unbiased and can contain antibodies to almost any target, while immune libraries are strongly biased toward the vaccine antigen [5]. Either form of library contains antibody genes that theoretically represent the entire polyclonal antibody repertoire of the donor(s). This is in contrast to methods such as hybridoma technology and single B-cell cloning that involve screening a relatively small proportion of the antibody repertoire due to their reliance on antibodies secreted by individual B-cells [6, 7]. Phage display technology is built upon the ability of the pIII coat protein of M13 filamentous bacteriophage to display antibody fragments, such as antigen-binding fragments (Fab), without compromising their ability to bind to a target of interest [8]. Antibody libraries are cloned into bacteriophage DNA, such that each phage particle produces a terminal pIII coat protein fused to a single antibody fragment [9]. Since the phage contains the genetic information of the antibody within its particle, the phenotype and genotype of the displayed fragment are physically linked. Antibodies with desirable binding properties can then be selected for and the genes encoding these antibodies can be recovered [10].

This chapter describes a general method for the isolation of antibody fragments that bind to GAS antigens. It has recently been used to identify Fab fragments that bind to GAS adhesins from naïve-human, immune-murine, and immune-rabbit Fab-phage libraries (Raynes et al., in preparation), and could be adapted for use with alternative GAS antigens and antibody fragment formats. This protocol uses biopanning to select for individual Fab-phage clones that bind to the target antigen (shown in Fig. 1). The highly diverse Fab-phage library is incubated with a target GAS antigen that has been biotinylated and immobilized onto streptavidin-coated beads. The nonbinding Fab-phage are then washed away to remove them from the selection process while the bound phage are eluted and amplified in *Escherichia coli* to enrich for binders in the polyclonal pool. This enriched pool is then used in the following round of biopanning. Typically, three to five rounds are sufficient to isolate Fab-phage that bind to the target protein.

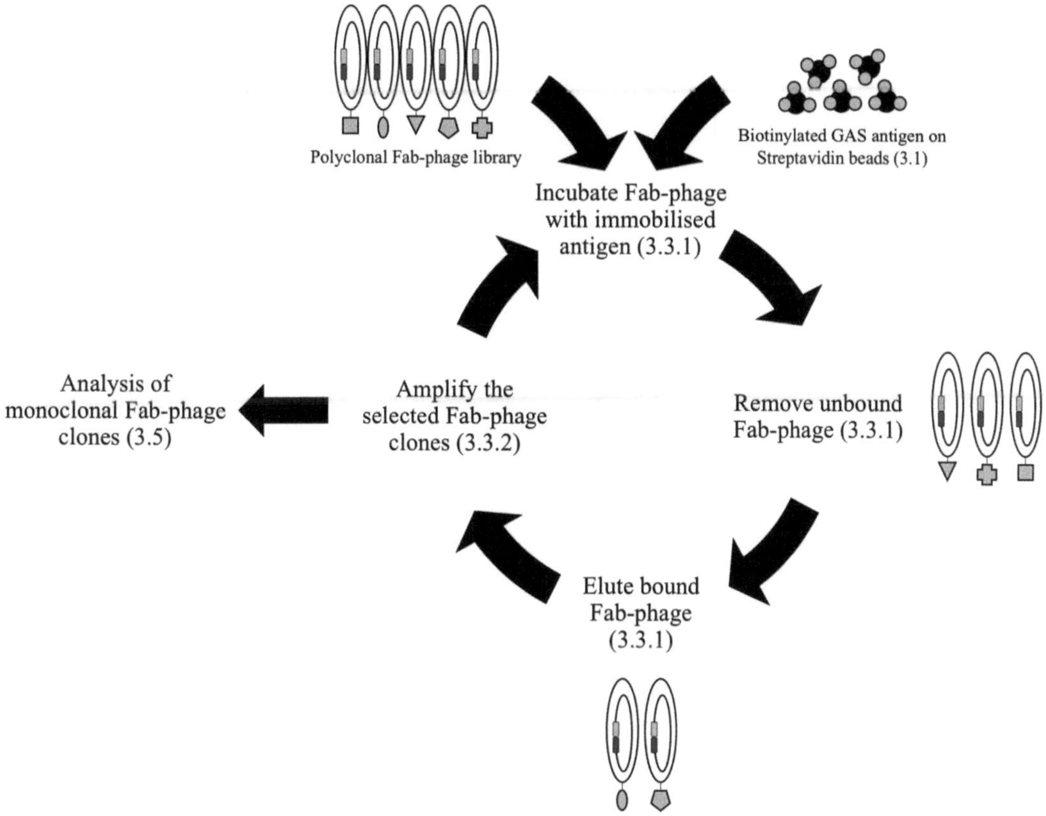

Fig. 1 Overview of the biopanning procedure to isolate Fab clones that bind to GAS antigens. A naïve-human or immune-animal Fab-phage library is incubated with a biotinylated GAS antigen immobilized onto streptavidin resin. Nonbinding phage are removed with multiple washing steps and the bound phage are eluted and amplified for further rounds of selection. Following three to five rounds of biopanning, individual Fab-phage clones can be isolated and analyzed. The numbers in brackets refer to the relevant steps detailed in this chapter

2 Materials

2.1 Biotinylation of GAS Antigens

1. GAS protein of interest at 1 mg/mL in phosphate-buffered saline (PBS), pH 7.4.
2. EZ-Link NHS-PEG4-Biotin (Thermo Fisher Scientific).
3. 100 mM Tris or 100 mM glycine solution
4. Pierce Biotin Quantitation Kit (Thermo Fisher Scientific).
5. Slide-A-Lyzer Dialysis Cassette (Thermo Fisher Scientific).

2.2 Biopanning

1. A Fab-phage library (*see* **Note 1**).
2. Filtered PBS, pH 7.4.
3. Biotinylated protein (from Subheading 3.1, **step 4**).
4. Casein blocking buffer: 0.1% (wt/vol) casein in PBS.

5. Milk blocking buffer (MB): 4% (wt/vol) low-fat milk powder in PBS, pH 7.4.
6. Dynabeads® M-280 Streptavidin (Thermo Fisher Scientific).
7. PBST: 0.1% Tween-20 in PBS.
8. M13K07 Helper Phage (*see* **Note 2**).
9. XL-1 Blue *Escherichia coli* (Agilent Technologies) grown in the presence of tetracycline to ensure the F episome is maintained.
10. Antibiotic stock solutions: 100 mg/mL ampicillin (AMP) (dissolved in Milli-Q water), 20 mg/mL tetracycline (TET) (in 70% ethanol), and 50 mg/mL kanamycin (KAN) (in Milli-Q water).
11. 40% (wt/vol) glucose solution (sterile).
12. Tryptone Yeast extract agar (2×TY agar): 16 g/L tryptone, 10 g/L yeast extract, 5 g/L NaCl, and 15 g/L bacto-agar adjusted to pH 7.0 before autoclaving.
13. 2×TYG agar: TY agar supplemented with antibiotics and glucose, as required. The final antibiotic concentrations are 100 μg/mL AMP, 10 μg/mL TET, and 50 μg/mL KAN. The final concentration of glucose is 2%.
14. Super Broth (SB) medium buffered with 3-(*N*-Morpholino) propanesulfonic acid (MOPS): 30 g/L tryptone, 20 g/L yeast extract, and 10 g/L MOPS in Milli-Q water, adjusted to pH 7.0 before autoclaving.
15. SB medium supplemented with antibiotics and glucose, as required. The final antibiotic concentrations are 100 μg/mL AMP, 10 μg/mL TET, and 50 μg/mL KAN. The final concentration of glucose is 2%.
16. 0.2 M glycine/HCl, pH 2.2.
17. 2 M Tris–HCl, pH 9.0.
18. PEG-6000 solution: 20% (wt/vol) PEG-6000 and 2.5 M NaCl.
19. 80% glycerol (sterile).
20. 2% sodium azide solution.
21. Magnetic rack that holds 1.5 mL tubes.

2.3 Screening Clones Using Monoclonal Phage ELISA

1. 96-well deep well plates for growth of cultures.
2. SB medium supplemented with antibiotics and glucose, as required (*see* Subheading 2.2, **item 15**).
3. AeraSeal™ Breathable film for cultures in 96-well deep well plates (Sigma-Aldrich).
4. Immunosorp flat-bottom microtiter plates (Nunc).
5. Filtered PBS, pH 7.4.

6. Milk blocking buffer (MB) (from Subheading 2.2, **item 5**).
7. Anti-M13-horseradish peroxidase (HRP) conjugate.
8. 3,3′,5,5′-Tetramethylbenzidine (TMB) solution.
9. 1 M HCl.

2.4 Fingerprinting to Identify Unique Clones

1. Polymerase chain reaction (PCR) machine.
2. PrimeSTAR HS DNA Polymerase with supplied buffer and deoxynucleotide triphosphates (dNTPs) (Takara Bio).
3. Primers suitable for PCR and sequencing of Fab clones from the Fab-phage library (*see* **Note 1**).
4. Agarose gels: 1% (wt/vol) and 4% (wt/vol) agarose.
5. Nucleic acid staining solution.
6. Mva I (BstN I) restriction enzyme.

3 Methods

This protocol describes the process of biopanning antibody-phage libraries against a biotinylated target protein immobilized on streptavidin resin. This method is preferred, as it allows libraries to be screened against the antigen in solution and avoids the partial denaturation of proteins that can occur when proteins are directly adsorbed onto plastic plates [11, 12]. However, the biopanning protocol could also be adapted for use with antigens that are directly coated onto immunoplates if desired.

3.1 Biotinylation of GAS Antigens and Immobilization on Streptavidin Resin

This section describes the in vitro chemical biotinylation of a GAS antigen, as it is a simple, highly efficient, and commonly used method to biotinylate any protein. This process utilizes a biotin analog that reacts with primary amine groups, resulting in the random biotinylation of multiple accessible lysine residues and the N-termini of polypeptides [13]. Alternatively, the targeted addition of a single biotin per protein can be achieved enzymatically. In this case, the protein is produced fused to a 15 amino acid AviTag peptide, which can be specifically biotinylated in vivo or in vitro by *E. coli* biotin ligase (BirA) (for full details, refer to [14]). This method has two major advantages, in that the protein remains in its native state and the orientation of the antigen can be controlled during biopanning and ELISA analysis (*see* **Note 3**).

1. Use 0.5–2 mg of GAS protein at 1 mg/mL in PBS. Add NHS-PEG$_4$-Biotin for a 20:1 molar ratio of biotin to protein and incubate on ice for 2 h.
2. Add 50 μL of 100 mM Tris or 100 mM glycine to inhibit the reaction.

3. Remove excess biotin by dialyzing overnight into PBS using a Slide-A-Lyzer Dialysis Cassette (*see* **Note 4**).
4. Confirm the antigen is biotinylated and quantify the average moles of biotin per mole of protein using the Pierce Biotin Quantitation Kit according to the manufacturer's instructions.
5. Wash 70 μL of magnetic streptavidin-coated beads twice with 1 mL PBS in a 1.5 mL tube. Place the tube containing the resin in a magnetic rack for 1 min and carefully remove the PBS, once the resin has collected on the magnet side of the tube.
6. Remove from the magnetic rack and add 500 μL of biotinylated GAS protein at a final concentration of 1400 nM in 0.1% casein.
7. Rock the mixture gently at room temperature (RT) for 1 h and then wash the beads three times with 1 mL of PBS (as in **step 5**).

3.2 Preparation of Phage Library

This protocol has been used to isolate Fab that bind to GAS antigens from naïve-human and immune-animal Fab-phage libraries but can be adapted for use with other libraries (*see* **Note 1**).

1. Libraries are typically stored in aliquots at −80 °C in 15% glycerol. Thaw an aliquot of the Fab-phage library containing 10^{12} pfu (*see* **Note 1**).
2. Add one-fifth volume of PEG-6000 solution and incubate on ice for 1 h.
3. Centrifuge for 10 min at 10,000 × *g*, 4 °C to pellet phage. The phage will appear as a white pellet.
4. Resuspend the phage pellet in 200 μL PBS and centrifuge for 2 min at 10,000 × *g* to pellet any debris.
5. Transfer the 200 μL of phage supernatant to a clean 1.5 mL tube.
6. Add 800 μL of MB to the phage supernatant and incubate for 1 h at RT, with gentle rocking, to block the library.

3.3 First Round of Biopanning

3.3.1 Phage Binding and Elution

1. Mix the blocked phage library (from Subheading 3.2, **step 6**) with the streptavidin beads coated with immobilized GAS protein (from Subheading 3.1, **step 7**) (*see* **Note 5**).
2. Rock the mixture gently for 2 h at RT.
3. Place the tube containing the mixture in the magnetic rack and remove the supernatant, once the resin has collected on the side of the tube. Wash the magnetic beads six times with 1 mL of PBST and twice with 1 mL of PBS.
4. Elute the bound phage by adding 350 μL of 0.2 M glycine/HCl, pH 2.2 to the washed beads. Incubate for 5 min at RT on a rotating wheel.

5. Separate supernatant and beads by applying to a magnetic rack. Transfer supernatant to a clean 1.5 mL tube containing 20 μL of 2 M Tris–HCl, pH 9.0 and mix to neutralize.

3.3.2 Infection of X-L1 Blue E. coli and Amplification of the Eluted Phage

1. Inoculate 4 mL of SB with a single colony of XL-1 Blue and grow to an optical density 600 nm (OD_{600}) of 0.6–0.8 (log phase). Infect the culture with the neutralized phage and incubate at RT without shaking for 15 min.
2. Add 4 mL of SB medium (prewarmed to 37 °C) supplemented with 20 μg/mL ampicillin and 10 μg/mL tetracycline.
3. Remove 2 μL of infected XL-1 Blue cells for output titering. Dilute this 2 μL in 200 μL of SB (1:100) and plate 10 μL and 100 μL volumes of this onto 2× TYG/AMP plates. Incubate plates overnight at 37 °C and calculate the output titer by multiplying the number of colonies by the culture volume and dividing by the plating volume (*see* **Note 6**).
4. Incubate the 8 mL culture from **step 2** for 1 h at 37 °C (200 rpm).
5. Add ampicillin to a final concentration of 50 μg/mL and continue shaking the culture at 37 °C (200 rpm) for 1 h.
6. Add M13KO7 helper phage at a multiplicity of infection (MOI) of 20:1 (phage-to-cells ratio; *see* **Note 7**) and transfer the culture to a sterile 500 mL disposable flask.
7. Add SB medium (prewarmed to 37 °C) supplemented with 100 μg/mL ampicillin and 10 μg/mL tetracycline to a final culture volume of 100 mL and shake at 37 °C (200 rpm) for 2 h.
8. Add kanamycin to a final concentration of 70 μg/mL and continue shaking the culture overnight at 37 °C (200 rpm).

3.3.3 Preparation of Phage for Subsequent Panning Rounds

1. Centrifuge the overnight culture for 15 min at 4000 × *g*, 4 °C.
2. Transfer the supernatant containing the phage to a clean centrifuge bottle and add one-fifth volume (20 mL) of PEG solution.
3. Mix and incubate on ice for 1 h to precipitate the phage particles.
4. Centrifuge for 15 min at 10,000 × *g*, 4 °C to pellet the phage.
5. Discard the supernatant and resuspend the phage pellet in 2 mL PBS, pH 7.4.
6. Centrifuge for 2 min at 10,000 × *g* to pellet any cell debris.
7. Transfer the supernatant containing the phage to a clean tube and repeat the PEG precipitation (**steps 2–5**).

8. Resuspend the pellet in 1 mL PBS, pH 7.4. Remove 2 μL of the phage suspension for titering. The remaining phage can be stored at 4 °C or −80 °C for subsequent panning (*see* **Note 8**).
9. To titer the amplified phage, make serial dilutions of the phage preparation at 10^{-3}, 10^{-6}, 10^{-8}, 10^{-9}, and 10^{-10} in SB media.
10. Add 1 μL of the 10^{-8}, 10^{-9}, and 10^{-10} dilutions to 50 μL of log-phase XL-1 Blue cells and incubate at RT without shaking for 15 min.
11. Plate onto 2× TYG/AMP plates and incubate overnight at 37 °C.
12. Count colonies and calculate the titer of the amplified phage. This titer will determine the volume of phage used in the next round of biopanning.

3.4 Subsequent Rounds of Panning to Enrich for Antibodies That Bind to GAS Antigens

Following the first round of biopanning and amplification, the pool of polyclonal phage should become enriched for clones that bind the GAS antigen of choice. Subsequent rounds of panning (normally three to five rounds) further enrich for high-affinity clones. The stringency of selection can be increased by decreasing the amount of target antigen and increasing the number of washing steps, as shown in Table 1. This promotes the selection of high-affinity Fab-phage clones that bind specifically to the target antigen.

Table 1
Biopanning conditions to enrich for high-affinity T-antigen binding clones

Biopanning round				
	1	2	3	4
Phage input (pfu)	Phage library (10^{12})	Amplified phage from first round (10^{11} to 10^{12})	Amplified phage from second round (10^{11} to 10^{12})	Amplified phage from third round (10^{11} to 10^{12})
GAS antigen concentration (nM)	1400	300	150	150
Streptavidin-coated magnetic bead volume (μL)	70	50	30	30
Blocking and deselection	0.1% casein 2% Milk	0.1% casein 2% Milk 5 μg M280 beads	0.1% casein 2% Milk 5 μg M280 beads	0.1% casein 2% Milk 5 μg M280 beads
Number of washes	6× PBST 2× PBS	14× PBST 2× PBS	14× PBST 2× PBS	14× PBST 2× PBS

3.5 Screening Individual Clones

After the desired number of biopanning rounds, individual colonies from the final round output titer plates (from Subheading 3.3.2, **step 3**) can be tested for antigen binding by phage ELISA (enzyme-linked immunosorbent assay (*see* **Note 9**)).

3.5.1 Small-Scale Phage Rescue

XL-1 Blue *E. coli* colonies containing Fab-phage clones are individually picked from output titer plates following the final round of biopanning and grown in 96-well deep well plates for small-scale phage rescue.

1. Add 500 μL of SB/AMP/TET-glucose medium to each well of the "master" 96-well deep well plate.
2. Seed each well with a unique XL-1 Blue colony using sterile pipette tips (*see* **Note 10**). Seal the plate with breathable film and incubate overnight at 37 °C, 800 rpm in a thermomixer.
3. Spin the plate for 3 min at 500 × *g* at RT to remove condensation from the underside of the film before opening.
4. Transfer 5 μL of each overnight culture to the corresponding wells of a new 96-well deep well plate containing 500 μL SB/AMP/TET-glucose medium. Seal the new plate with breathable film and incubate for 3 h at 37 °C, 800 rpm in a thermomixer until the OD_{600} ~0.5.
5. Add M13KO7 helper phage at a multiplicity of infection (MOI) of 20:1. Dilute helper phage in an appropriate amount of SB and add 50 μL of this dilution to each well.
6. Seal the plate with breathable film and shake for 30 s at 800 rpm to mix.
7. Stop the shaking and incubate the plate undisturbed for 30 min at 37 °C.
8. Pellet the cells by centrifuging the plate for 5 min at 4000 × *g*. Remove the supernatant by inverting the plate and resuspend each cell pellet with 500 μL SB/AMP/KAN/TET.
9. Seal the plate with breathable film and shake overnight at 30 °C, 800 rpm in a thermomixer.
10. Centrifuge the plate for 5 min at 4000 × *g* to pellet the cells and remove 400 μL of the phage-containing supernatant to a new 96-well deep well plate. Keep the plate on ice until required or store at 4 °C.

3.5.2 Monoclonal Phage ELISA

1. Dilute the desired GAS protein in PBS to 5 μg/mL and add 100 μL to each well of a 96-well immunoplate. Incubate for 1 h at RT.
2. Wash the plate three times with PBST.
3. To block, add 200 μL of MB to each well and incubate overnight at 4 °C.

4. Wash the plate three times with PBST.
5. Dilute the phage-containing supernatant (from Subheading 3.5.1, **step 10**) 1 in 4 with MB and add 100 μL of the diluted phage to the corresponding well on the immunoplate. Incubate for 1 h at RT.
6. Wash the plate five times with PBST.
7. Add 100 μL of anti-M13-HRP antibody (*see* **Note 11**) diluted 1: 5000 in MB to each well and incubate the plate for 1 h at RT.
8. Wash three times with PBST.
9. Add 50 μL of TMB substrate to each well.
10. Cover plate with aluminum foil and incubate at RT for 1–30 min until a blue color develops (*see* **Note 12**).
11. Stop the reaction with 50 μL of 1 M HCl. The color in the well will be yellow after the addition of HCl.
12. Read the absorbance at 450 nm on an optical density plate reader. Fab-phage clones should specifically bind to the target antigen and not to unrelated antigens such as bovine serum albumin (BSA), biotin, or streptavidin.

3.5.3 Restriction Digestion Fingerprinting to Identify Unique Clones

Following the identification of clones that bind to the target antigen, each clone can immediately be sequenced to identify unique clones. However, it is not uncommon for a small number of clones to dominate the selection process or for multiple copies of one clone to be present. Where a large number of binding clones are identified by monoclonal phage ELISA, it may be useful to fingerprint each clone prior to sequencing in order to identify identical fingerprints and reduce the number of clones sequenced.

1. Centrifuge the "master" 96-well deep well plate (from Subheading 3.5.1, **step 2**) for 3 min at 500 × *g* to remove condensation from the underside of the film before opening.
2. Transfer 0.1 μL–5 μL of culture from the wells containing antigen-binding clones (identified in Subheading 3.5.2, **step 12**) into individual PCR tubes.
3. Make PCR mixes as shown in Table 2 and perform PCR as shown in Table 3.
4. Analyze 5 μL of each PCR product on a 1% agarose gel (*see* **Note 13**).
5. Make the Mva I restriction digest reaction mix shown in Table 4 and incubate for 1 h at 37 °C (*see* **Note 14**).
6. Analyze the entire reaction volume from each clone on a 4% agarose gel.
7. Visually inspect the banding patterns produced by the digestion of each Fab clone to identify unique clones.

Table 2
Reaction mixture for PCR amplification of Fab inserts

	Per reaction
Clone culture (volume may require optimization)	0.1 μL
Forward primer	0.45 μM
Reverse primer	0.45 μM
5× PrimeSTAR buffer (Mg^{2+})	1×
dNTP mix	0.2 mM
PrimeSTAR HS DNA polymerase	0.2 μL
Sterilized distilled water	Up to 20 μL

Table 3
Conditions for PCR amplification of Fab inserts

	Temperature	Time
Initial denaturation	98 °C	3 min
30 cycles of:		
Denaturation	98 °C	30 s
Annealing	Primer dependent	15 s
Extension	72 °C	2 min
Final extension	72 °C	5 min

Table 4
Reaction mixture for Mva I digestion of amplified Fab fragments

	Per reaction
PCR product	10 μL
Buffer H	1.2 μL
Mva I	0.2 μL
Sterilized distilled water	Up to 12 μL

4 Notes

1. Our laboratory has used a naïve-human Fab-phage library called HX02 [15, 16] to isolate T-antigen binding Fab. We have also constructed and biopanned immune Fab-phage libraries from mice and rabbits vaccinated with GAS T-antigens and M proteins using established protocols

[17, 18]. Alternatively, commercial antibody libraries are available from SourceBioscience (the Human Domain Antibody (DAb) Library and the ATLAS scFv library) or can be custom-built by a number of companies including Creative Biolabs and Oak Biosciences. Matching primers that enable PCR amplification of antibody fragment gene inserts are generally available (or can be designed based on the phagemid sequence) for downstream clone screening (*see* also **Note 13**).

2. Helper phage are usually stored as glycerol stocks at −80 °C with 15% glycerol. They can be readily amplified (*see* [19, 20]) from these frozen stocks as required and are stable at 4 °C for up to 2 months.
3. Our laboratory has used chemical biotinylation methods to biotinylate GAS T-antigens and, following biopanning, has successfully isolated T-antigen binding Fab. However, random biotinylation may be unsuitable for use with lysine-rich antigens such as the GAS M protein. The M proteins that we have worked with are approximately 15% lysine residues, and random biotinylation causes denaturation of the M protein and the partial loss of epitopes. When working with lysine-rich proteins, specific enzymatic biotinylation of AviTag-fusion proteins is recommended.
4. Other methods can be used to remove excess biotin such as a desalting column or performing size-exclusion chromatography with PBS.
5. Biopanning with proteins immobilized on streptavidin resin can result in the selection and enrichment of streptavidin binders. To prevent this, subsequent rounds of panning deselect streptavidin binders from the library prior to biopanning. Alternatively, other biotin-binding resin such as Neutravidin can be used in place of streptavidin.
6. The input and output titers should be monitored at each round of biopanning. With an input of 10^{11} to 10^{12} phage, 10^5 to 10^7 clones would be expected after the first round of selection. Ideally, after each round of biopanning the ratio of output pfu to input pfu will increase, indicating enrichment of binders. However, if this does not occur it does not necessarily mean that selection and enrichment have been unsuccessful, and clones from the output plates should always be checked by phage ELISA after the final round of panning.
7. The number of helper phage to add for an MOI 20 is calculated as follows:

   culture volume (mL) $\times$ OD_{600} $\times$ 3×10^8 $\times$ 20, where OD_{600} is the optical density (OD) of a suspension of cells at 600 nm (OD_{600}) and 3×10^8 is the concentration of bacterial cells in 1 mL culture when $OD_{600} = 1$ and 20 is MOI.

8. Purified phage can be stored for several weeks at 4 °C with a final concentration of 0.02% sodium azide. Phage can be stored long term at −80 °C with 15% glycerol.
9. Clones can also be screened against an unrelated antigen (e.g., BSA) and streptavidin resin to ensure that binding is specific for the target antigen.
10. Well A1 can be left empty as a sterile control well. The master plate can be sealed with airtight sealing tape and be stored at 4 °C for up to 2 weeks. For longer storage, the plate can be stored at −80 °C with 15% glycerol in each well.
11. The anti-M13-HRP antibody binds to the pVIII protein; multiple copies of which are assembled into the M13 bacteriophage coat.
12. Generally, an incubation time of 1–5 min is sufficient for signal to develop while allowing the distinction between stronger and weaker binding clones.
13. This PCR step may also have some use as a screening tool. We have found that while most clones are readily amplified by PCR, those that are resistant to amplification tend to have significant deleterious mutations when sequenced.
14. The Mva I restriction enzyme cleaves the sequence CC/WGG, which occurs frequently in Fab sequences. When Fab clones with unique sequences are digested with Mva I, different banding patterns are observed, while digestion of identical Fab clones results in identical banding patterns. Mva I restriction digestion is suitable for fingerprinting Fab clones from human and rabbit libraries. However, Fab from murine libraries tend to be cleaved into fragments that are too small for visual analysis and, as such, proceeding directly to sequencing for murine libraries is recommended.

References

1. Carapetis JR, Steer AC, Mulholland EK, Weber M (2005) The global burden of group A streptococcal diseases. Lancet Infect Dis 5:685–694
2. Carapetis JR, Beaton A, Cunningham MW et al (2016) Acute rheumatic fever and rheumatic heart disease. Nat Rev Dis Primers 1:1–24
3. Steer AC, Carapetis JR, Dale JB et al (2016) Status of research and development of vaccines for Streptococcus pyogenes. Vaccine 34:2953–2958
4. Walker MJ, Barnett TC, McArthur JD et al (2014) Disease manifestations and pathogenic mechanisms of group A Streptococcus. Clin Microbiol Rev 27:264–301
5. Bradbury ARM, Marks JD (2004) Antibodies from phage antibody libraries. J Immunol Methods 290:29–49
6. Freysd'ottir J (2000) Production of monoclonal antibodies. Methods Mol Med 40:267–279
7. Hofmann D, Lai JR (2017) Exploring human antimicrobial antibody responses on a single B cell level. Clin Vaccine Immunol 24:1147
8. McCafferty J, Griffiths AD, Winter G, Chiswell DJ (1990) Phage antibodies: filamentous phage displaying antibody variable domains. Nature 348:552–554
9. Barbas CF, Kang AS, Lerner RA, Benkovic SJ (1991) Assembly of combinatorial antibody

libraries on phage surfaces: the gene III site. Proc Natl Acad Sci U S A 88:7978–7982

10. Hoogenboom HR (2005) Selecting and screening recombinant antibody libraries. Nat Biotechnol 23:1105–1116
11. Davies J, Dawkes AC, Haymes AG et al (1994) A scanning tunnelling microscopy comparison of passive antibody adsorption and biotinylated antibody linkage to streptavidin on microtiter wells. J Immunol Methods 167:263–269
12. Tay MYF, Lee CC, Vasudevan SG, Moreland NJ (2014) Identification of dengue-specific human antibody fragments using phage display. Methods Mol Biol 1138:161–173
13. Chames P, Hoogenboom HR, Henderikx P (2002) Selection of antibodies against biotinylated antigens. Methods Mol Biol 178:147–157
14. Kay BK, Thai S, Volgina VV (2009) High-throughput biotinylation of proteins. Methods Mol Biol 498:185–196
15. Moreland NJ, Tay MYF, Lim E et al (2010) High affinity human antibody fragments to dengue virus non-structural protein 3. PLoS Negl Trop Dis 4:e881
16. Zhao Y, Moreland NJ, Tay MYF et al (2014) Identification and molecular characterization of human antibody fragments specific for dengue NS5 protein. Virus Res 179:225–230
17. Andris-Widhopf J, Steinberger P, Fuller R et al (2011) Generation of human fab antibody libraries: PCR amplification and assembly of light- and heavy-chain coding sequences. Cold Spring Harb Protoc 2011(9):pdb.prot065565. https://doi.org/10.1101/pdb.prot065565
18. Rader C (2009) Generation and selection of rabbit antibody libraries by phage display. Methods Mol Biol 525:101–128
19. Lee CMY, Iorno N, Sierro F, Christ D (2007) Selection of human antibody fragments by phage display. Nat Protoc 2:3001–3008
20. Coomber DWJ (2002) Panning of antibody phage-display libraries. Standard protocols. Methods Mol Biol 178:133–145

Part IV

Group A Streptococcus Killing Assays and Infection Models

Chapter 20

Assays to Analyze Adhesion of Group A Streptococcus to Host Cells

Adrina H. J. Khemlani, Thomas Proft, and Jacelyn M. S. Loh

Abstract

The critical first step of Group A Streptococcus (GAS) pathogenesis is adhesion to the host pharyngeal and skin epithelial cell surfaces (Brouwer et al., FEBS Lett 590:3739–3757, 2016). Host-cell adhesion assays provide a straightforward model to study these host–pathogen interactions. Here, we describe the culturing of immortalized cell lines into monolayers to mimic host epithelia. Various GAS strains can then be added to study their adhesion properties. In addition, we describe the use of antibodies raised against the cell-surface components of GAS to study if these are able to neutralize the binding of GAS to the cell lines. This provides an indication if these cell-surface components are involved in adhesion and if antibodies generated against them function through neutralization.

Key words Host-cell adhesion, Immortalized cell lines, Antibodies

1 Introduction

A wide variety of GAS cell-surface components have been reported to mediate adhesion to human cells. These were characterized commonly through studies employing human-derived cells and tissues including primary cultures of tonsil epithelium and keratinocytes as well as immortalized cell lines.

Various GAS adhesins, such as lipoteichoic acid, hyaluronic acid (HA) capsule, certain M proteins, laminin- and collagen-binding proteins, various different fibronectin-binding proteins and pili, have been characterized in studies utilizing immortalized cell lines [2–4]. These studies have provided in-depth insights into the crucial first step of GAS pathogenesis. We have used immortalized cell lines to study GAS tissue tropism and to examine if antibodies raised against pili are able to neutralize the adhesive properties of GAS, hence preventing the onset of a GAS infection [5–7].

Here, we describe a host-cell adhesion assay for measuring the adherence of GAS to a human immortalized keratinocyte cell line (HaCaT) and a human-derived nasopharyngeal carcinoma cell line

Thomas Proft and Jacelyn M. S. Loh (eds.), *Group A Streptococcus: Methods and Protocols*, Methods in Molecular Biology, vol. 2136, https://doi.org/10.1007/978-1-0716-0467-0_20,

(Detroit 562), which represent the two main sites of infection by this human-specific pathogen. Wild-type or bioluminescent-labeled bacteria are incubated with these cell monolayers cultured on a multiwell plate, then washed to remove unbound bacteria before enumeration by plating or measuring bioluminescence. For neutralization assays, antibodies or recombinant proteins can be added to the wells.

2 Materials

Prepare all solutions using sterile distilled water (prepared by purifying deionized water, to attain a sensitivity of 18 MΩ-cm at 25 °C) and analytical grade reagents or tissue culture (TC) grade media. All cell culture work must be performed in a class II biological safety cabinet. Sterilize all media for cell culture by passing it through a membrane filter, 0.22 μm pore size. Store all media and reagents at 4 °C up to a month (unless indicated otherwise). Before use, prewarm the media for cells in a 37 °C water bath for 20 min. Follow waste disposal regulations for all materials as per your laboratory facility.

2.1 Cell Culture

1. HaCaT or Detroit 562 cell lines.
2. TC grade phosphate-buffered saline (PBS) without Ca^{2+} and Mg^{2+}.
3. HaCaT medium: Dulbecco's Modified Eagle Medium (DMEM) with 4.5 g/L glucose, 4 mM L-glutamine (*see* **Note 1**), 0.04 mM phenol red without sodium pyruvate, 10% fetal bovine serum (FBS) (*see* **Note 2**), 0.02 units/ml penicillin, and 0.02 μg/ml streptomycin (*see* **Note 3**). For the adhesion assay, prepare the media without penicillin and streptomycin.
4. Detroit 562 medium: DMEM with 1 g/L glucose, 4 mM L-glutamine, 0.04 mM phenol red, 1 mM sodium pyruvate, 10% FBS, 0.02 units/ml penicillin, and 0.02 μg/ml streptomycin (*see* **Note 3**). For the adhesion assay, prepare the media without penicillin and streptomycin.
5. 0.05% trypsin/0.48 mM ethylenediaminetetraacetic acid (EDTA) solution prepared in PBS without Ca^{2+} and Mg^{2+} (*see* **Note 4**).
6. 96-well TC-treated sterile plates.
7. 96-well TC-treated sterile white plates with clear bottom.
8. T25 and T75 tissue culture flasks.
9. Hemocytometer.
10. Brightfield microscope.

2.2 Bacteria

1. Selected GAS strains (*see* **Note 5**).
2. Brain-Heart Infusion (BHI) Medium (*see* **Note 6**).
3. Blood agar plates or BHI agar plates supplemented with appropriate antibiotics if using bioluminescent GAS.
4. PBS.

2.3 Adhesion Assay

1. Hank's Balanced Salt solution (HBSS) with 1% FBS.
2. 50 μg/ml luciferin (if bioluminescent bacteria expressing luciferase are used): Prepare the stock at 30 mg/ml in water and store frozen. Prepare the working stock concentration by diluting in HBSS.
3. 1.5 ml Eppendorf tubes
4. Plate reader.

3 Methods

3.1 Preparation of Cell Monolayer

1. Culture HaCaT and Detroit 562 cells in their appropriate medium with antibiotics at 37 °C and 5% carbon dioxide in a T75 flask.
2. Aspirate off culture medium and rinse the monolayer with 10 ml PBS without Ca^{2+} and Mg^{2+}.
3. Aspirate off PBS and add 2.5 ml trypsin/EDTA solution to detach the cells. Incubate at 37 °C for approximately 10 min for HaCaT and 5 min for Detroit 562. Observe the cells under a microscope, detached cells appear rounded. If not, incubate further until most of them appear rounded.
4. Tap the sides of the flask and add 2.5 ml of culture medium to neutralize the action of trypsin. Using a serological pipette, flush the surface of the flask with the medium that is in the flask, to further dislodge the cells.
5. Transfer cells into a sterile 50 ml tube and centrifuge at 300 × *g* for 5 min.
6. Aspirate off the supernatant and resuspend the cell pellet in 5 ml of culture medium without antibiotics. Count the cells with a hemocytometer and adjust the cell concentration using culture medium without antibiotics. In our laboratory, the HaCaT cell pellet is resuspended to a final concentration of 4×10^5 cells/ml and the Detroit cell pellet is resuspended to a final concentration of 1×10^6 cells/ml (*see* **Note 7**).
7. Dispense 100 μl of the cell suspension into each well of a 96-well plate. For bioluminescent strains, we use white plates with a clear bottom, as they enhance the signal and lower signal spreadover. Set up three wells for each sample.

8. Incubate cells overnight at 37 °C with 5% carbon dioxide.
9. The cells are ready for use in adhesion experiments when a confluent monolayer is formed, usually within 24 h, by observing under a microscope.

3.2 Determine the Number of GAS Colony-Forming Units with Respect to OD_{600}

1. Streak GAS on BHI agar plates (or blood agar plates). Use agar plates with appropriate antibiotics, if using bioluminescent strains carrying a plasmid with the antibiotic selection marker.
2. Incubate the agar plates at 37 °C overnight.
3. Inoculate a single colony in 15 ml BHI medium with appropriate antibiotics (if required) in a 15 ml tube at 37 °C overnight.
4. Dilute the overnight culture in BHI medium at a 1:10 dilution, that is, add 1 ml of overnight culture to 9 ml of fresh medium.
5. Culture bacteria until the required growth phase/optical density is reached (*see* **Note 8**).
6. Centrifuge at 5000 × *g* for 5 min and wash by resuspending in 10 ml PBS.
7. Centrifuge at 5000 × *g* for 5 min, aspirate the supernatant, and resuspend the bacterial pellet in 10 ml HBSS/1% FBS.
8. Perform a tenfold dilution series by adding 100 μl of bacterial culture to 900 μl of HBSS/1% FBS and vortex tube between each dilution (*see* **Note 9**).
9. For each dilution, pipette 20 μl droplets onto an agar plate in triplicate.
10. Allow the drops to dry before incubating at 37 °C overnight.
11. The next day, enumerate the bacteria to establish the colony-forming unit (CFU) obtained at the required growth phase/optical density.

3.3 Host-Cell Adhesion Assay

1. Culture bacteria as described in Subheading 3.2, **steps 1–7**. Based on the enumeration determined in Subheading 3.2, dilute the bacterial culture to a CFU of ten times more than the concentration of monolayer cells used to seed the wells, to achieve a multiplicity of infection (MOI) of 10 (*see* **Note 10**). In this case, GAS is diluted to 4×10^6/ml when using HaCaT or 1×10^7/ml when using Detroit 562 in HBSS/1% FBS.
2. Remove medium from the cell monolayer and add 100 μl of PBS to wash.
3. Aspirate off the PBS and add 100 μl of the bacteria to each well, achieving a MOI of 10:1. Each different strain will require triplicate inoculum control wells, and triplicate experiential wells (*see* **Note 11**).

4. For cell only control wells, add 100 μl of HBSS/1% serum without bacteria (*see* **Note 12**). Set up additional triplicate wells for counting the actual cells used in the assay.
5. Incubate the plate for 1 h at 37 °C/5% CO_2.
6. Titrate and drop plate GAS inoculum, so you know how many bacteria are added each time as in Subheading 3.2, **steps 8–11**.
7. After 1 h incubation, the experimental wells are washed five times with HBSS/1% FBS. A wash is done by adding 200 μl of HBSS/1% FBS, followed by carefully aspirating the liquid without disturbing the monolayers. Inoculum control wells are left unwashed.

3.3.1 Adhesion Assay with Bioluminescent GAS Expressing Luciferase

1. After performing the washes for experimental wells, add 100 μl HBSS/1% FBS to each well.
2. Add 100 μl of 50 μg/ml luciferin to each well and determine the bioluminescence intensity (bioluminescent light unit, BLU) on a plate reader.
3. The percentage of adherence can be calculated as follows:

$$\frac{\text{BLU of experiment well} - \text{average background}}{\text{BLU of inoculum control well} - \text{average background}} \times 100$$

4. To count the actual cells used in the assay, remove the medium from the three wells set up for this purpose (Subheading 3.3, **step 4**). To wash, add 100 μl of PBS to each well and aspirate the PBS. Add 50 μl of trypsin/EDTA solution to each well. Incubate for 5–10 min at 37 °C or until cells appear rounded under a microscope. Dislodge the cells by using a pipette tip to scrape the bottom of the wells. Count the cells with an hemocytometer.
5. In order to minimize inaccuracy raised from the variation of confluency, the adherence percentage can be further divided by the MOI determined from the actual inoculum and cell numbers counted during the assay to give an adherence index number.

3.3.2 Adhesion Assay with Nonbioluminescent GAS

1. After performing the washes for experimental wells, add 50 μl of trypsin/EDTA solution to each well.
2. Incubate at 37 °C for approximately 10 min for HaCaT and 5 min for Detroit 562.
3. Add 50 μl of HBSS/1% FBS to the well to neutralize the action of trypsin.
4. Scrape the bottom of a well using a pipette tip to dislodge the cells and transfer the contents of each well into Eppendorf tubes.

5. Centrifuge at 300 × *g* for 5 min.
6. Aspirate the medium and resuspend the pellet in 200 μl of PBS.
7. Do a series of tenfold dilutions in PBS, by adding 100 μl of cells to 900 μl of PBS for each dilution.
8. For each dilution, pipette 20 μl droplets in triplicate onto BHI agar plates with the appropriate antibiotics for bioluminescent strains or blood agar plates.
9. Allow the drops to dry before incubating the agar plates at 37 °C overnight.
10. The next day, enumerate the GAS CFU. The percentage of adherence is calculated as follows:

$$\frac{\text{CFU of experiment well}}{\text{CFU of GAS inoculum}} \times 100$$

11. To count the actual cells used in the assay, remove the medium from the three wells set up for this purpose (Subheading 3.3, **step 4**). To wash, add 100 μl of PBS to each well and aspirate the PBS. Add 50 μl of trypsin/EDTA solution to each well. Incubate for 5–10 min at 37 °C or until cells appear rounded under a microscope. Dislodge the cells by using a pipette tip to scrape the bottom of the wells. Count the cells with an hemocytometer.
12. In order to minimize inaccuracy raised from the variation of confluency, the adherence percentage can be further divided by the MOI determined from the actual inoculum and cell numbers counted during the assay to give an adherence index number.

4 Notes

1. DMEM for these endothelial cell lines should contain L-glutamine to facilitate attachment to the cell culture flasks or 96-well plates. L-Glutamine spontaneously degrades in aqueous solution; hence, this medium should be prepared fresh and stored at 4 °C in the dark for up to a month. A more stable form of L-glutamine is the dipeptide L-alanine-L-glutamine that is also contained in certain commercially available DMEMs.
2. The fetal bovine serum does not need to be heat-inactivated. In our laboratory, we purchase serum from a single lot number to reduce variations in the experiments. In order to minimize freeze-thaw cycles of the serum, we aliquot the serum into 50 ml aliquots and store at −20 °C for long-term storage. Once thawed, the serum is stored at 4 °C for up to a month.

3. For general cell maintenance, antibiotics can be added to the medium. For setting up assay plates described in this chapter, it is important that the medium is free of antibiotics, as this will kill the bacteria when included.
4. Trypsin/EDTA activity is inhibited by the presence of cations and serum. Hence, the PBS used should be free from cations. Suitable alternatives can be used to dislodge the cells from the wells. After treatment with trypsin/EDTA solution, always observe cells under a microscope. Detached cells appear rounded. Incubating cells with too high a trypsin concentration for a long period can damage cell membranes and kill the cells.
5. Plasmids to generate bioluminescent strains can be obtained from Addgene (https://www.addgene.org). Refer to Chapter 6 on how to generate the strains.
6. We have found that the bacteria grow well in fresh BHI medium stored on the bench for no longer than 2 weeks. For every experiment, streak bacteria from a glycerol stock on blood agar plates to confirm hemolysis and inoculate single colony from a plate in BHI medium. Plates with GAS growing on it can be kept at 4 °C for up to a month.
7. The seeding densities would be different for different cell lines. The seeding densities provided for HaCaT and Detroit 562 are expected to generate confluent monolayers within 24 h.
8. The expression of GAS virulence factors or adhesins is variable at different growth phases/optical density. Determine the optimal growth phase/optical density at which expression of the virulence factor or adhesin of interest is maximal. In addition, different GAS strains give different CFUs at the same O.D.; hence, steps in Subheading 3.2 have to be done on each GAS strain analyzed.
9. Dispense the bacterial culture on the walls of the tube. Do not immerse the tip in the diluent (HBSS/1% FBS). Change tips between each dilution. A series of six dilutions should cover the expected range, that is, a dilution of 10^6 of the overnight culture.
10. The MOI (multiplicity of infection) has been tested previously in our laboratory and a MOI of 10 was found to be optimal. You can test a range of MOIs (usually 10:1 up to 1000:1 can be tested) as required.
11. The host-cell adhesion assay can be extended to study the effect of antibodies on the binding of GAS to the cell monolayer. In the case of a neutralizing effect, a reduced percentage of adherence will be observed. Some GAS strains have surface proteins with high binding capacity for immunoglobulins in a

nonspecific manner and inhibition of adherence caused by this can mask specific antibodies. To incorporate this into the background activity, bacteria can be preincubated with 10 μg/ml of naïve immunoglobulin G (IgG) at 37 °C for 30 min, prior to the addition of specific test antibodies. Specific antibodies to GAS virulence factors raised in rabbits, mouse, etc., are added at the same time when bacteria are added to the monolayer cells at a desired concentration (titration is recommended). Likewise, a range of competitive inhibitors (e.g., recombinant proteins) can be added as an alternative. Resuspension volumes of bacteria need to be adjusted accordingly, so that the total volume of bacteria + antibody added to each well is 100 μl.

12. These triplicate wells are for background bioluminescent reading.

Acknowledgments

This work was supported by the Health Research Council (HRC) New Zealand and the Maurice Wilkins Centre for Molecular Biodiscovery. J.M.S.L. is a Heart Foundation NZ research fellow.

References

1. Brouwer S, Barnett TC, Rivera-Hernandez T et al (2016) Streptococcus pyogenes adhesion and colonization. FEBS Lett 590 (21):3739–3757. https://doi.org/10.1002/1873-3468.12254
2. Walker MJ, Barnett TC, McArthur JD et al (2014) Disease manifestations and pathogenic mechanisms of group A Streptococcus. Clin Microbiol Rev 27(2):264–301. https://doi.org/10.1128/CMR.00101-13
3. Ryan PA, Juncosa B (2016) Group A streptococcal adherence. In: Ferretti JJ, Stevens DL, Fischetti VA (eds) Streptococcus pyogenes: basic biology to clinical manifestations. University of Oklahoma Health Sciences Center, Oklahoma City (OK)
4. Courtney HS, Hasty DL, Dale JB (2002) Molecular mechanisms of adhesion, colonization, and invasion of group a streptococci. Ann Med 34(2):77–87
5. Loh JM, Tsai JC, Proft T (2017) The ability of group a streptococcus to adhere to immortalised human skin versus throat cell lines does not reflect their predicted tissue tropism. Clin Microbiol Infect 23(9):677. https://doi.org/10.1016/j.cmi.2017.03.011
6. Loh JMS, Lorenz N, Tsai CJY et al (2017) Mucosal vaccination with pili from group a streptococcus expressed on Lactococcus lactis generates protective immune responses. Sci Rep 7(1):7174. https://doi.org/10.1038/s41598-017-07602-0
7. Tsai JC, Loh JM, Clow F et al (2016) The group a streptococcus serotype M2 pilus plays a role in host cell adhesion and immune evasion. Mol Microbiol 103(2):282–298. https://doi.org/10.1111/mmi.13556

Chapter 21

The Use of *Galleria mellonella* (Wax Moth) as an Infection Model for Group A Streptococcus

Catherine (Jia-Yun) Tsai, Jacelyn M. S. Loh, and Thomas Proft

Abstract

Recently, the use of *Galleria mellonella* larvae as a nonmammalian model to simulate bacterial infectious diseases has shown to be a rapid, simple, and cost-effective alternative. The insect's innate immune response is remarkably similar to that of the vertebrates, and consists of both the cellular and the humoral immune response. Here, we provide a protocol for using *G. mellonella* larvae to study virulence of GAS, including the use of a health score system for quantitative analysis and the methods for assessing post-infection bacterial burden in vivo.

Key words Group A Streptococcus, *Streptococcus pyogenes*, Wax worms, *Galleria mellonella* larvae, Infection model

1 Introduction

Modeling disease in animals provides evident understanding for pathogenicity and virulence mechanisms in the study of infectious disease. However, as GAS is a human exclusive pathogen, it has always been problematic to establish reliable animal infection models for in vivo analyses. Rodent and primate animal models have been important tools for investigating molecular and cellular events during GAS pathogenesis, as well as for dissecting the complex interplay between host and pathogen. Nonetheless, increasing costs and ethical requirements involved in mammalian animal experiments further hinder the development of these models to become a robust platform.

Recently, the larvae of the wax moth (also known as honeycomb moth) *Galleria mellonella*, or colloquially known as the wax worm, has gained attention and interest in its use as an infection model for pathogenic bacteria and fungi. The wax worm shares several advantages with the classic nonmammalian model animals such as the nematode *Caenorhabditis elegans* and the fruit fly *Drosophila melanogaster*, including being cost-effective, easy to

Thomas Proft and Jacelyn M. S. Loh (eds.), *Group A Streptococcus: Methods and Protocols*, Methods in Molecular Biology, vol. 2136, https://doi.org/10.1007/978-1-0716-0467-0_21,

obtain and maintain, more amenable, and more ethically acceptable than mammalian models. In addition, the innate immune system of wax worms is functionally and structurally similar to mammals, and contains both humoral as well as cellular immunity [1, 2]. The insect's humoral defense mechanisms include melanization, hemolymph coagulation, and antimicrobial peptide production. Furthermore, the wax worm hemocytes can perform cellular defenses akin to phagocytosis, nodulization, and large-scale encapsulation found in human system [3]. On top of these features that make the wax worms particularly useful for studying human pathogens, the key advantage of the wax worm over *C. elegans* and *D. melanogaster* is that it can survive at 37 °C. This is highly advantageous, as many human pathogens undergo significant transcriptomic changes at this human body temperature [4].

Increasing numbers of research labs have utilized the wax worm model for studying bacterial diseases and testing antimicrobial drugs [5]. Recently a few studies have utilized wax worms to model infections of multiple GAS strains [6, 7]. These studies showed that the virulence demonstrated in wax worms could be correlated, at least to some degree, with both molecular virulence characteristics and pathogenicity shown in mammalian models. Loh et al. validated the use of wax worm model for studying individual virulence factors, in this case the well-established M protein and pilus produced by the M1 serotype GAS [6]. Since then, a number of researchers have employed the wax worm model to investigate putative virulence factors or virulence modulators, such as the pilus produced by the M2 serotype [8], nuclease SpnA [9], 5′-nucleotidase S5nA [10], and L-lactate dehydrogenase LDH [11]. Furthermore, Afshar et al. established a disease model using *G. mellonella* to study an epidemic lineage of invasive emm3 GAS [12].

When used as an infection model in the context of bacterial pathogens, the last instar larvae about 5 weeks post hatching are usually ideal. Healthy larvae of the size between 1.5 and 2.5 cm and presenting with a cream color and no markings should be chosen. The most efficient infection route is through intrahemocoelic injection via the lower left proleg or through the skin. Apart from using survival rate of the infected larvae to evaluate the severity of infection, a health index scoring system summarized in Table 1 has been developed by Loh et al. for quantifiable assessment of the infection outcome [6]. This scoring system takes into account four major observations to evaluate the health and viability of the larvae: survival, mobility, cocoon formation, and melanization. An example of the appearances of larvae with different health scores is shown in Fig. 1. Post-infection bacterial burden can also be quantified by homogenizing the larvae, plating the homogenate on selective agar and enumerating the recovered colonies.

Table 1
Wax worm health index scoring system

Category	Description	Score
Activity	Active without stimulation	3
	Active when stimulated	2
	Minimal activity upon stimulation	1
	None	0
Cocoon formation	Full cocoon	1
	Partial cocoon	0.5
	None	0
Melanization	None	4
	<3 spots on the tail/no dotted line	3
	≥3 spots on beige wax worm	2
	Brown or complete melanization (black)	0
Survival	Alive	2
	Dead	0

Table adapted from [5] and reproduced with permission from Taylor & Francis under the Creative Commons Public Licence

2 Materials

All materials should be prepared under sterile condition, and be handled with aseptic techniques. Media and buffers should be made up with ultrapure water and sterilized by autoclaving or 0.22-μm membrane filtration.

2.1 Bacterial Culture

1. THY: Todd Hewitt broth supplemented with 0.2% yeast. Brain Heart Infusion (BHI) broth can also be used.
2. 1.5-ml sterile plastic tubes.
3. Spectrometer.
4. Selected Group A Streptococcus strains.

2.2 *G. mellonella* Larvae

1. Antibiotic-free *G. mellonella* larvae from a preferred supplier (*see* **Note 1**).

Choose larvae of weight between 100 and 150 mg, and of length between 15 and 25 mm at time of infection. Ideally, at least 10 larvae per inoculation group should be prepared.

2.3 Materials for Injection and Monitoring

1. Phosphate-buffered saline (PBS): 2.7 mM KCl, 150 mM NaCl, 10 mM phosphate salts pH 7.4.
2. Petri dishes (9-cm diameter).
3. 29G Microfine insulin syringe.

2.4 Materials for Assessing Bacteria Burden by Homogenization

1. Phosphate-buffered saline (PBS) *see* Subheading 2.3, **item 1**.
2. 1.5-ml sterile plastic tubes.
3. Conical homogenizer or micro pestle.
4. Columbia horse blood agar plates.

3 Methods

All procedures should be undertaken according to the containment regulations applicable to the laboratory. Used materials and *G. mellonella* larvae should be decontaminated using appropriate methods (such as 10% bleach of commercial disinfectant), and be disposed of according to the biohazard waste regulations.

3.1 Preparing Bacterial Suspension for Injection

1. Grow 1.5 ml GAS strains in THY or BHI at 37 °C under static conditions until the culture reaches mid-log phase (judged by an optical density at 600 nm [OD_{600nm}] of 0.4–0.6 (*see* **Note 2**).
2. Spin down the bacteria at 5000 × *g* for 10 min and remove the culture media.
3. Wash bacterial pellet once with phosphate-buffered saline (PBS), then re-suspend the bacteria in appropriate volume of PBS to obtain desired concentration (*see* **Note 2**). For consistency across repetitions of experiments, frozen aliquots can be prepared (*see* **Note 3**).

3.2 Preparation and Infection of G. mellonella Larvae

1. Maintain the larvae at controlled room temperature (~25 °C or 77 °F) in the dark with food, and use within 2 weeks. Before the experiments, inspect the larvae and select healthy-looking insects of similar size. When necessary, use forceps to remove larvae from their cocoons. Groups of larvae can be kept in 9-cm petri dishes. Employ containment measures to prevent the larvae from escaping, such as using tapes to secure the lids of the petri dish, or keeping the petri dishes in a plastic container with a lid (but not airtight).
2. Hold one larva between the index finger and the thumb. Arch the larva toward its back and secure the position with fingers. Inject 20 μl of the bacterial inoculum (*see* **Note 2**) to the lower left proleg (the second to last pair of legs) of a *G. mellonella* larva using a 29G microfine insulin syringe. Alternatively, a restraint device can be used to secure the larvae during injection, to minimize the risk of accidental needle stick injury [13].
3. Confirm the infecting dose by preparing drop plates of serial dilutions for colony enumeration. Perform a serial dilution (for example, 1:10) of the inoculum until the expected concentration ranging between 1×10^4 and 1×10^5 CFU/ml is reached.

4. In duplicate or triplicate, drop 10 μl of the diluted bacterial suspension (of the target concentration, plus one dilution above and one dilution below) on an agar plate.
5. Incubate at 37 °C overnight and count the bacterial colonies.

3.3 Monitoring of G. mellonella Larvae

1. Incubate each group of 10 larvae at 37 °C in a lidded 9-cm petri dish without food (*see* **Note 4**) and monitor daily for up to 5 days post-infection.
2. Assess responses to the infection by survival analysis.
3. Alternatively, a health index scoring system, which comprises four categories of phenotypes related to overall health (Table 1), can be used to record the health status of each infected wax worm. A healthier wax worm demonstrates higher activity and cocoon formation, while a less healthy animal would be more sedentary and produce melanin as an immune response against infection. If the larvae are enclosed in a cocoon, it is necessary to remove the cocoon completely before assessing the level of activity. The scoring system correlates higher scores with healthier phenotype, thus a healthy, uninfected wax worm typically scores between 9 and 10, while a seriously infected, dead *G. mellonella* larvae would have a score of 0.

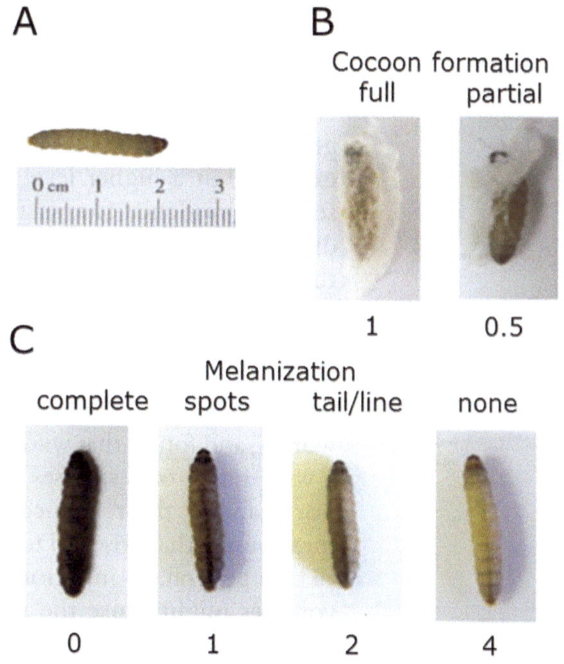

Fig. 1 Photographic images of wax worm larvae. (**a**) Healthy last instar larvae with a cream color and size of 1.5–2.5 cm. (**b** and **c**) Image examples of different levels of cocoon formation and melanization, respectively, with corresponding health index scores below each image. (Figure reproduced from [5] with permission from Taylor & Francis under the Creative Commons Public Licence)

3.4 Assessment of Post-infection Bacterial Burden

1. Submerge each wax worm in 0.5 ml PBS and homogenize by mechanical disruption using a small pestle and mortar, such as a micro pestle or a glass conical homogenizer.
2. Prepare serial dilutions of the homogenate in PBS and plate onto horse blood agar or BHI agar supplemented with antibiotics (where applicable). With several recently developed reporter plasmids for labelling GAS strains, kinetics of bacterial growth in vivo and endpoint bacterial burden can also be assessed without the need for homogenization (*see* **Note 5**).

4 Notes

1. The breeding and growing conditions of different *G. mellonella* larvae suppliers may cause inconsistencies in experiment results and difficulties in obtaining reproducible results. Use of antibiotics, growth temperature, natural flora, exposure to hormones or genetic makeup may contribute to these variations. Therefore, it is important to source larvae from a supplier with standardized procedures in breeding and housing the insects. Commercial larvae intended as feed for reptiles and birds and as a fishing bait should be avoided. An example of the source for standardized *G. mellonella* larvae is the TruLarv breeding line produced by Biosystems Technology Ltd. (www.biosystemstechnology.com).
2. Infectious dose and growth phase of the GAS inoculum play a critical role in the infection outcome. The expression of many virulence factors in GAS is tightly regulated during different growth phase. In general, cell wall-associated factors are expressed at a higher level during early growth stages, while secreted virulence factors tend to be have higher expression during later growth stages. If genetically modified GAS is used, the induction and expression profile of the introduced gene (or complemented gene in a knockout strain) should also be considered when choosing the growth phase of the bacterial inoculum. As for the infectious dose, Loh et al. [4] examined the virulence of 7 different M types of GAS in infecting wax worms, and found that wild-type GAS of different M types gave a wide range of responses in the wax worm infection model. For example, the reference M1 strain (SF370) exhibits a half maximum lethal dose (LD_{50}) of 6×10^6 at day 1 post-infection [4]. In general, an injection dose of 10–20 μl is ideal, as larger volumes might cause the risk of leaking or imbalance of the larvae's body fluids.
3. Frozen GAS stock can be prepared to ensure that the dose and growth phase of GAS inoculums across different experiments are consistent. Culture GAS strains in BHI or THY at 37 °C

under static conditions until reaching desired growth phase. Spin down the bacteria at 5000 × *g* and remove the growth media. Resuspend bacterial pellet in PBS containing 10% glycerol at an OD_{600nm} equivalent to 10. Aliquots of 1 ml should be stored at −80 °C until required. It is recommended to perform a viability test before the actual experiment, by slowly thawing one aliquot on ice and preparing drop plates of serial dilutions for colony enumeration. To start the experiment, thaw one aliquot of the frozen bacterial stock on ice, and centrifuge 5000 × *g* for 5 min to remove the PBS/glycerol solution. Wash the pellets once with 1 ml PBS by centrifuging at 5000 × *g* for 5 min. Dilute the bacterial suspension to required concentration with PBS.

4. There is a need to conduct standardized *G. mellonella* laboratory handling that strictly determines the testing conditions. It was observed by Pelega et al. [14] that incubation temperature after infection of *G. mellonella* larvae by *Acinetobacter baumannii* (ATCC 17978) affected the survival of the larvae. On the other hand, Benville et al. showed that nutritional deprivation of the infected *G. mellonella* larvae led to suppression of immune response and increased susceptibility to infection [15]. Therefore, starving the larvae post-infection may favor the infection progress, and facilitate the observation of signs of illness in the larvae.

5. An ideal reporter labelling system should provide strong signals that correlate linearly with viable bacterial cell counts. Loh et al. reported a novel strategy using a toxin-antitoxin-stabilized plasmid to generate a wide range of bioluminescent GAS strains [16] (*See* Chapter 6). Post-infection bacterial burden can be easily assessed by biophotonic imaging (BPI), using light intensity (in bioluminescence unit) for quantification. Figure 2 shows an example of *G. mellonella* larvae infected with bioluminescence-labelled M2 GAS strain (MG10270).

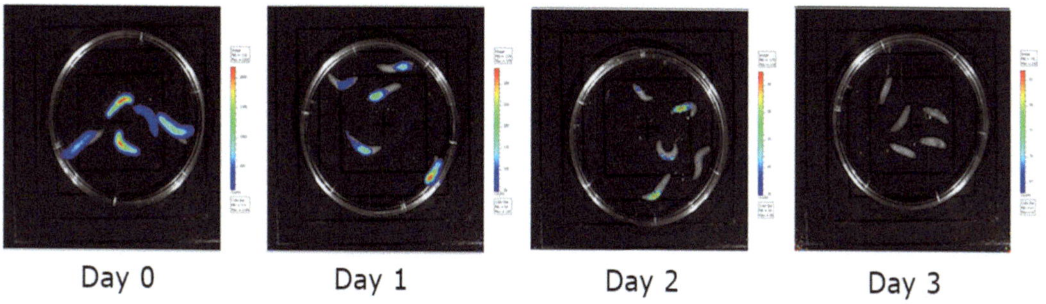

Fig. 2 A group of 5 wax worms were inoculated with 7 × 10^5 CFUs of GAS M2 strain MG10270. In vivo live bacteria were detected using an in vivo imaging system (IVIS Spectrum, Perkin Elmer). Exposure times were extended each day as required for the detection of the signal (Day 0 = 0.5 s, Day 1 = 2 s, Day 2 = 1 min, and Day 3 = 5 min). Complete clearance of live GAS was observed on Day 3 post-infection

References

1. Hoffmann JA (1995) Innate immunity of insects. Curr Opin Immunol 7(1):4–10
2. Lionakis MS (2011) Drosophila and Galleria insect model hosts: new tools for the study of fungal virulence, pharmacology and immunology. Virulence 2(6):521–527
3. Kavanagh K, Reeves EP (2004) Exploiting the potential of insects for in vivo pathogenicity testing of microbial pathogens. FEMS Microbiol Rev 28(1):101–112
4. Smoot LM, Smoot JC, Graham MR et al (2001) Global differential gene expression in response to growth temperature alteration in group A Streptococcus. Proc Natl Acad Sci U S A 98(18):10416–10421
5. Tsai CJ, Loh JM, Proft T (2016) Galleria mellonella infection models for the study of bacterial diseases and for antimicrobial drug testing. Virulence 7(3):214–229
6. Loh JMS, Adenwalla N, Wiles S, Proft T (2013) Galleria mellonella larvae as an infection model for group A streptococcus. Virulence 4(5):419–428. https://doi.org/10.4161/viru.24930
7. Olsen RJ, Watkins ME, Cantu CC et al (2011) Virulence of serotype M3 Group A Streptococcus strains in wax worms (Galleria mellonella-larvae). Virulence 2(2):111–119. https://doi.org/10.4161/viru.2.2.14338
8. Tsai JC, Loh JM, Clow F et al (2017) The Group A Streptococcus serotype M2 pilus plays a role in host cell adhesion and immune evasion. Mol Microbiol 103(2):282–298. https://doi.org/10.1111/mmi.13556
9. Chalmers CKA, Sohn CR, Loh JMS et al (2017) Streptococcus pyogenes nuclease A (SpnA) mediated virulence does not exclusively depend on nuclease activity. J Microbiol Immunol Infect pii: S1684-1182(17)30236-0. https://doi.org/10.1016/j.jmii.2017.09.006. [Epub ahead of print]
10. Dangel MLDJ, Haßelbarth S, Krogull M et al (2019) The 5′-nucleotidase S5nA is dispensable for evasion of phagocytosis and biofilm formation in Streptococcus pyogenes. PLoS One 14(1):e0211074
11. Oehmcke-Hecht S, Nass LE, Wichura JB et al (2017) Deletion of the L-lactate dehydrogenase gene ldh in Streptococcus pyogenes leads to a loss of SpeB activity and a hypovirulent phenotype. Front Microbiol 8:1841. https://doi.org/10.3389/fmicb.2017.01841
12. Afshar B, Turner CE, Lamagni TL et al (2017) Enhanced nasopharyngeal infection and shedding associated with an epidemic lineage of emm3 group A Streptococcus. Virulence 8 (7):1390–1400. https://doi.org/10.1080/21505594.2017.1325070
13. Dalton JPUB, Swift S, Wiles S (2017) A novel restraint device for injection of Galleria mellonella larvae that minimizes the risk of accidental operator needle stick injury. Front Cell Infect Microbiol 7(99) https://doi.org/10.3389/fcimb.2017.00099 [online]
14. Peleg AY, Jara S, Monga D et al (2009) Galleria mellonella as a model system to study acinetobacter baumannii pathogenesis and therapeutics. Antimicrob Agents Chemother 53 (6):2605–2609. https://doi.org/10.1128/aac.01533-08
15. Banville N, Fallon J, McLoughlin K, Kavanagh K (2011) Disruption of haemocyte function by exposure to cytochalasin b or nocodazole increases the susceptibility of Galleria mellonella larvae to infection. Microbes Infect 13 (14–15):1191–1198. https://doi.org/10.1016/j.micinf.2011.07.001
16. Loh JM, Proft T (2013) Toxin-antitoxin-stabilized reporter plasmids for biophotonic imaging of Group A streptococcus. Appl Microbiol Biotechnol 97(22):9737–9745. https://doi.org/10.1007/s00253-013-5200-7

Chapter 22

A Superficial Skin Scarification Method in Mice to Mimic *Streptococcus pyogenes* Skin Infection in Humans

Manisha Pandey and Michael F. Good

Abstract

Skin infections caused by *Streptococcus pyogenes* are a significant health concern in the tropics and among Indigenous populations of developed countries. To study the immunobiology of skin infections and to develop preventative and therapeutic measures to target these infections, an in vivo model is vital. We developed a mouse model to investigate immunity to skin infections and for testing the efficacy of several vaccine candidates. This article describes the method of induction of superficial skin infections, their progression and regression, and processing of samples to obtain critical readouts and analyses. This model could be valuable to inform clinical studies.

Key words *Streptococcus pyogenes*, Pyoderma, Superficial skin infection, Bacterial burden, Scarification

1 Introduction

Streptococcal skin infection is a serious health burden in the tropics, in developing countries and among Indigenous populations of developed countries. A superficial skin infection (pyoderma or impetigo) is a primary site of infection and a reservoir for transmission. From this primary site, infections can lead to serious invasive disease including septicemia, streptococcal toxic shock syndrome (STSS), and necrotizing fasciitis, all of which can have devastating outcomes [1, 2]. Moreover, pyoderma can lead to post streptococcal glomerulonephritis and has been linked with acute rheumatic fever (ARF) and rheumatic heart disease (RHD) [3–5].

Disease pathogenesis and translational research leading to vaccines and therapeutics cannot be modeled adequately with in vitro or in silico approaches, which do not account for the myriad host defense mechanisms and host tissue complexity that a pathogen will encounter in vivo. Despite some limitations, animal models provide a multifactorial environment to model a human disease. There is a paucity of models that mimic the human situation of *S. pyogenes*

Thomas Proft and Jacelyn M. S. Loh (eds.), *Group A Streptococcus: Methods and Protocols*, Methods in Molecular Biology, vol. 2136, https://doi.org/10.1007/978-1-0716-0467-0_22, © Springer Science+Business Media, LLC, part of Springer Nature 2020

pyoderma. The models that involve either subcutaneous or intradermal injection of *S. pyogenes* involve the needle bypassing the skin layers and not reflecting the superficial aspect of *S. pyogenes* infection where the skin must be broken to allow the bacteria to enter [6, 7]. A humanized mouse (hu-mouse) model developed by engrafting neonatal foreskin onto the flanks of the SCID mouse has also been described [8]. The absence of adaptive immune response in SCID mice limits the utility of this model for testing the efficacy of vaccine or antimicrobials.

The murine superficial skin infection model described here closely mimics the physiological conditions encountered by human hosts under natural conditions. Damage to the stratum corneum is a prominent feature in natural disease and provides *S. pyogenes* access to sub corneal keratinocytes [8]. In our model, infection leads to massive influx of inflammatory infiltrates (Fig. 1), a finding that parallels the pathophysiology of streptococcal skin infection in humans as shown in the humanized mouse (Hu-SCID) model (Fig. 2). Despite innate host defenses, mice develop local skin infection and septicemia which gradually decrease from day 6 onward and which could completely clear by day 9 post infection (Fig. 3a and b). However, clearance by day 9 is dependent on the virulence of the infecting *S. pyogenes* strain. The anatomical differences in the human and mouse skin limit the generation of pyoderma like pathology in our model. This limitation has previously been reported for subcutaneous infection model [9]. In addition, the lack of human-specific targets for virulence factors such as streptokinase and superantigens in regular mice may also restrict the assessment of infection outcome in human-like situations [9]. To overcome the issue of superantigen non-responsiveness in regular mice, we have expanded our superficial infection model to include HLA humanized mice. Following superficial skin infection with a lethal clinical isolate, the HLA humanized mice developed septicemia and streptococcal toxic shock syndrome (STSS) like pathology with detectable levels of toxin in their blood [10].

In summary, the strengths of this model, which include its simplicity, adaptability (to various mouse strains), and high throughput outweigh the limitations listed above and therefore advocate its tremendous utility. This superficial skin infection model should be applicable to the evaluation of novel antimicrobials and vaccine candidates in the treatment and prevention of superficial infections caused by *S. pyogenes.*

2 Materials

2.1 Bacterial Preparation and Mice Procedures

1. *S. pyogenes* glycerol stocks (stored in −80 °C freezer).
2. 4- to 6-week-old female BALB/c mice or 4- to 6-week- old male and/or female HLA-B6 mice.

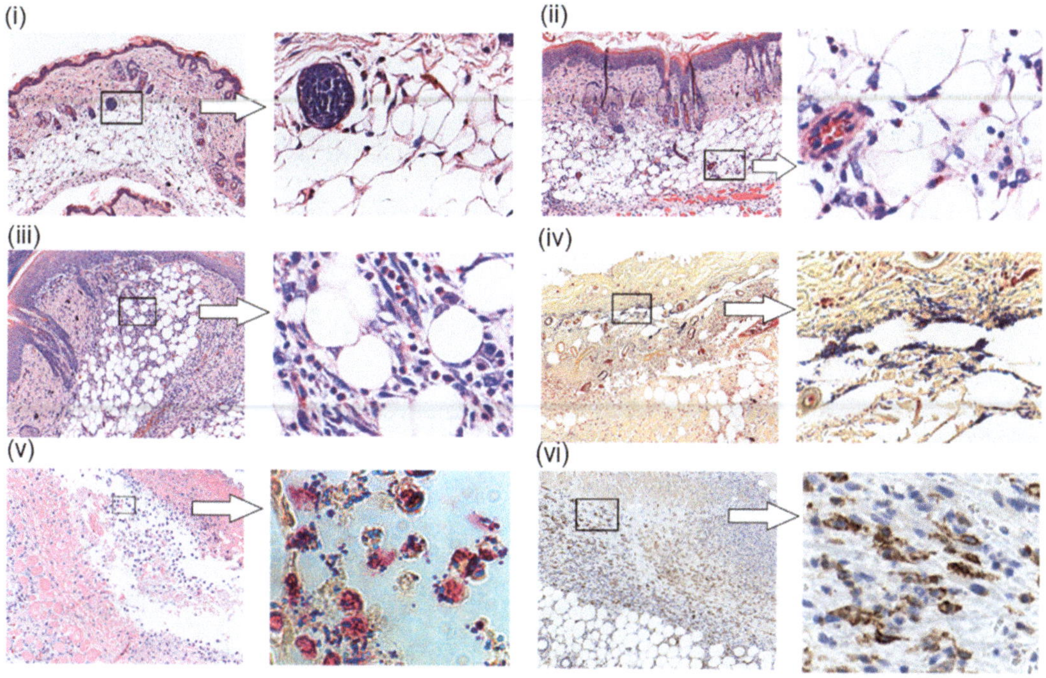

Fig. 1 Development of superficial skin infection model for GAS: BALB/c mice were infected with *S. pyogenes* M1 using a newly established method of superficial skin infection. Histopathology of skin tissue using the superficial skin infection model is illustrated. Skin biopsy samples from naive (i), naïve-scarified-uninfected (ii), and naïve-scarified-GAS infected (iii) mice were collected on day 3 post infection. The samples were fixed in 1% *para*-formaldehyde and embedded in paraffin or frozen in OCT for immunohistochemistry. The samples were stained with Giemsa. Post infection, the inflammatory cell infiltrates consisting of mononuclear cells and neutrophils are shown (iii). A Gram-stained section of skin biopsy sample from GAS-infected mice demonstrating the presence of GAS in dermal layer (iv). A 5× magnification of skin section demonstrating neutrophils engulfing GAS and undergoing the process of degranulation (v) and a 100× magnification of inset is shown (v). A F4/80 stained skin section demonstrating the ingress of macrophages in naïve-scarified-GAS-infected mice (vi). Original magnification is 5×. The 50× magnification of inset is shown next to each image and 100× for inset in image (v) at the arrowhead. (Figure reproduced from [14] with permission from the American Association of Immunologists, Inc)

3. Syringes (1 mL and 5 mL) and needles (26.5 gauge).
4. Anesthetic: Ketamine: Xylazyl: water-1:1:10. 2 mL of Ketamine (100 mg/mL) and 2 mL of Xylaxil-20 (20 mg/mL); make up to 20 mL with double-distilled water (DDW) (*see* **Note 1**).
5. Clipper (Wahl).
6. Ethanol (80%): Add 200 mL of DDW to 800 mL of Ethanol.
7. Sterile surgical equipment (pairs of scissors, tweezers, and metal file for skin scarification).
8. Band-aid (plastic).

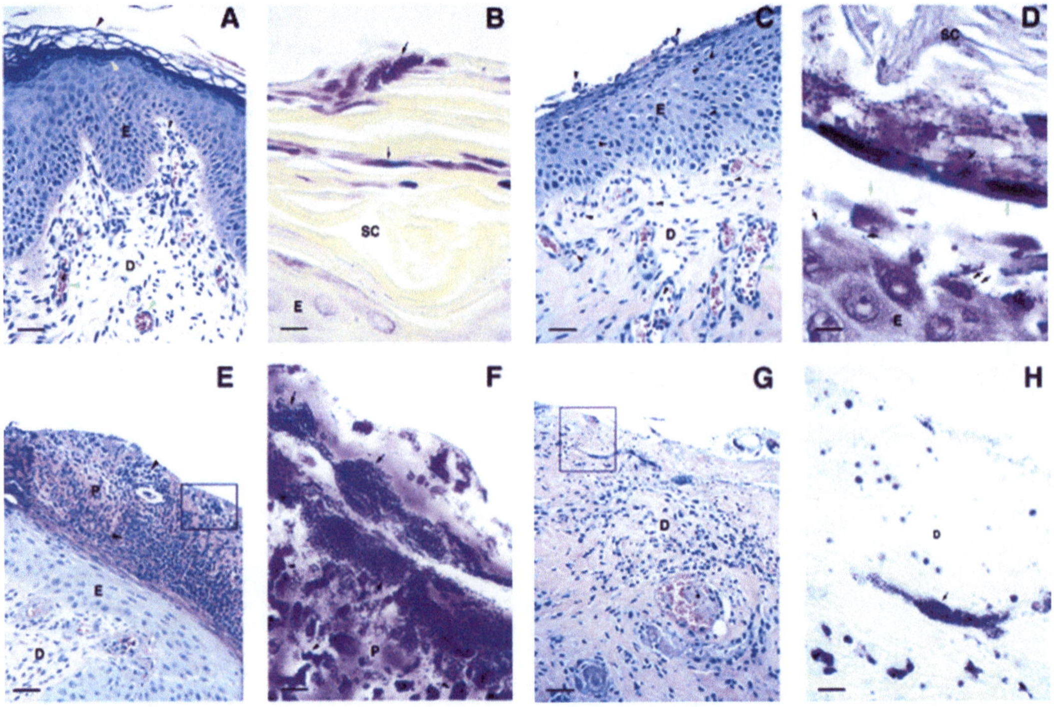

Fig. 2 Histopathology of human skin grafts. Photomicrographs of formalin-fixed tissue sections of human skin grafts inoculated with strain D471 (**a** and **b**) or 29,487 (**c** through **h**) and biopsied at 48 h (**a** through **d**), 96 h (**e** and **f**), or 7 days (**g** and **h**) are shown. Tissue sections are stained with hematoxylin-eosin (**a**, **c**, **e**, and **g**) or tissue Gram stain (**b**, **d**, **f**, and **h**). Panels (**a** and **b**) are from the same graft. The tissue section in panel (**f**) is adjacent to that shown in panel (**e**) and corresponds to the boxed area of panel (**e**). The tissue section in panel **h** is adjacent to that shown in panel (**g**) and corresponds to the boxed area of panel (**g**). Highlighted features are PMNs (black arrowheads), dermal blood vessels (green arrowheads), stratum corneum (red arrowhead), healthy granular keratinocyte layer (yellow arrowhead), Gram-positive cocci (black arrows), parakeratosis (red arrow), damaged keratinocytes (green arrows), dermal ulcer (green asterisk), thrombotic vessel (black asterisk), cell nuclei (blue arrowheads). *SC* stratum corneum, *E* epidermis, *D* dermis, *P* purulent exudate. Bars, 35 μm (**a**, **e**, and **g**), 28 μm (**c**), 7 μm (**b**, **f**, and **h**), and 5.6 μm (**d**). (Figure reproduced from [8] with permission from the American Society for Microbiology)

9. Beads: 0.5 mm zirconium oxide beads, 0.9–2 mm stainless steel blend and 3.2 mm stainless steel balls, Next Advance, USA.
10. Safe lock microfuge tubes, 1.7 mL.
11. Bullet blender tissue homogenizer (Next Advance, USA).
12. Light box.

2.2 Media and Solutions

1. THBYN: THB with 1% yeast extract and 1% Neopeptone. Dissolve 14.6 g THB, 4 g Neopeptone, and 4 g yeast in 400 mL deionized distilled water (DDW). Autoclave the solution (121 °C for 20 min, keep sterile at all times).
2. THBYN with 15% glycerol: add 15 mL glycerol to 85 mL THBYN.

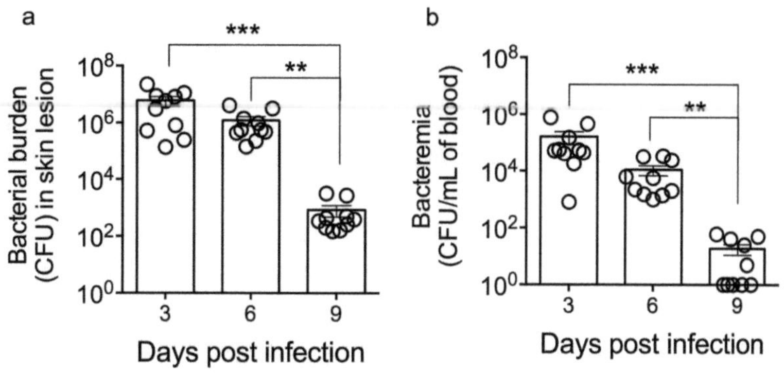

Fig. 3 Bacterial burden assessment in mice infected via superficial scarification method: BALB/c mice were infected with *S. pyogenes* using superficial scarification method of skin infection. Mice were infected via the skin route and on days 3,6, and 9 post-infection, 10 mice per group were culled and skin and blood samples collected. The samples were processed, serially diluted, and bacterial-burden was reported as the number of CFU in skin lesions (**a**) or CFU/mL of blood (**b**)

3. Columbia Blood Agar (CBA): Dissolve 15.6 g CBA in 400 mL DDW. This solution will be enough for 20 plates. Autoclave the solution (121 °C for 20 min), keep sterile at all times. Allow it to cool down to 40–50 °C and add 8 mL of defibrinated horse blood. Mix by gently inverting the schott bottle. Take extreme care not to generate air bubbles during mixing, as they will cause obstruction during CFU enumeration.
4. Streptococcal Selective Medium: Dissolve the contents of one vial of streptococcal selective supplement (Oxoid, UK) in 2 mL of 1:1 mix of ethanol: sterile distilled water following the manufacturer's instructions. Add the content of the vial to 500 mL of CBA + 2% horse blood at 50 °C (prepared as above) and mix well.
5. DPBS (Dulbecco's phosphate-buffered saline).
6. DPBS with 10 mM EDTA: Add 2 mL of EDTA (0.5 M, pH 8.0) to 98 mL of DPBS.

3 Methods

3.1 Bacterial Inoculum

1. Retrieve *S. pyogenes* glycerol stocks from the −80 °C freezer.
2. Take a loopful of frozen culture and perform a 16-streak on CBA plates and grow for 18 h at 37 °C.
3. Confirm the homogeneity of the culture by visual appearance and pick a few (3–4) single colonies with a loop and inoculate 5 mL of THBYN broth. Grow at 37 °C for 18 h.

4. Following overnight incubation, centrifuge the bacterial culture at 1800 × *g* for 10 min. Remove the supernatant and resuspend the pellet in 10 mL of cold THBYN. Following a second round of centrifugation at 1800 × *g* for 10 min resuspend the pellet in 5 mL of cold THBYN.
5. Take 1 mL of bacterial suspension and perform tenfold serial dilutions (10^{-1} to 10^{-6}) in THBYN. Perform pour plating to count the numbers of CFU in the bacterial suspension.
6. For pour plating, place 50 μL inoculum of each dilution in the center of a sterile petri dish. Pour molten cooled CBA (approx. 15 mL) into the petri dish containing the inoculum and mix well with gentle swirling (up and down, left to right, clock-wise, anti-clock wise, ten times each) of blood agar (*see* **Note 2**).
7. Allow the plates to cool down in the laminar flow hood. Once the CBA is solidified, cover the plates, invert them, and transfer the inverted plates to a 37 °C incubator for 18 h.
8. After 18 h of incubation, remove the plates and count the individual colony-forming units (CFUs) on each plate.
9. While counting, ensure that all the colonies demonstrate the same phenotype. The zone of beta-hemolysis (characteristic of *S. pyogenes*) should be easy to identify (*see* **Note 3**).
10. Calculate the bacterial concentration (CFU/mL) in the original bacterial suspension using the following formula: number of CFU on plate X reciprocal of dilution of sample X fraction of 1 mL of sample plated = number of bacteria/mL. For example, if 38 CFU are on the plate with 50 μL of 10^{-4} dilution, then the count is $38 \times 10{,}000 \times 20 = 7.6 \times 10^6$ CFU/mL of original bacterial suspension. This is the ***neat stock***.
11. Dilute the original bacterial suspension (stored at 4 °C) in fresh THBN with 15% glycerol to get to the desired concentration.
12. Prepare ***working stock*** using the calculation: $C_1V_1 = C_2V_2$, *where* C_1 = *bacterial concentration in neat stock;* V_1 = *the volume required from the neat stock (unknown),*
    C_2 = *aimed concentration for the working stock;* V_2 = *final volume of resuspended working stock.*
13. Aliquot working stock into multiple cryovials (0.5–1 mL volumes) and store at −80 °C (*see* **Note 4**).

3.2 Animal Adaptation of S. pyogenes Isolates

S. pyogenes is a human-specific pathogen and hence passaging thorough animals is a prerequisite to enhance its virulence in a murine model. This step is not needed when humanized (e.g., HLA transgenic) mice are used.

1. Inoculate 2 × 8 mL of sterile THBYN with 100 μL of bacterial glycerol stock (from the −80 °C). Incubate overnight at 37 °C.

2. Centrifuge the combined bacterial cultures at 1800 × *g* for 10 min. Remove the supernatant and resuspend the pellet in 20 mL of cold THBYN. Following a second round of centrifugation (at 1800 × *g* for 10 min), resuspend the pellet in 5 mL of cold THBYN. This is the neat suspension.
3. Take 1 mL of the neat suspension to read absorbance at 600 nm. This will give an estimate of bacterial concentration in the broth. OD_{600} reading of 0.65 equals 10^8 CFU/mL (*see* **Note 5**).
4. Perform tenfold serial dilutions of neat suspension (10^{-1} to 10^{-6}) in THBYN and follow **steps 5–7** as in Subheading 3.1. The CFU counts on the CBA plates will define the concentration of live bacteria in the inoculum.
5. Store plate at 4 °C until animal adaptation is achieved.
6. Intraperitoneally administer 2 mice with 400 μL of the neat bacterial suspension at a concentration ranging between 10^7 and 10^8 CFU/mL using a 1-mL syringe.
7. Monitor mice closely in accordance with the score sheet (refer to Table 1 for clinical scoring details).
8. After 24 h of infection, cull mice via CO_2 asphyxiation. Harvest the spleen aseptically and place in 1 mL of sterile DPBS in a petri dish. Press the spleen gently using the flat end of a sterile 5-mL syringe to obtain a homogenous suspension.
9. Collect 500 μL of suspension with a 1-mL syringe carefully, while avoiding the cell debris. Intraperitoneally administer a naive mouse with 400 μL of the suspension.
10. Perform a 16-streak and tenfold serial dilutions with the remaining spleen suspension (*see* **steps 6–9**) to confirm the bacterial concentration in the injected bacterial suspension.
11. Repeat **steps 7–10** for at least five to six times.

    Save CBA plates after each round of passaging should there be a need to go back to the previous passage of bacteria, due to low or no recovery from the spleen of the subsequently infected mouse.
12. The passaging (animal adaptation) of *S. pyogenes* isolate will be confirmed when 3 mice consecutively show signs of illness (*see* **Note 6**).

3.3 Superficial Skin Infection

For a schematic representation of the methodology, refer to Fig. 4.

3.3.1 Anesthetization of Mice

1. Anesthetize mice via an intraperitoneal injection using 100–150 μL of a ketamine: xylaxyl: water cocktail (*see* **Note 7**).

Table 1
Clinical score determination of mice to assess disease severity

Parameter	Score 0	1	2	3
Coat condition/ Grooming	No ruffling/ Normal grooming	Sight ruffling/ Reduced grooming	Moderate ruffling/ Poor grooming	Highly ruffled/No grooming
Skin lesion/ purulent exudate	Small wound (<2 mm) with no purulent drainage	Medium wound (3–4 mm) with small purulent drainage	Medium wound (3–4 mm) with medium purulent drainage	Large wound (>5 mm) with excessive purulent drainage
Reaction/ Movements	Normal reaction to stimuli/ Normal movement	Reduced reaction to stimuli/Lethargic movement	Slow response to stimuli/Reduced movement	Lack of response to stimuli/Impaired movement
Posture	Normal gait/No hunching	Slight hunching while at rest	Moderate hunching while at rest	Extreme hunching while at rest
Pallor/Eye(s) Morphology	Normal pallor/ Normal eye (s) morphology	Fair pallor/Enlarged eye(s)	Reduced pallor/ Slight discharge from eye(s)	Significant change in pallor/ Discharge around eyes
Weight loss	<5%	5–10%	11-15%	>15%
Other Physiological signs	Normal breathing	Increased breathing	Rapid breathing/ Minor Shivering	Uncontrolled breathing/ Shivering

Clinical score definitions	Clinical score	Status
	0–3	Healthy
	4–6	Mid Disease
	7–10	Moderate Disease
	11–30	Severe

2. Once fully anesthetized, apply the eye cream to prevent drying of mouse eyes while they are under anesthesia and maintain the boy temperature (*see* **Note 8**).
3. Shave the fur at the nape of the neck using a clipper, ensuring a close shave.
4. Disinfect the exposed skin by spraying 80% ethanol while covering their face to make sure that ethanol does not get on the face and especially eyes.

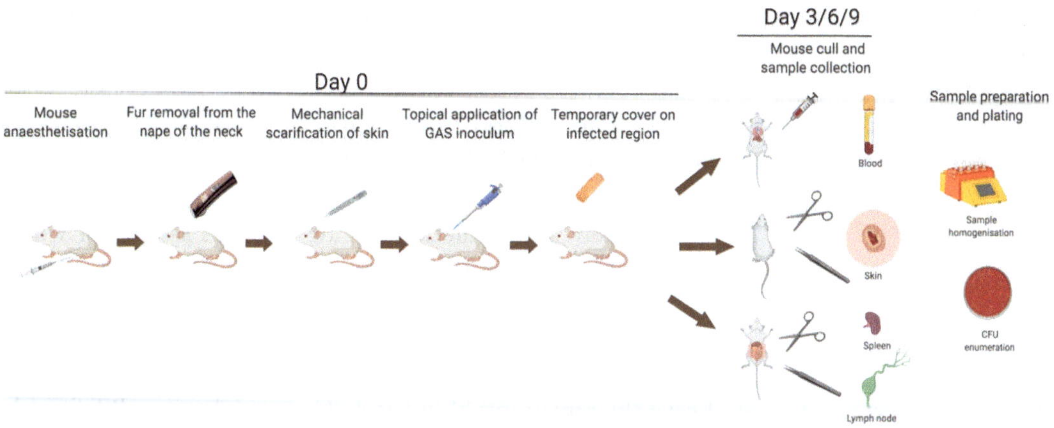

Fig. 4 A schematic summarizing the process of superficial skin infection in mice, sample collection, and processing

3.3.2 Skin Site Scarification and Infection

1. Allow the ethanol to completely evaporate, then lift a small fold of shaved skin away from the body using a pair for fine sterile tweezers.
2. Use the other hand with a second set of tweezers to grip a very small section of folded skin. The first hand with tweezers is then used to secure mouse's head and ears away from the folded skin, to avoid any injury during the scarification process.
3. Create a superficial skin wound by gently running a sterile metal file over the raised skin in the same direction for a total of five times (*see* **Note 9**). This procedure requires two pairs of hands. Only a superficial abrasion of the skin is needed. The scarified area should show a broken skin layer with some redness.
4. Using a P10 and a sterile pipette tip, inoculate the wounded site with 10 μL of bacterial inoculum containing optimal dose for bacterial infection (*see* **Note 10**).
5. Allow the inoculum to absorb completely on the abraded site (*see* **Note 11**). The reddening at of the infection site will become more pronounced.
6. Once the area is visibly dry, apply a temporary cover (band-aid) over the site for protection. A glistening appearance of the infection site is an indication that the inoculum is completely absorbed.
7. House each mouse separately to prevent mouse-to-mouse grooming and transmission of infection.
8. Monitor all mice on a daily basis. A trained investigator who is blinded to the experimental setup should score the mice. The blinded scorer will subjectively score each mouse (refer to Table 1) and record clinical scores.

3.4 Sample Collection and Preparation

1. Sacrifice a designated number of mice from each group on days 3, 6, and 9 post infection via CO_2 asphyxiation for sample collection (*see* **Note 12**).
2. Collect whole blood from each mouse via cardiac punctures immediately after CO_2 asphyxiation, using a 1-mL syringe and 26.5-gauge needle.
3. Immediately after collection, draw a 50-μL sample of whole blood and instantly dilute 1:2 in DPBS containing 10 mM EDTA. These samples will be used for blood bacterial burden determination. Store the diluted blood sample at 4 °C until plating.
4. Leave the remaining blood at 4 °C overnight or at RT for 4 h. Remove blood clots using sterile toothpicks or pipette tips and centrifuge at 845 × *g* for 10 min to separate the serum from the RBCs. Aspirate the serum and store at −20 °C for future use. These samples can be used in ELISA or Western blot to assess antigen-specific antibodies post infection.
5. Excise skin tissue from the site of infection aseptically (measuring approximately 5 mm × 5 mm) and place in 500 μL sterile cold DPBS. Theses samples will be used to assess streptococcal burden at the infection site.
6. Harvest the spleen of infected mice aseptically and store in 500 μL sterile ice-cold PBS. Theses samples will be used to assess streptococcal burden at sterile systemic sites.
7. Harvest lymph nodes (cervical and axillary) to assess the bacterial dissemination into draining lymph nodes. It is important to keep all the tissue samples cold (on ice or in the refrigerator) until the processing and plating are done. This will stop bacteria growing in samples post collection.

3.5 Sample Processing and Bacterial Burden Determination

1. Wash skin sample three times with by replacing fresh DPBS in the tube. Trim the skin samples into thin strips for faster homogenization (*see* **Note 13**).
2. Place the samples into a fresh safe-lock tube containing 300 μL of PBS with small beads (SSB14B—0.9–2 mm stainless steel blend), volume equivalent to 1× the volume of the sample and 2 large silver balls (SSB32—3.2 mm stainless steel ball). Homogenize the samples at a setting for Speed 10 and Time 5 using the bullet blender tissue homogenizer.
3. Follow **steps 1–2** for spleen and lymph node samples.
4. For spleen samples, use 0.5 mm zirconium oxide beads. Use a volume equivalent to 1× the volume of the sample and homogenize at a setting for Speed 10 and Time 3.

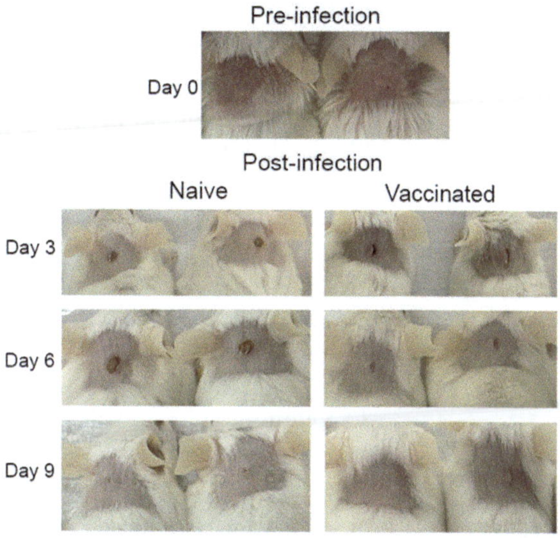

Fig. 5 Gross pathology at the skin site of infection: Images of naïve and vaccinated mice showing gross pathology at the skin site of infection. Mice were vaccinated with J8-DT or left unvaccinated. Two weeks post last immunization, mice were superficially infected with *S. pyogenes.* Designated numbers of mice were culled on days 3, 6, and 9 post *S. pyogenes* infection. Images of mice pre- (day 0) and post infection (days 3, 6, and 9) are shown

5. For lymph node samples, use 0.5 mm zirconium oxide beads and homogenize the samples at a setting for Speed 10 and Time 3 (*see* **Note 14**).
6. Serially dilute skin, spleen, and lymph node homogenates up to 10^{-6} in DPBS while on ice.

   For gross pathology at the skin site of infection at each cull time point (day 3, 6, and 9), refer to Fig. 5. *The figure also demonstrates the differences in the gross pathology between vaccinated and naïve mice.*

3.6 Quantification of Bacterial Burden

1. For enumeration of bacteria, plate the serially diluted homogenates on plates with Streptococcus selective medium (*see* **Note 15**).
2. *Blood*: Plate 50 μL of diluted blood samples in duplicate using the pour plate method with *Streptococcus* selective medium (*see* Subheading 3.1, **step 5** and **6**).
3. *Skin and Spleen*: Plate neat and serially diluted skin and spleen samples in duplicate using the pour plate method as described above.
4. *Lymph Nodes*: Plate neat and 10^{-2} diluted lymph node samples in duplicate using the pour plate method as described above.

5. Incubate plates at 37 °C overnight and count individual beta-hemolytic streptococcal colonies on the next day.
6. Bacterial burden in the skin is calculated as CFU in the excised biopsy sample (as per lesion). In spleen and lymph nodes, it is calculated in the total volume of homogenates and reported as bacterial burden per tissue. For blood samples, the CFU counts are calculated for a total of 1 mL volume and reported as CFU/mL blood.

4 Notes

1. Ketamine is a controlled S8 drug and must be stored and used following regulatory requirements.
2. It is critical to swirl the blood agar plates consistently to get an even spread of CFU and accurate CFU counts. This step is crucial, as the accuracy of CFU counts in infection dose will impact on the overall outcome of the experiment.
3. Depending upon the initial bacterial concentration in the THBYN broth, it is possible that the CFU numbers on plates with lower dilutions (10^{-1} to 10^{-3}) are too high to count accurately. In that case, move to the plates with higher dilutions. As a general guide, the CFU counts between 50 and 200 on a 100-cm plate are considered reliable.
4. Working stock prepared at a concentration of 1×10^8 CFU/mL would deliver 1×10^6 CFU when 10-μL volume is inoculated at the infection site.
5. It is to be noted that the OD reading will represent both live and dead bacteria and therefore it is necessary to confirm the live bacterial counts by plating on CBA plates. The correlation between OD reading and bacterial concentration are dependent on the *S. pyogenes* strain and therefore should be defined for each strain used.
6. *S. pyogenes* strains vary significantly in their ability to attain animal adaption. If the spleen bacterial burden during initial passaging is low, it is recommended that the spleen homogenate is inoculated in fresh THBYN broth and grown overnight prior to its administration into subsequent mouse for passaging. This additional step may be required for initial 3–4 passages.
7. The volume of anesthetic will depend on the body weight of mouse. Please refer to published guidelines [11]. It is critical that mice receive the optimal dose of anesthetic to keep them sedated for at least 15–20 min. This will allow time for adequate scarification, bacterial inoculum application, and its adsorption, as well as time for application of the band-aid. In

addition, adequate anesthetization will ensure that mice will not experience pain during these procedures. Inadequate dose of anesthetic might result in mice being mobile during the procedure, causing disruption to the procedures. Moreover, if mice do not remain unconscious during the procedure, their handling will be challenging.

8. Application of eye cream is recommended to prevent mouse eye drying under the anesthesia [12]. It is important to maintain the body temperature of the mice during this procedure, as both the anesthetic and spraying the ethanol will greatly reduce the body temperature of the mice, which can then inadvertently slow the recovery of the mice and increase their risk of dying under the anesthetic. Use of a heating pad/heat lamp or a similar device is recommended.
9. The depth of scarification/abrasion ultimately defines the severity of infection. The systemic invasion of bacteria relies on the appropriate damage to the skin to allow bacterial dissemination to the systemic organs and blood. Therefore, consistency of this process is critical for reproducibility. Superficial scarification should damage the epidermal layer, where the bacterial inoculum will be applied. A consistent process will allow a comparison between various strains concerning their virulence potential. Also, the intensity of scarification dictates the formation of a scab on the infected site which plays an important role in the quantification of bacteria in skin lesion. The scab may fall off at different times post infection (usually after day 3 post infection), leading to significant variations in bacterial counts in skin lesions. Consistency in the scarification procedure will reduce this variation.
10. One of the common oversights relates to application of bacterial inoculum on the abraded site without complete removal of ethanol. It is important that ethanol is completely evaporated from the shaved site before a superficial scar is created and prior to application of bacterial inoculum. Failure to do so will result in bacterial killing at the infection site, leading to reduced infection dose and inconsistency in infection outcome. Typically, an inoculum dose of 1–2 × 10^6 CFU/10 μL is required for infectivity.
11. It is important to ensure that the bacterial inoculum is completely absorbed on the scarified surface prior to the application of the band-aid. Application of the band-aid on a wet surface will lead to the bacteria being absorbed on to the band-aid, resulting in their removal from the infection site and thus reduced infection dose.
12. A cohort size of 8–10 mice/time point is critical to generate statistical power.

13. Washing of skin biopsy samples prior to their transfer into fresh tubes is an important step, as it removes surface contaminants and also removes any loose/unadhered bacteria on the skin.
14. A specific combination of speed and time is required for optimal homogenization of different types of samples. For additional information regarding the homogenization speed and time, refer to the manufacturer's instructions [13].
15. Streptococcus selective medium is commonly used for enumeration of *S. pyogenes.* Alternatively, Columbia blood agar medium containing 2% defibrinated horse blood with specific antibiotics can also be used for selection of the antibiotic-resistant strains used for challenge. One of the most common antibiotics used is streptomycin sulfate at 200 μg/mL. Please note that this option will require use of an antibiotic-resistant strain for infection of mice. An antibiotic-resistant strain can be generated by serial passaging of *S. pyogenes* in vitro with increasing concentration of an antibiotic.

References

1. Carapetis JR, Steer AC, Mulholland EK, Weber M (2005) The global burden of group A streptococcal diseases. Lancet Infect Dis 5 (11):685–694
2. Gardiner DL, Sriprakash KS (1996) Molecular epidemiology of impetiginous group A streptococcal infections in aboriginal communities of northern Australia. J Clin Microbiol 34 (6):1448–1452
3. Norton R, Smith HV, Wood N et al (2004) Invasive group A streptococcal disease in North Queensland (1996-2001). Indian J Med Res 119:148–151
4. Carapetis JR, Walker AM, Hibble M et al (1999) Clinical and epidemiological features of group A streptococcal bacteraemia in a region with hyperendemic superficial streptococcal infection. Epidemiol Infect 122 (1):59–65. https://doi.org/10.1017/S0950268898001952
5. McDonald M, Currie BJ, Carapetis JR (2004) Acute rheumatic fever: a chink in the chain that links the heart to the throat? Lancet Infect Dis 4(4):240–245. https://doi.org/10.1016/S1473-3099(04)00975-2
6. Raeder RH, Barker-Merrill L, Lester T et al (2000) A pivotal role for interferon-gamma in protection against group A streptococcal skin infection. J Infect Dis 181(2):639–645
7. Mishalian I, Ordan M, Peled A et al (2011) Recruited macrophages control dissemination of group A Streptococcus from infected soft tissues. J Immunol 187(11):6022–6031. https://doi.org/10.4049/jimmunol.1101385
8. Scaramuzzino DA, McNiff JM, Bessen DE (2000) Humanized in vivo model for streptococcal impetigo. Infect Immun 68 (5):2880–2887. https://doi.org/10.1128/Iai.68.5.2880-2887.2000
9. Ferretti JJ, Stevens DL, Fischetti VA (2017) Streptococcus pyogenes: Basic Biology to Clinical Manifestations. University of Oklahoma Healthe Sciences Center, Oklahoma City (OK). https://www.ncbi.nlm.nih.gov/books/NBK333424/pdf/Bookshelf_NBK333424.pdf
10. Pandey M, Ozberk V, Chen Z et al (2019) Antibodies to the conserved region of the M protein and a streptococcal superantigen cooperatively resolve toxic shock-like syndrome in HLA humanized mice. Sci Adv 5(9):eaax3013
11. Flecknell P (2015) Laboratory Animal Anaesthesia. Academic Press, Cambridge. https://www.elsevier.com/books/laboratory-animal-anaesthesia/flecknell/978-0-12-800036-6. eBook ISBN: 9780128005781

12. https://allergan-web-cdn-prod.azureedge.net/allerganaustralia/allerganaustralia/media/allergan-australia/products/pdfs/pi/agprcfni20916.pdf
13. https://www.nextadvance.com/homogenizer-tissue-cell-culture-bullet-blender-support/homogenizer-cell-disrupter-protocols/
14. Pandey M, Langshaw E, Hartas J et al (2015) A synthetic M protein peptide synergizes with a CXC chemokine protease to induce vaccine-mediated protection against virulent streptococcal pyoderma and bacteremia. J Immunol 194(12):5915–5925. https://doi.org/10.4049/jimmunol.1500157

Chapter 23

A Mouse Nasopharyngeal Colonization Model for Group A Streptococcus

Adrina H. J. Khemlani, Thomas Proft, and Jacelyn M. S. Loh

Abstract

Group A Streptococcus (GAS) often exists as an asymptomatic colonizer of the upper respiratory tract in humans. Unsurprisingly, a high proportion of symptomatic infections caused by GAS include pharyngitis. While not usually life-threatening, these infections cause significant morbidity and economic burden/loss of productivity, and can have downstream life-threatening autoimmune consequences. Modeling asymptomatic colonization in animals is, therefore, a useful tool to dissect host–bacteria interactions and to evaluate efficacy of vaccines aimed at reducing the burden of carriage. Here we describe a mouse model of nasopharyngeal colonization using nasal challenge of susceptible mice and the evaluation of subsequent bacterial burden.

Key words Mouse model, Nasopharyngeal colonization, FVB/n mouse

1 Introduction

Although Group A Streptococcus (GAS) is a human-restricted pathogen, many have successfully modeled infection in animals to answer important questions about the interplay between bacteria and host. The pharynx is a common reservoir and infection site for GAS in humans, and the most representative animal models used to recapitulate human pharyngeal infection are non-human primates [1–3]. High cost and accessibility of non-human primates are, however, prohibitive in many settings. Thus, a number of murine models of infection have been developed to best mimic GAS colonization and infection in vivo (reviewed in [4]).

Here, we provide detailed methodology for a mouse nasopharyngeal model that was originally described by Alam et al. This model uses the FVB/n mouse strain which was shown to be more susceptible to GAS colonization [5]. GAS is introduced through the nares in a low volume under isoflurane anesthesia preventing entry into the lungs. This can result in an asymptomatic carriage of bacteria for at least 10 days depending on the bacterial strain used.

Thomas Proft and Jacelyn M. S. Loh (eds.), *Group A Streptococcus: Methods and Protocols*, Methods in Molecular Biology, vol. 2136, https://doi.org/10.1007/978-1-0716-0467-0_23,

Bacterial colonization can be monitored daily by detecting bacteria shed from the nares through a simple process of gently pressing the nose onto agar and looking for outgrowth of β-hemolytic bacteria. Bacterial colonization can also be monitored by biophotonic imaging as an alternative if a suitable bioluminescent strain is available. Overall bacterial burden in the nasopharynx can also be assessed at the experimental endpoint by homogenizing the tissue and plating for enumeration. In our hands, two clinical isolates belonging to the M89T11 and M49T18.2 types were the best colonizers, with typically >80% colonization rates in male mice after 10 days [6].

2 Materials

2.1 Bacteria Preparation

1. Selected GAS strains (*see* **Note 1**).
2. Brain–heart infusion (BHI) medium prepared according to manufacturer's instructions (*see* **Note 2**).
3. Blood agar plates.
4. BHI agar plates (optional, *see* **Note 3**): add 15 g agar per liter of BHI medium prior to autoclaving. Add antibiotics at appropriate concentrations prior to pouring plates.
5. Phosphate-buffered saline (PBS).

2.2 Mouse Challenge and Monitoring

1. Isoflurane and appropriate delivery device (*see* **Note 4**).
2. FVB/n mice (*see* **Note 5**).
3. Individually ventilated cages (IVCs).
4. D-luciferin (optional, *see* **Note 6**).
5. Biophotonic imager (optional, *see* **Note 6**).
6. 70% Ethanol.
7. Mouse dissection equipment.
8. Bead homogenizer tubes containing 2.8 mm ceramic beads (*see* **Note 7**).
9. Bead homogenizing device (*see* **Note 7**).

3 Methods

3.1 GAS Preparation

1. Streak GAS strain onto a blood agar plate (or appropriate alternative) using standard microbiology techniques to obtain single colonies. Incubate overnight at 37 °C.
2. Inoculate a single colony of GAS in 1.5 ml BHI medium overnight at 37 °C under static conditions.
3. Add the 1.5 ml overnight culture to 13.5 ml fresh pre-warmed BHI medium (*see* **Note 8**) in a 15-ml tube and incubate at

37 °C under static conditions until $OD_{600} = 0.4–0.6$ (*see* **Note 9**).

4. Centrifuge at 5000 × *g* for 10 min at room temperature (RT) and discard the supernatant.
5. Resuspend the pellet in 10-ml sterile PBS.
6. Centrifuge at 5000 × *g* for 10 min at RT and discard the supernatant.
7. Resuspend the pellet in appropriate volume of PBS to obtain 2×10^{10} CFU/ml (*see* **Note 10**).
8. Enumerate the CFU by drop plating serial dilutions onto blood agar plates to determine exact inoculum used for each experiment.

3.2 Nasopharyngeal Challenge

Mice are infected under isoflurane anesthesia intranasally with 10^8 CFU GAS in exponential phase (prepared as in Subheading 3.1) or PBS control in a 5 μl volume. Mice are then monitored daily.

1. Set up the anesthetic equipment in accordance to manufacturer's instructions and your institutional requirements.
2. Place the mice into the induction chamber (*see* **Note 11**) and induce anesthesia at 5% isoflurane. The level of light anesthesia can be maintained at 2–2.5% isoflurane as required (*see* **Note 12**).
3. Pick up each mouse by the scruff and turn it such that it is facing you in the supine position with the head elevated.
4. Using a micropipette, deposit the bacteria onto the external nares (*see* **Note 13**) allowing the 5 μl of bacteria to be inhaled by the mouse (*see* **Note 14**).
5. Return the mouse to its cage and monitor for recovery. Regular welfare monitoring should be performed throughout experimentation in accordance with institutional ethical requirements.

3.3 Monitoring of Bacterial Shedding

This method can be performed daily until experimental endpoint (*see* **Note 15**).

1. Hold the mouse by the scruff of the neck and restrain in one hand.
2. Gently press the nares of each mouse onto the surface of a blood agar or alternative antibiotic selection plate (*see* **Note 3**) ten times.
3. Incubate the plate at 37 °C overnight.
4. Record the number of presses that were positive for GAS on the plates (*see* **Note 16**).

3.4 Monitoring by Biophotonic Imaging

An alternative or additional monitoring procedure to that in Subheading 3.3 can be performed if a bioluminescent reporter strain and imaging equipment are available (*see* **Note 6**).

1. Mice are anesthetized as outlined in Subheading 3.2, **steps 1** and **2**.
2. Substrate (if required) for imaging can be introduced through the nares as described in Subheading 3.2, **steps 3** and **4**. If the pLZ12Km2-P23R:TA:ffluc reporter is used (*see* Chapter 6), we recommend a D-luciferin substrate concentration of 15 mg/ml delivered in a 25 μl volume (12.5 μl per nare).
3. A suitable imaging device with heated chamber can be used to monitor bacterial load/localization in mice under anesthesia.

3.5 Enumeration of Bacterial Load

1. Sacrifice the mice by an approved method in accordance with institutional ethical requirements.
2. Spray surface of mouse with 70% ethanol.
3. Strip back skin of mouse head to remove fur.
4. Dissect complete nasal turbinates (*see* **Note 17**) and place into pre-weighed bead homogenizer tube containing 2.8 mm ceramic beads and 1 ml PBS.
5. Homogenize the sample with an appropriate bead homogenizing device (*see* **Note** 7).
6. Measure the weight of the sample (*see* **Note 18**).
7. Titrate the homogenate in PBS and drop plate on blood agar or antibiotic selection plates (depending on GAS strain used, as required) for enumeration.

4 Notes

1. There is a wide variation in the ability of different GAS strains to colonize mice. Alam et al. reported that the M75 strain had improved rates of colonization compared to other strains tested [5]. In our hands, the M89T11 and M49T18.2 clinical isolates were the best colonizers among the 19 strains tested [6].
2. This medium can be kept at RT for up to 2 weeks. We have found that some strains grow slower when medium has been stored for longer than 2 weeks in our lab. Alternative growth media may also be substituted for BHI, e.g., Todd–Hewitt–yeast (THY) broth. We used BHI when screening GAS strains for good colonizers as some of our clinical isolates failed to grow in THY.
3. BHI plates with antibiotics can be used when enumerating or monitoring bacterial shedding of strains that carry antibiotic

resistance genes. This will lower or eliminate visualization of the background flora in the animal colony, if present. Blood agar plates are used if no antibiotic selection is available and GAS can be distinguished from background flora by β-hemolysis.

4. Check the anesthetic equipment before using. Make sure there is adequate anesthetic agent prior to use. Check that there is an adequate supply of oxygen and it is properly attached to the anesthetic machine.
5. Male mice were found to be more susceptible to colonization by GAS than female mice and are typically used in our experiments. If using female mice, larger sample sizes would need to be considered.
6. D-luciferin is the substrate needed for visualization of bioluminescence if using a firefly luciferase reporter stain. Appropriate imaging equipment with heated chamber is needed for in vivo imaging. We use the IVIS kinetic imager by PerkinElmer.
7. Other suitable homogenizing method could also be used. We used the Bead Rupter by Omni International.
8. It is important to have the medium pre-warmed to 37 °C. Failure to do so may result in inconsistent time required to reach the optical density required.
9. We used exponential phase cultures in our experiments. Use of alternative growth phases may be possible, but would need to be tested by the user.
10. Prior to procedure, the appropriate volume needed to obtain a 2×10^{10} CFU/ml suspension would need to be determined by plating. We found that resuspension of the bacterial pellet in 1/300 of the original culture volume (at $OD_{600} = 0.5$) was appropriate.
11. Depending on the size of your induction chamber, whole groups of mice can be anesthetized at once and removed from chamber individually for inoculation.
12. A moderate depth and consistent respiratory rate works best for this procedure, i.e., breathing not too rapid or too deep.
13. We found a more consistent result if the full 5 μl was delivered through a single nare (we used the left). Avoidance of directly touching the nare with the pipette tip will prevent damaging the nasal membranes.
14. Watch for the droplet to be inhaled by the mouse before returning to its recovery cage.
15. We have not observed beyond 10 days and, therefore, cannot confirm if colonization is maintained beyond this time point.

16. Mice were scored as having cleared the infection if no β-hemolytic colonies were cultured for at least 2 consecutive days.
17. This was performed by cutting between the orbits, and between each orbit and the upper jaw.
18. The weight of the tubes with and without the tissue sample is recorded to enable calculation of the tissue weight for reporting bacterial load as CFU/g tissue. We use a fine scale balance to do this accurately.

Acknowledgments

This work was supported by the Health Research Council (HRC) New Zealand and the Maurice Wilkins Centre for Molecular Biodiscovery. J.M.S.L. is a Heart Foundation NZ research fellow.

References

1. Rivera-Hernandez T, Carnathan DG, Jones S et al (2019) An experimental group A streptococcus vaccine that reduces pharyngitis and tonsillitis in a nonhuman primate model. MBio 10 (2):e00693–e00619. https://doi.org/10.1128/mBio.00693-19
2. Skinner JM, Caro-Aguilar IC, Payne AM et al (2011) Comparison of rhesus and cynomolgus macaques in a Streptococcus pyogenes infection model for vaccine evaluation. Microb Pathog 50 (1):39–47. https://doi.org/10.1016/j.micpath.2010.10.004
3. Sumby P, Tart AH, Musser JM (2008) A non-human primate model of acute group a Streptococcus pharyngitis. Methods Mol Biol 431:255–267. https://doi.org/10.1007/978-1-60327-032-8_20
4. Gogos A, Federle MJ (2019) Modeling Streptococcus pyogenes pharyngeal colonization in the mouse. Front Cell Infect Microbiol 9:137. https://doi.org/10.3389/fcimb.2019.00137
5. Alam FM, Turner CE, Smith K et al (2013) Inactivation of the CovR/S virulence regulator impairs infection in an improved murine model of Streptococcus pyogenes naso-pharyngeal infection. PLoS One 8(4):e61655. https://doi.org/10.1371/journal.pone.
6. Loh JMS, Lorenz N, Tsai CJY et al (2017) Mucosal vaccination with pili from group a streptococcus expressed on Lactococcus lactis generates protective immune responses. Sci Rep 7(1):7174. https://doi.org/10.1038/s41598-017-07602-0.

Chapter 24

Humanized Plasminogen Mouse Model to Study Group A Streptococcus Invasive Disease

Tania Rivera-Hernandez and Mark J. Walker

Abstract

This chapter presents the methodology to carry out infection of humanized plasminogen mice with Group A Streptococcus (GAS). This model of invasive disease has been widely used within the field to study the virulence of different GAS strains, host–pathogen interactions, the importance of particular virulence factors, and preclinical evaluation of novel treatments and vaccines. The model has shown to be highly reproducible and therefore represents an invaluable tool for GAS research.

Key words Group A Streptococcus, Invasive disease, Humanized plasminogen mouse, Bacterial dissemination and survival

1 Introduction

Group A Streptococcus (GAS) is a strictly human pathogen which causes a significant disease burden worldwide [1]. The use of animal models in GAS research has made valuable contributions to improve our understanding of disease mechanisms and host–pathogen interactions. GAS has the ability to infect different tissues such as the tonsils, skin, lungs, and the fascia, and can also cause systemic infections [2]. This wide spectrum of tissue tropisms, severity of infection, and host restriction has resulted in the lack of a single standardized animal model and instead has led to the use of multiple animal models, which include mice, rats, and nonhuman primates, among others [3]. One of the most reliable models to mimic invasive disease is the infection of transgenic mice expressing human plasminogen under the control of the murine albumin gene promoter (*Alb*PLG1) [4]. The model accounts for the production of GAS streptokinase (Ska), a plasminogen activator enzyme with high specificity for human plasminogen [5]. Production of Ska leads to the conversion of human plasminogen into plasmin at the bacterial surface, which degrades components of the extracellular matrix [6], thus facilitating bacterial dissemination

Thomas Proft and Jacelyn M. S. Loh (eds.), *Group A Streptococcus: Methods and Protocols*, Methods in Molecular Biology, vol. 2136, https://doi.org/10.1007/978-1-0716-0467-0_24,

[7, 8]. The use of the humanized plasminogen mouse model has proven useful for investigating the role of GAS virulence factors important for bacterial dissemination and systemic disease [9–11] and preclinical development of new treatments [12]. In addition, the model has been used successfully to show protective vaccine-mediated immunity using homologous M protein as an antigen [13], making the model suitable for vaccine studies.

Mice are typically infected via subcutaneous injection. At 3 days post-infection, bacteria spread systemically and can be found in sterile organs such as blood, liver, and spleen [14]. From this point onward, the disease progresses rapidly and symptoms become more severe. The model allows quantification of mouse survival following infection or estimation of bacterial burden in the lesion site as well as sterile organs, thus demonstrating the versatility for different applications within GAS research.

2 Materials

2.1 Preparation of Frozen Stock GAS Inoculum

1. GAS strains of interest (*see* **Note 1**).
2. Horse blood agar plates.
3. Todd–Hewitt broth supplemented with 1% (w/v) yeast extract (THYB).
4. THYB supplemented with 30% (v/v) glycerol.
5. THY agar (THYA) plates: THYB supplemented with 0.2% (w/v) yeast extract and 1.5% (w/v) agar.
6. 15-mL screw cap tubes, sterile.
7. Liquid nitrogen.

2.2 Preparation of Fresh GAS Inoculum

1. GAS strains of interest.
2. Horse blood agar plates.
3. THYB.
4. THYA plates.
5. Phosphate-buffered saline (PBS), sterile.
6. 15-mL screw cap tubes, sterile.
7. 50-mL screw cap tubes, sterile.

2.3 Mouse Infection

1. Mice heterozygous for the human plasminogen gene (*Alb*PLG1) [4]. Mice are available from The Jackson Laboratory (B6.Cg-Tg(Alb-PLG)1Dgi/J stock No: 007238).
2. Mouse fur clippers.
3. Hair removal cream for sensitive skin.
4. Soft tissues.
5. Inhalable anesthetic such as methoxyflurane or isoflurane.

6. Sealed chamber.
7. Ophthalmic veterinary ointment.
8. Isopropyl alcohol pads/wipes.
9. 1-mL syringes.
10. 26–27 gauge needles.
11. Score cards for disease monitoring.
12. CO_2 chamber (or other suitable method) for humane euthanasia.

2.4 Determination of Bacterial Burden

1. CO_2 chamber (or other suitable method) for humane euthanasia.
2. 1-mL syringes.
3. 26–27 gauge needles.
4. 1.5-mL tubes.
5. EDTA-treated blood collection tubes.
6. Sterile dissection kit.
7. PBS, sterile.
8. Tissue homogenizer.
9. THYA plates.

3 Methods

3.1 Preparation of Frozen Stock GAS Inoculum

A GAS frozen stock can be aliquoted and stored at −80 °C for extended periods of time, maintaining cell viability and capacity to cause infection in mice (*see* **Note 2**).

1. Inoculate GAS strain on a horse blood agar plate, and incubate overnight at 37 °C.
2. Use a single colony to inoculate 10 mL of pre-warmed THYB, and incubate ON at 37 °C static.
3. Measure the optical density at 600 nm (OD_{600}) of the overnight culture and inoculate 200 mL of pre-warmed THYB so that initial OD_{600} reaches 0.05. Incubate at 37 °C static until OD_{600} reaches 0.5–0.6.
4. Pellet the bacterial cells by centrifugation at 5000 × *g* for 10 min at 4 °C.
5. Wash the pellet with 50 mL of ice-cold fresh THYB.
6. Repeat wash with ice-cold THYB.
7. Fully resuspend bacterial pellet in 50 mL of ice-cold THYB.
8. Add the same volume of ice-cold THYB 30% glycerol (for a final glycerol concentration of 15%). Aliquot culture in 4-mL aliquots.

9. Snap-freeze aliquots in liquid nitrogen and store at −80 °C.
10. To determine the colony-forming unit per mL (CFU/mL) of the bacterial stock, place one of the frozen aliquots on ice and incubate until thawed. Prepare bacterial dilutions and plate on THYA. Incubate overnight at 37 °C and determine bacterial concentration. Repeat this step for five individual aliquots to obtain a mean concentration of the stock. Calculate the volume required for infection based on the CFU/mL of the stock and the desired infecting dose.
11. On the day of infection, thaw on ice the number of aliquots needed for infection and confirm the CFU/mL by plating.

3.2 Preparation of Fresh GAS Inoculum

The use of a fresh GAS inoculum for infection requires careful characterization of the growth characteristics of the bacterial strain. Given that chain length varies between strains, it is common to see differences in the correlation between OD_{600} and the number of CFUs. We recommend to carry out the following protocol to determine the OD_{600} required to obtain the desired number of CFUs and repeat the protocol at least three times to ensure consistency. When comparing two strains, it is important to obtain bacterial suspensions with similar CFU/mL concentration and infect using the same dose.

1. Inoculate strain on a horse blood agar plate, and incubate overnight at 37 °C.
2. Use a single colony to inoculate 10 mL of pre-warmed THYB, and incubate ON at 37 °C static.
3. Measure OD_{600} of the overnight culture and inoculate 40 mL of pre-warmed THYB so that initial OD_{600} reaches 0.05. Incubate at 37 °C static until OD_{600} reaches 0.5–0.6.
4. Pellet the bacterial cells by centrifugation at 5000 × *g* for 10 min at 4 °C.
5. Wash the pellet with 40 mL of PBS.
6. Repeat wash with PBS.
7. Resuspend the bacterial pellet to the pre-determined desired OD_{600} and store at 4 °C until used for infection.
8. Prepare bacterial dilutions and plate on THYA. Incubate overnight at 37 °C to determine bacterial concentration and infecting dose.

3.3 Mouse Infection

Mice are infected on day 0 via subcutaneous injection of a bacterial suspension from a frozen stock (*see* Subheading 3.1) or prepared fresh on the day of infection (*see* Subheading 3.2). Standard experiments require groups of 10 mice heterozygous for the human plasminogen gene (*Alb*PLG1); five males and five females are used. We recommend to use mice between 6 and 14 weeks of age.

1. On day −1, tag mice to facilitate identification and follow disease progression.
2. Trim the mouse fur on the lower back area above the base of the tail.
3. Apply a thin layer of hair removal cream to the trimmed area. House the mouse individually for 2 min. Remove the cream by wiping the area using a dampened tissue. Make sure to remove all excess cream to avoid irritation of the skin.
4. On infection day (day 0), load 1-mL syringes with the bacterial suspension (*see* **Note 3**).
5. Place mouse in the sealed chamber for anesthesia. Wait until mouse is unconscious and monitor breathing until it is slow and steady. Take the mouse out of the chamber and assess the level of anesthesia by applying a firm toe pinch (pedal reflex). If the reflex is still present, provide further exposure to anesthetic.
6. Once under anesthesia, apply ophthalmic ointment to prevent eye desiccation.
7. Clean the shaved area in the lower back of the mouse using an isopropyl alcohol pad.
8. Inject subcutaneously a maximum of 100 μL of the bacterial suspension into the shaved and clean area. A visible bubble should form under the skin once the liquid is injected.
9. Place the mouse in a suitable warm area for recovery before returning it to its cage.
10. Score mice twice a day (morning and afternoon) following the criteria listed in Table 1. An individual score is assigned to each

Table 1
Mouse scoring system

Indicators	Independent variables	Score
Locomotion	Walking normally	*0*
	Limping, stiffness, slowed movement	*1*
	Visible inflammation (swollen limbs, tail, or face)	*2*
	No movement	*3*
Behavior	Normal, active, and responsive to stimuli	*0*
	Away from littermates	*1*
	Aggressive or huddled in corner	*2*
	Lack of response to external stimuli	*3*
Appearance	Normal, groomed	*0*
	Ruffled fur	*1*
	Animal is hunched, shivering, visibly dehydrated	*2*
	Reluctant to move	*3*

mouse in order to follow disease progression (*see* **Note 4**). Any animal euthanized due to the presence of severe symptoms is considered as an event for Logrank (Mantel–Cox test) survival analysis.

11. Ten days after infection, euthanize surviving animals. These animals are considered as censored subjects for Logrank (Mantel–Cox test) survival analysis.

3.4 Determination of Bacterial Burden

An alternative to survival analysis is to evaluate bacterial burden and dissemination after infection. Days 3 and 5 post-infection are suitable time points to quantify bacteria in the lesion (skin), blood, liver, and spleen. If the aim of the experiment is the determination of bacterial burden, it is important to use a sub-lethal infection dose to prevent animals progressing to severe symptoms of infection and avoid having to euthanize animals before the established collection time points.

1. At days 3 and 5 post-infection, use a suitable method for euthanasia and sacrifice mice for organ collection.
2. Perform a heart puncture to collect a blood sample and place the blood in an EDTA-treated collection tube to avoid coagulation.
3. Using a sterile dissection kit, excise the skin lesion area and collect it in a 1.5-mL tube containing 1 mL of sterile PBS (*see* **Note 5**).
4. Aseptically remove the spleen and liver and collect it in a 1.5-mL tube containing 1 mL of sterile PBS (*see* **Note 5**).
5. Using a tissue homogenizer, process the skin, liver, and spleen samples in order to obtain homogeneous suspensions.
6. Make tenfold serial dilutions of homogenized tissues and blood, and plate on TYHA. Incubate plates at 37 °C overnight and count colonies to determine CFU/mL.

4 Notes

1. This protocol has been optimized using GAS M1T1 strain 5448 [15]. An infecting dose of approximately 1×10^8 CFU delivered in a volume of 100 μL results in survival rates from 0% to 30%.
2. On the day of infection, a single-use aliquot is thawed and used for infection. The main advantage of preparing a frozen stock is that it allows use of the same bacterial preparation for different experiments, improving reproducibility. Before preparing a master stock, we recommend to characterize the growth of the strain of interest in order to determine the final

resuspension volume based on the desired CFU/mL concentration. The culture volumes used in this particular protocol were optimized for GAS M1T1 strain 5448 [15] in order to obtain a stock at approximately 1×10^9 CFU/mL.

3. Special care should be taken when handling needles and syringes loaded with GAS to prevent needle-stick injuries. In case of an incident, attend a physician and receive antibiotic therapy for GAS infection as soon as possible.
4. Once infected, mice should be monitored twice a day for 10 days. A score sheet should be kept to keep track of symptoms and disease progression in order to know when to humanely euthanize a mouse when condition is worsening. At days 1 and 2 post-infection, animals will develop a skin lesion that can progress or improve within the following days. Most animals will show severe symptoms of infection between days 3 and 5; however, occasionally, some animals will show severe symptoms at later time points. Any mouse with a score of *3* on any of the indicators or a score of *2* on two or more indicators should be euthanized immediately and considered as succumbed to infection. These animals are considered as an event for Logrank (Mantel–Cox test) survival analysis. Any mouse with a score of *2* on one indicator should be monitored closely to determine if the disease progresses in subsequent check-ups.
5. If desired, skin, liver, and spleen tissues can be weighed to normalize the CFU per g of tissue.

References

1. Carapetis JR, Steer AC, Mulholland EK et al (2005) The global burden of group A streptococcal diseases. Lancet Infect Dis 5 (11):685–694. https://doi.org/10.1016/S1473-3099(05)70267-X
2. Walker MJ, Barnett TC, McArthur JD et al (2014) Disease manifestations and pathogenic mechanisms of group A *Streptococcus*. Clin Microbiol Rev 27(2):264–301. https://doi.org/10.1128/CMR.00101-13
3. Watson ME Jr, Neely MN, Caparon MG (2016) Animal models of *Streptococcus pyogenes* infection. In: Ferretti JJ, Stevens DL, Fischetti VA (eds) *Streptococcus pyogenes*: basic biology to clinical manifestations. University of Oklahoma Health Sciences Center, Oklahoma City (OK), pp 629–660
4. Sun H, Ringdahl U, Homeister JW et al (2004) Plasminogen is a critical host pathogenicity factor for group A streptococcal infection. Science 305(5688):1283–1286. https://doi.org/10.1126/science.1101245
5. Yakovlev SA, Rublenko MV, Izdepsky VI et al (1995) Activating effect of the plasminogen activators on plasminogens of different mammalia species. Thromb Res 79(4):423–428. https://doi.org/10.1016/0049-3848(95)00131-A
6. Coleman JL, Benach JL (1999) Use of the plasminogen activation system by microorganisms. J Lab Clin Med 134(6):567–576. https://doi.org/10.1016/S0022-2143(99)90095-1
7. Cole JN, Barnett TC, Nizet V et al (2011) Molecular insight into invasive group A streptococcal disease. Nat Rev Microbiol 9 (10):724–736. https://doi.org/10.1038/nrmicro2648
8. Cole JN, McArthur JD, McKay FC et al (2006) Trigger for group A streptococcal M1T1

invasive disease. FASEB J 20(10):1745–1747. https://doi.org/10.1096/fj.06-5804fje

9. Cook SM, Skora A, Gillen CM et al (2012) Streptokinase variants from *Streptococcus pyogenes* isolates display altered plasminogen activation characteristics—implications for pathogenesis. Mol Microbiol 86 (5):1052–1062. https://doi.org/10.1111/mmi.12037

10. Maamary PG, Sanderson-Smith ML, Aziz RK et al (2010) Parameters governing invasive disease propensity of non-M1 serotype group A streptococci. J Innate Immun 2(6):596–606. https://doi.org/10.1159/000317640

11. Brouwer S, Cork AJ, Ong CLY et al (2018) Endopeptidase PepO regulates the SpeB cysteine protease and is essential for the virulence of invasive M1T1 *Streptococcus pyogenes*. J Bacteriol 200(8). https://doi.org/10.1128/JB.00654-17

12. Sun H, Xu Y, Sitkiewicz I et al (2012) Inhibitor of streptokinase gene expression improves survival after group A *Streptococcus* infection in mice. Proc Natl Acad Sci U S A 109 (9):3469–3474. https://doi.org/10.1073/pnas.1201031109

13. Rivera-Hernandez T, Pandey M, Henningham A et al (2016) Differing efficacies of lead group A streptococcal vaccine candidates and full-length M protein in cutaneous and invasive disease models. MBio 7(3):e00618–e00616. https://doi.org/10.1128/mBio.00618-16

14. Walker MJ, Hollands A, Sanderson-Smith ML et al (2007) DNase Sda1 provides selection pressure for a switch to invasive group A streptococcal infection. Nat Med 13:981–985. https://doi.org/10.1038/nm1612

15. Chatellier S, Ihendyane N, Kansal RG et al (2000) Genetic relatedness and superantigen expression in group A *Streptococcus* serotype M1 isolates from patients with severe and nonsevere invasive diseases. Infect Immun 68 (6):3523–3534. https://doi.org/10.1128/IAI.68.6.3523-3534.2000

Chapter 25

Lancefield Whole Blood Killing Assay to Evaluate Vaccine Efficacy

Mark Reglinski

Abstract

While the Lancefield whole blood killing assay is named after the renowned streptococcal researcher Rebecca Lancefield, the protocol was first described by Todd in 1927 (Br J Exp Pathol 8:1–5, 1927). Initially, the assay was used to identify novel Group A Streptococcal (GAS) serotypes through the supplementation of non-immune human blood (often from infants) with type-specific antisera prepared in rabbits (Lancefield, J Exp Med 106:525–544, 1957; Maxted, Br J Exp Pathol 37:415–422, 1956) and to demonstrate the impressive longevity of type-specific immunity in patients following invasive GAS infection (Lancefield, J Exp Med 110:271–292, 1959). The modern assay is routinely used to screen defined GAS mutants (Wessels, Bronze, Proc Natl Acad Sci U S A 91:12238–12242, 1994; Zinkernagel et al., Cell Host Microbe 4:170–178, 2008) or transposon libraries (Le Breton et al., Infect Immun 81:862-875, 2013) for enhanced susceptibility to opsonophagocytic killing or to screen vaccine antisera (Salehi et al., mSphere 3: e00617–e00618, 2018) or other serological preparations (Reglinski et al., Sci Rep 5:15825, 2015) for anti-streptococcal activity.

Key words Lancefield assay, Whole blood, Bacterial killing, Opsonophagocytosis, Antiserum, Mutant screening, Multiplication factor

1 Introduction

The fundamental whole blood killing assay protocol has scarcely changed since Todd's first description [1]. Whole blood from healthy human donors is aseptically collected into tubes containing a suitable anti-coagulant and supplemented with 50–100 colony forming units (CFU) of freshly prepared GAS. The samples are incubated at 37 °C for 3 h under constant agitation and the bacterial density at the beginning and at the end of the experiment are used to calculate the resulting degree of bacterial growth (denoted the multiplication factor). Multiplication factors calculated under different reaction conditions can be compared to determine, for example, the bactericidal activity of serological

Thomas Proft and Jacelyn M. S. Loh (eds.), *Group A Streptococcus: Methods and Protocols*, Methods in Molecular Biology, vol. 2136, https://doi.org/10.1007/978-1-0716-0467-0_25,

preparations or the role of certain genes in GAS resistance to phagocytosis.

Despite several attempts in the mid-twentieth century, it is clear that the human whole blood used in the assay cannot be reliably substituted with animal blood which remains only weakly bactericidal even when supplemented with human immune factors [10]. The only exception is non-human primate blood which readily kills GAS in the presence of bactericidal human antibodies [10, 3]. The specific differences between human and animal blood that are responsible for this disparity remain unclear; however, Stollerman and others have provided evidence for a human-specific co-opsonin that is essential for GAS killing [11–13]. Thus, the use of freshly isolated human whole blood remains an absolute requirement for a successful assay.

2 Materials

Prepare all solutions using distilled water and research-grade reagents. Ethical approval covering the use of human blood from healthy donors must be obtained prior to conducting this protocol. Blood collections should be performed by a licensed physician or trained phlebotomist.

1. Frozen stocks of GAS strains of interest stored at −80 °C in 20% glycerol.
2. Columbia blood agar (CBA) plates (*see* **Note 1**): Prepare 400 ml of Columbia blood agar base in a 500-ml Duran bottle containing an autoclave safe magnetic stir bar according to the manufacturer's instructions and sterilize by autoclaving at 121 °C for 15 min. Cool the base medium to 50 °C and add 5% sterile defibrinated blood stirring continuously. Pour plates immediately and allow to set under the flame or in a laminar flow hood with the lids ajar.
3. Todd–Hewitt broth (THB): Prepare 400 ml of Todd–Hewitt broth in a 500-ml Duran bottle according to the manufacturer's instructions and sterilize by autoclaving at 121 °C for 15 min.
4. Phosphate-buffered saline (PBS): Dissolve 8 g of sodium chloride (NaCl), 0.2 g of potassium chloride (KCl), 1.44 g of dibasic sodium phosphate (Na_2HPO_4), and 0.24 g of monobasic potassium phosphate (KH_2PO_4) in 800 ml H_2O and adjust the pH to 7.4 with HCl as required. Add H_2O to 1 L and sterilize by autoclaving at 121 °C for 15 min.
5. BD Vacutainer® Safety-Lok™ blood collection sets with pre-attached holder (21-gauge needle).

6. 6 ml BD Vacutainer® blood collection tubes, lithium heparin spray coated (*see* **Note 2**).
7. Alcohol swabs.
8. Tourniquet.
9. 2-ml microcentrifuge tubes (*see* **Note 3**).
10. End-over-end rotator capable of securely holding the reaction tubes.
11. 5% CO_2 static incubator set at 37 °C.

3 Method

1. Streak GAS strain of interest onto a CBA plate and culture overnight at 37 °C in 5% CO_2.
2. Inoculate a single colony into 5 ml of THB and culture overnight at 37 °C in 5% CO_2 in a static incubator.
3. Inoculate 500 μl of the resulting overnight culture 1:10 into fresh, pre-warmed THB and incubate at 37 °C in 5% CO_2 in a static incubator until an A_{600} of 0.15–0.2 is reached (*see* **Note 4**).
4. Dilute culture 1:10,000 in sterile PBS. Plate 30 μl of the diluted GAS onto a CBA plate and culture overnight to determine the starting CFU/reaction.
5. Aseptically collect heparinized whole blood from at least three healthy donors (*see* **Note 5**) and, if appropriate, supplement with test reagents of interest; for example, rabbit antiserum (*see* **Note 6**), pooled human immunoglobulin, or putative bacteriostatic agents.
6. Inoculate 30 μl of the diluted GAS into 270 μl of heparinized blood and incubate for 3 h (*see* **Note 7**) at 37 °C in 5% CO_2 with constant end-over-end rotation (*see* **Note 8**).
7. Place reactions on ice and serially dilute 1:10, 1:100, and 1:1000 in PBS (*see* **Note 9**).
8. Plate 15 μl of a neat, 1:10, 1:100, and 1:1000 dilution onto CBA plates and culture overnight to determine the final CFU/reaction.
9. Determine the CFU added to the reactions at 0 h and the CFU present in the reactions at 3 h (*see* **Note 10**), and then divide the CFU/reaction (3 h) by the CFU/reaction (0 h) to give the multiplication factor.

4 Notes

1. Pre-poured Columbia blood agar plates are available for purchase from several commercial suppliers. In addition to being more convenient, purchased plates tend to be higher quality and more consistent than plates prepared in-house.
2. Rothbard reported a marked reduction in bacteriostatic activity when the anticoagulant heparin was replaced with the calcium-neutralizing agents, potassium, and ammonium oxalate or sodium citrate [10]. Defibrination is not recommended due to the risk of decreasing the number of leukocytes below the level required for effective phagocytosis.
3. 1.5 ml microcentrifuge tubes can also be used; however, it is important to ensure that the blood and streptococci mix effectively while the reactions are agitated.
4. The optical density of 0.15–0.2 is given for guidance and is based on growth of the invasive GAS isolate H305 (National Collection of Type Cultures number: NCTC8198) [14] in freshly prepared THB. The optical density that results in a starting inoculum of 50–100 CFU must be determined empirically for each strain used. Freshly prepared cultures should be used, and not frozen culture stocks.
5. The biggest hurdle that must be overcome when running a whole blood killing assays is inter-donor variation in bacterial multiplication. We and others have found that this affect can be partially ablated through prescreening of donors for their ability to support GAS multiplication, or digestion of endogenous immunoglobulin prior to addition of bacteria [15, 16]. However, as discussed above, the bactericidal power possessed by the blood of some individuals cannot be explained satisfactorily by the presence of anti-GAS antibodies. If the difference in bacterial growth in whole blood is subtle between test and control conditions, there may be no alternative but to increase the number of donor blood samples until the null hypothesis can be rejected (or confirmed) with adequate statistical stringency. Three technical replicates of each donor blood sample should be set up to ensure accurate intra-donor multiplication factors are obtained. It is recommended that the assay is set up within 2 h of blood withdrawal.
6. Serum from vaccinated rabbits is most commonly used to assess the efficacy of the humoral immune response to specific antigens [17, 18]. The dilution of antiserum required to promote killing should be determined empirically and will depend upon the concentration and functionality of the test antibodies. Reactions containing pre-immune serum should be included as a control. Reactions can also be supplemented with patient

serum or purified human antibodies [9]. Only scant references could be found where mouse antiserum was used in place of rabbit [19, 20].

7. While a 3 h incubation was employed in Rebecca Lancefield's seminal studies [4] and is generally accepted as the standard protocol, published incubation times range from 30 min to 24 h [6, 21, 22]. Thus, the incubation time may have to be optimized for specific applications. Fuller and colleagues noted that after 3 h, the surviving streptococci multiply rapidly to reach densities of "many millions per C.C." within 24 h. At such densities, subtle difference between immune and non-immune reactions may be difficult to detect [22].
8. Reactions must be agitated to ensure that the leukocytes do not sediment, as this will substantially reduce the apparent bactericidal activity of the blood samples. In his original publication, Todd demonstrated that the bactericidal activity of paired blood samples was reduced if incubated under static conditions, and postulated that this resulted from a reduction in contact time between the leukocytes and streptococci [1]. Sealing of tubes and/or use of a 5% CO_2 incubator is essential. If the blood is agitated in an open tube, the rapid loss of CO_2 may affect the survival and growth of the GAS isolate [22].
9. Serial tenfold dilutions in PBS can be prepared in a 96-well plate, allowing three columns of 5-μl dots to be carefully spotted onto a prewarmed blood agar plate using a 1–10-μl multichannel pipette. The PBS is allowed to evaporate and the plates are cultured overnight at 37 °C. The next day one of the resulting bacterial dilutions will contain a countable number of colonies. As undiluted blood will not evaporate, the lowest dilution that can be assayed in this manner is 1:10.
10. The starting CFU is equivalent to the number of colonies present in 30 μl of the initial GAS dilution (plated during **step 4**). The final CFU is equivalent to the number of colonies present in the 300 μl reaction following 3 h of incubation (an aliquot of which is plated during **step 9**).

References

1. Todd EW (1927) A method of measuring the increase or decrease of the population of Hæmolytic streptococci in blood. Br J Exp Pathol 8(1):1–5
2. Lancefield RC (1957) Differentiation of group a streptococci with a common R antigen into three serological types, with special reference to the bactericidal test. J Exp Med 106 (4):525–544
3. Maxted WR (1956) The indirect bactericidal test as a means of identifying antibody to the M antigen of Streptococcus pyogenes. Br J Exp Pathol 37(4):415–422
4. Lancefield RC (1959) Persistence of type-specific antibodies in man following infection with group A streptococci. J Exp Med 110 (2):271–292
5. Wessels MR, Bronze MS (1994) Critical role of the group A streptococcal capsule in pharyngeal colonization and infection in mice. Proc Natl Acad Sci U S A 91(25):12238–12242

6. Zinkernagel AS, Timmer AM, Pence MA et al (2008) The IL-8 protease SpyCEP/ScpC of group A Streptococcus promotes resistance to neutrophil killing. Cell Host Microbe 4 (2):170–178. https://doi.org/10.1016/j.chom.2008.07.002
7. Le Breton Y, Mistry P, Valdes KM et al (2013) Genome-wide identification of genes required for fitness of group A Streptococcus in human blood. Infect Immun 81(3):862–875. https://doi.org/10.1128/iai.00837-12
8. Salehi S, Hohn CM, Penfound TA, Dale JB (2018) Development of an Opsonophagocytic killing assay using HL-60 cells for detection of functional antibodies against Streptococcus pyogenes. mSphere 3(6):e00617–e00618. https://doi.org/10.1128/mSphere.00617-18
9. Reglinski M, Gierula M, Lynskey NN et al (2015) Identification of the Streptococcus pyogenes surface antigens recognised by pooled human immunoglobulin. Sci Rep 5:15825. https://doi.org/10.1038/srep15825srep15825
10. Rothbard S (1945) Bacteriostatic effect of human sera on group A Streptococci: II. Comparative bacteriostatic effect of Normal whole blood from different animal species in the presence of human convalescent sera. J Exp Med 82(2):107–118
11. Stollerman GH, Kantor FS, Gordon BD (1958) Accessory plasma factors involved in the bactericidal test for type-specific antibody to group A Streptococci. I. A typical behavior of some human and rabbit bloods. J Exp Med 108(4):475–491
12. Stollerman GH, Rytel M, Ortiz J (1963) Accessory plasma factors involved in the bactericidal test for type-specific antibody to group A Streptococci. II. Human plasma cofactor (s) enhancing opsonization of encapsulated organisms. J Exp Med 117:1–17
13. Kahlich R, Svec M, Prochazka O (1983) Complex surveillance of streptococcus pyogenes. VI. Relationship between hypersensitivity to zymosan and indicators of immunity against group A streptococci. J Hyg Epidemiol Microbiol Immunol 27(2):219–228
14. Unnikrishnan M, Cohen J, Sriskandan S (2001) Complementation of a speA negative Streptococcus pyogenes with speA: effects on virulence and production of streptococcal pyrogenic exotoxin A. Microb Pathog 31 (3):109–114. https://doi.org/10.1006/mpat.2001.0453
15. Bauer MJ, Georgousakis MM, Vu T et al (2012) Evaluation of novel Streptococcus pyogenes vaccine candidates incorporating multiple conserved sequences from the C-repeat region of the M-protein. Vaccine 30 (12):2197–2205. https://doi.org/10.1016/j.vaccine.2011.12.115
16. Reglinski M, Lynskey NN, Sriskandan S (2016) Modification of the classical Lancefield assay of group A streptococcal killing to reduce inter-donor variation. J Microbiol Methods 124:69–71. https://doi.org/10.1016/j.mimet.2016.03.015
17. Courtney HS, Hasty DL, Dale JB (2006) Anti-phagocytic mechanisms of Streptococcus pyogenes: binding of fibrinogen to M-related protein. Mol Microbiol 59(3):936–947. https://doi.org/10.1111/j.1365-2958.2005.04977.x
18. Lannergard J, Gustafsson MC, Waldemarsson J et al (2011) The Hypervariable region of Streptococcus pyogenes M protein escapes antibody attack by antigenic variation and weak immunogenicity. Cell Host Microbe 10(2):147–157. https://doi.org/10.1016/j.chom.2011.06.011
19. Wittner MK, Fox EN (1977) Homologous and heterologous protection of mice with group A streptococcal M protein vaccines. Infect Immun 15(1):104–108
20. Bronze MS, McKinsey DS, Beachey EH, Dale JB (1988) Protective immunity evoked by locally administered group A streptococcal vaccines in mice. J Immunol 141(8):2767–2770
21. Keefer CS, Spink WW (1936) Studies of hemolytic streptococcal infection I. Factors influencing the outcome of erysipelas. J Clin Invest 15 (1):17–19. https://doi.org/10.1172/JCI100754
22. Fuller AT, Colebrook L, Maxted WR (1939) Factors which determine the fate of hæmolytic streptococci (group A) in shed blood and in serum. J Pathol Bacteriol 48(2):443–464. https://doi.org/10.1002/path.1700480217

Chapter 26

An Opsonophagocytic Killing Assay for the Evaluation of Group A Streptococcus Vaccine Antisera

Reuben McGregor, Scott Jones, Raynes M Jeremy, David Goldblatt, and Nicole J. Moreland

Abstract

Group A Streptococcus (GAS) is a major cause of global mortality, yet there are no licensed GAS vaccines. Vaccine progress has been hampered, in part, by a lack of standardized assays able to quantify antibody function in test antisera. The most widely used assay was developed over 50 years ago by Rebecca Lancefield and relies on human whole blood as a source of complement and neutrophils. Recently, an opsonophagocytic killing (OPK) assay has been developed for GAS by adapting the OPK methods utilized in *Streptococcus pneumoniae* vaccine testing. This assay uses dimethylformamide (DMF)-differentiated human promyelocytic leukemia cells (HL-60 cells) as a source of neutrophils and baby rabbit complement, thus removing the major sources of variation in the Lancefield assays. This protocol outlines methods for performing a GAS OPK assay including titering test sera to generate an opsonic index. This in vitro assay could aid in selecting vaccine candidates by demonstrating whether candidate-induced antibodies lead to complement deposition and opsonophagocytic killing.

Key words Complement, HL-60 cells, Phagocytosis, Group A Streptococcus, Vaccine, Antibody

1 Introduction

Group A Streptococcus (GAS) is a Gram-positive, human only, pathogen that causes a wide spectrum of disease, including superficial infections (such as pyoderma and pharyngitis), severe invasive GAS disease, acute rheumatic fever (ARF), and associated rheumatic heart disease (RHD) [1]. Despite significant global morbidity and mortality caused by these clinical syndromes, currently there is no licensed GAS vaccine [2]. Development of a vaccine to GAS has been hampered, in part, by the lack of a reliable, standardized assay to measure a functional immune response. The most widely used assays, developed in the 1950s by Rebecca Lancefield (known as the "Lancefield" assay), measure the survival of GAS in fresh human blood [3]. However, interindividual differences in complement and neutrophils result in large data variability making

Thomas Proft and Jacelyn M. S. Loh (eds.), *Group A Streptococcus: Methods and Protocols*, Methods in Molecular Biology, vol. 2136, https://doi.org/10.1007/978-1-0716-0467-0_26,

comparisons between laboratories difficult [4]. The need for improved assays capable of measuring functional GAS antibody responses was recently identified as an international priority for vaccine development [2, 5, 6] .

The development of an opsonophagocytic killing (OPK) assay for testing *Streptococcus pneumoniae* antisera [7] has facilitated the advancement of vaccines for this bacterial pathogen [8]. The assay readout is an "opsonic index," which is the reciprocal value of the serum dilution at which 50% killing is obtained. These values have been shown to correlate with the ability of vaccine candidates to generate protection in clinical trials and are now routinely used to evaluate the *Streptococcus pneumoniae* vaccines [9]. Recently, OPK assays have been developed for GAS by ourselves and others, with proof of principle using rabbit antisera and human intravenous immunoglobulin [10, 11]. In order to overcome variability associated with traditional Lancefield assays, the GAS OPK uses an exogenous source of both complement (baby rabbit complement) and phagocytes (human promyelocytic leukemia cells (HL-60 cells)), in line with protocols developed for *S. pneumoniae*. The HL-60 cells are differentiated to a neutrophil-like cell by adding dimethylformamide (DMF) and the complement is sourced from baby rabbits, meaning it has low heterophile antibody activity. When antibodies specific for GAS components on the bacterial surface are mixed with GAS bacteria, in conjunction with differentiated HL-60 cells and complement, the anti-GAS antibodies activate the classical complement pathway to stimulate opsonization of the bacteria with C4b and C3b complement factors. This leads to opsonophagocytosis and killing by the HL-60 phagocytic cell line.

This protocol describes a method for performing an OPK assay for GAS largely based on the *S. pneumoniae* OPK [10]. It makes use of the assay buffer utilized in the qualified *S. pneumoniae* assays, and which we have found to be sufficient for measuring opsonophagocytosis against selected GAS strains. In a recent report, Salehi et al. [11] describe a modified GAS OPK assay, in which additional reagents are included in the assay buffer. Given the growing momentum in GAS vaccine development, harmonizing OPK assay protocols for selected GAS strain types (*emm* types) between laboratories should now be a priority. Over and above vaccine development, the GAS OPK assay can be utilized to test natural immunity against GAS in human donors, as well as the efficacy of GAS specific monoclonal antibodies.

2 Materials

2.1 HL-60 Cell Culture

1. HL-60 cells (from American Type Culture Collection (ATCC)).
2. T-75 tissue culture flasks, filtered-cap.

3. Trypan blue cell viability stain.
4. M1 medium: Roswell Park Memorial Institute (RPMI) 1640, 10% fetal bovine serum (FBS), 1% L-glutamine, 100 U/ml penicillin, 100 U/ml streptomycin, 0.5 μg/ml amphotericin-B.
5. M2 medium: RPMI 1640, 10% FBS, 1% L-glutamine.
6. Freezing medium: 9 ml FBS, 1 ml dimethyl sulfoxide (DMSO).
7. Dimethylformamide (DMF).

2.2 Materials for Bacterial Growth, Optimum Dilution, and OPK Assay

1. GAS glycerol stock.
2. Horse blood agar plate.
3. Baby Rabbit Complement (3–4-week baby rabbit complement, Pel-Freez Biologicals) (*see* **Note 1**).
4. Hank's balanced salt solution (HBSS) containing calcium chloride and magnesium chloride without phenol red (HBSS++) (1×).
5. Hank's balanced salt solution without calcium chloride, magnesium chloride or phenol red (HBSS−−) (1×).
6. 100 × 100 mm square petri dishes.
7. Automated colony counter (ProtoCOL3, Synbiosis).

2.3 Plates and Medium Preparation

1. Five percent of 2,3,5-tetraphenyltetrazolium chloride (TTC): make up in sterile pyrogen-free water and sterile filter, store at 4 °C.
2. One percent gelatin: prepare in sterile pyrogen-free water and autoclave.
3. Todd-Hewitt-Yeast (THY) agar plates: 3% Todd-Hewitt broth, 0.5% yeast, 1.5% agar. Prepare in sterile pyrogen-free water. Autoclave and pour into 100 × 100 mm square plates.
4. Todd-Hewitt-Yeast (THY) overlay agar: 3% Todd-Hewitt broth, 0.5% yeast, 0.75% agar. Make up on day of assay, autoclave, and keep at 55 °C.
5. Opsonization buffer (OPS buffer): 5% heat-inactivated FBS, 10% 10× HBSS (++), 0.1% gelatin. Make up in sterile pyrogen-free water, prepare on day of assay.
6. Todd-Hewitt Broth (THB). Autoclave.
7. Tryptone glucose glycerol (TGG) medium: 3% tryptic soy broth, 0.5% glucose, 10% glycerol. Make up in sterile pyrogen-free water and autoclave.

3 Methods

3.1 Preparation of GAS TGG Working Stocks

1. Streak from GAS glycerol stock onto a horse blood agar plate.
2. Incubate plate overnight at 37 °C, 5% CO_2 or until colonies appear (*see* **Note 2**).
3. Inoculate a single GAS colony into 10 ml of THB.
4. Incubate overnight at 37 °C in 5% CO_2 without agitation.
5. Dilute GAS overnight culture tenfold in THB, prewarmed to 37 °C, in a 50 ml Falcon tube and incubate at 37 °C in 5% CO_2 without agitation.
6. Once an absorbance at 600 nm of 0.5–0.7 is reached, place the tube on ice and mix the bacterial suspension with TGG medium at 1:1 ratio, vortex and aliquot 500 μl per vial and store at −80 °C.

3.2 Preparation of Bacteria for Use in Optimum Dilution and OPK Assays

1. Remove one vial of bacteria from the −80 °C freezer and defrost at 37 °C.
2. Centrifuge the tube for 2 min at 13,000–15,000 × *g* at room temperature (RT) and remove the supernatant, being careful not to disturb the pellet.
3. Wash once with 1 ml of OPS buffer and centrifuge the tube at 13,000–15,000 × *g* at RT, remove the supernatant, and resuspend the pellet in 0.5 ml OPS buffer.
4. Proceed making the predetermined dilution (*see* Subheading 3.6).

3.3 HL-60 Cell Expansion from ATCC Master Stock

1. Seed the HL-60 cells at a concentration of 2×10^5/ml in a T-75 culture flask in M1 medium (*see* **Notes 3** and **4**).
2. Grow the HL-60 cells in M1 medium suspension without agitation at 37 °C with 5% CO_2, until they reach a density of 5×10^5 cells/ml. Maintain cells between 2×10^5 cells/ml and 5×10^5 cells/ml with M1 medium. The typical growth culture schedule requires feeding every 2–3 days (*see* **Note 5**).
3. When the concentration reaches 5×10^5 cells/ml in the desired number of flasks, pellet the HL-60 cells by centrifugation at 350 × *g* for 5 min and add freezing medium to obtain a concentration of 1×10^7 cell/ml.
4. Aliquot 1 ml, of HL-60 cells in freezing medium, into cryovials and transfer cryovials to liquid nitrogen.

3.4 HL-60 Working Stock Preparation

1. Seed a HL-60 aliquot, prepared in subheading 3.3, at 2×10^5/ml in M1 medium in a T-75 culture flask (*see* **Note 6**).
2. Grow the HL-60 cells in suspension without agitation at 37 °C with 5% CO_2, maintaining a density of 2×10^5 to 1×10^6/ml by splitting every 3–4 days in M1 medium.

3. Once the cell growth accelerates (*see* **Note** 7), centrifuge at $350 \times g$ for 5 min at RT, discard supernatant, and resuspend pellet in warm M2 medium maintaining cells at 2×10^5 to 1×10^6/ml.
4. After approximately 2 weeks of maintenance in M2 medium, HL-60 cells can be differentiated for use in OPK assays (*see* **Note 8**).

3.5 HL-60 Cell Differentiation and Preparation for Use in Assays

1. To differentiate HL-60 cells, centrifuge at $350 \times g$ for 5 min at RT and discard the supernatant (*see* **Note 9**).
2. Resuspend HL-60 cells in 0.8% DMF prepared in M2 medium prewarmed to 37 °C.
3. Incubate the HL-60 cells at 37 °C with 5% CO_2 for 3–4 days (*see* **Note 10**). Do not feed the cells during this time (*see* **Note 11**).
4. After 3- or 4-days differentiation, decant HL-60 cells into 50 ml Falcon tubes and centrifuge tubes for 5 min at $350 \times g$ at RT.
5. Discard the supernatant and resuspend HL-60 cells in equal volume HBSS−−. Centrifuge for 5 min at $350 \times g$ at RT.
6. Remove the supernatant and resuspend cells in equal volume HBSS++.
7. Centrifuge for 5 min at $350 \times g$ at RT.
8. Remove the supernatant and resuspend HL-60 cells at 1×10^7 cells/ml in OPS buffer. Count the cells and assess the viability (*see* **Notes 12** and **13**).
9. Store at RT until required (maximum of 6 h).

3.6 Optimum Dilution Assay

This assay will determine the dilution of bacteria required for input in the OPK assay, *see* Fig. 1 for diagrammatic representation.

1. Remove the required number of baby rabbit complement aliquots from the −80 °C freezer and thaw on ice.
2. Thaw one GAS TGG stock (prepared as in Subheading 3.1) by placing a tube at 37 °C.
3. Following wash (*see* Subheading 3.2), dilute the bacteria in OPS buffer tenfold in well A1 of a 96-well plate (Fig. 1, **step 1**).
4. Serially dilute the bacteria fivefold down a column of the 96-well plate (Fig. 1, **step 2**).
5. In a second 96-well plate, add 20 μl of OPS buffer to two columns for each serotype of bacteria to be tested (Fig. 1, **step 3**).
6. Transfer 10 μl of diluted bacteria prepared in **steps 1–4** to the appropriate wells in the second plate (in duplicate) (Fig. 1, **step 4**).

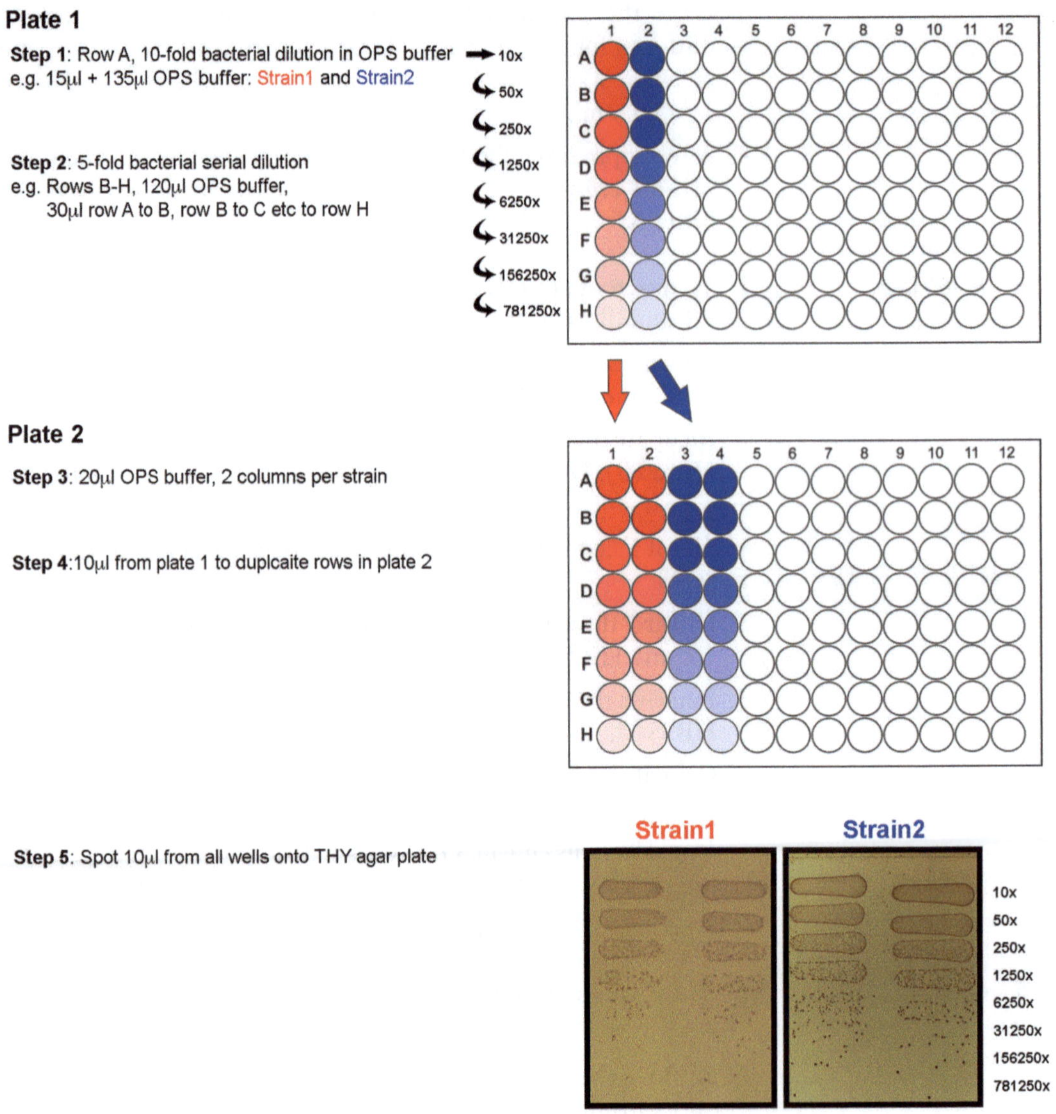

Fig. 1 Optimum dilution assay procedure. In this example, two GAS strain types are assessed for optimum dilution represented in red and blue. *OPS* opsonization, *THY* Todd-Hewitt Yeast

7. Incubate a 96-well plate at RT on a miniorbital shaker for 30 min at 700 rpm.
8. Following the incubation period, add 10 μl of active complement prediluted twofold in OPS buffer to each well in use in the 96-well plate (*see* **Note 14**).
9. Add 40 μl of differentiated HL-60 cells (prepared as in Subheading 3.5) to each well in the 96-well plate, covered with a lid that allows gas exchange.
10. Incubate on a miniorbital shaker at 37 °C/5% CO_2 for 45–90 min at 700 rpm.

11. Keep the plates on ice for 20 min to halt the phagocytic process.
12. Spot 10 μl from each column of 96-well plate onto a THY plate. Tilt the plate so the spots run and measure approximately 1 cm across. Allow the spot to dry at RT (Fig. 1, **step 5**).
13. Add 20 ml THY overlay agar (with TTC added at 1000-fold dilution) to each THY agar plate and wait for the overlay agar to solidify.
14. Incubate plates for 16–18 h in a 37 °C/5% CO_2 incubator, in this time bacteria should form red-colored colonies.
15. Read plates using an automated colony counter.
16. Determine the dilution of bacteria that yields 50–200 colony-forming units (CFUs) for input into OPK assay (*see* Subheading 3.7).

3.7 Opsonophagocytic Killing Assay (OPK Assay)

See Fig. 2 for a diagrammatic representation of the OPK assay using one GAS strain and five serum samples. Columns 1 and 2 are used as controls to calculate nonspecific killing (NSK) induced by complement alone and maximum bacterial growth in the absence of test sera. These values are crucial for the assay acceptance criteria and subsequent calculation of the "opsonic index" (*see* **Note 16**).

1. Remove the required number of complement aliquots from the −80 °C freezer and thaw on ice.
2. Add 20 μl of OPS buffer to columns 1 and 2, rows A–H of the 96-well plate (Fig. 2, **step 1**).
3. Add 20 μl of OPS buffer to columns 3–12, rows A–G (not row H) of the 96-well plate (Fig. 2, **step 2**).
4. Heat-inactivate the required volume of test serum by incubating at 56 °C for 30 min.
5. Make an initial dilution of test serum (twofold to tenfold) in OPS buffer and add 30 μl of this diluted serum to row H, columns 3–12 of plate in duplicate. Continue adding up to five samples in row H (Fig. 2, **step 3**).
6. Perform a threefold serial dilution of serum by transferring 10 μl from row H columns 3–12, through row A (Fig. 2, **step 4**).
7. Thaw one GAS TGG stock (prepared as in Subheading 3.1) by placing a tube at 37 °C.
8. Following wash (*see* Subheading 3.2) and resuspension in the appropriate volume of OPS buffer, add 10 μl of bacterial mixture to all wells (*see* **Note 15**) (Fig. 2, **step 5**).
9. Incubate the 96-well plates for 30 min at RT on a miniorbital shaker at 700 rpm (Fig. 2, **step 6**). This allows time for antibodies to opsonize the bacteria.

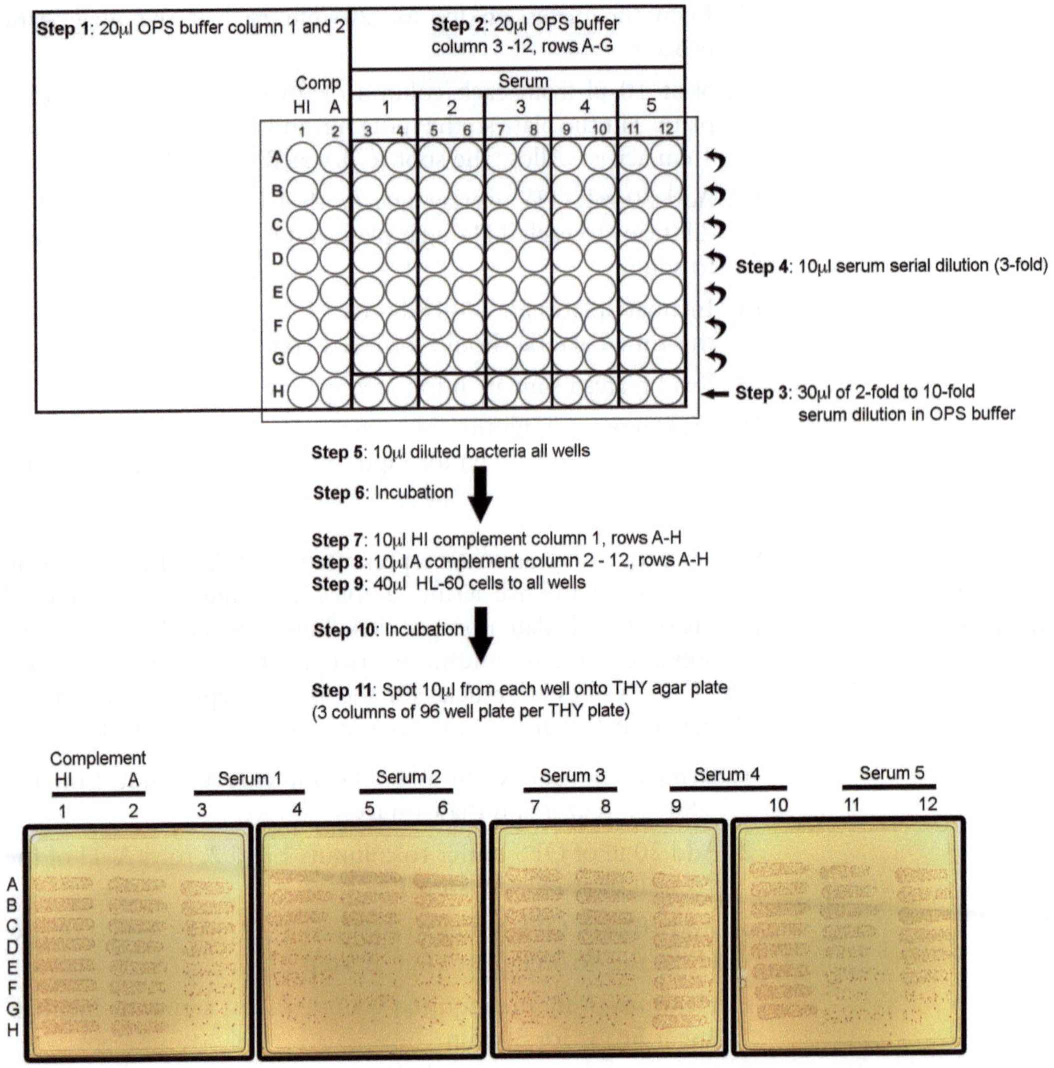

Fig. 2 OPK assay procedure setup. In this setup, one 96-well plate is spotted onto four THY plates. In this example, killing can be observed for all sera, except serum 4. Note the difference in bacterial number between column 1 and 2 is used to calculate the nonspecific killing. *Comp* Complement, *A* Active complement, *HI* Heat-Inactivated Complement, *OPS* opsonization, *THY* Todd-Hewitt Yeast

10. Heat-inactivate the required volume of complement by incubating at 56 °C for 30 min.
11. Following the 30-min incubation period, add 10 μl of heat-inactivated complement prediluted twofold in OPS buffer to, column 1 (Fig. 2, **step 7**).
12. Add 10 μl of active complement prediluted twofold in OPS buffer to all other wells (columns 2–12) (*see* **Note 16**) (Fig. 2, **step 8**).

13. Add 40 μl of differentiated HL-60 cells (prepared as in Subheading 3.5) to all wells and cover with a lid that allows gas exchange (Fig. 2, **step 9**).
14. Incubate on a miniorbital shaker at 700 rpm at 37 °C/5% CO_2 for 45–90 min (*see* **Note 17**) (Fig. 2, **step 10**).
15. Put plates on ice for 20 min to halt the phagocytosis process.
16. Spot 10 μl from each column of 96-well plate onto a THY plate. Tilt the plate so the spots run and measure approximately 1 cm across. Allow to dry at RT (Fig. 2, **step 11**).
17. Add 20 ml of overlay agar (with added TTC at 1000-fold dilution) to each THY agar plate and wait for overlay agar to solidify.
18. Incubate plates for 16–18 h in a 37 °C/5% CO_2 incubator, in this time bacteria should form red-colored colonies.
19. Read plates using an automated colony-counting software (*see* **Note 18**).

4 Notes

1. Baby rabbit complement is extremely sensitive to temperature fluctuations. When received from supplier, the complement should be thawed on ice and aliquoted into 1–5 ml aliquots for use in assay. Once frozen, aliquots should only be thawed once for use in assays and the remainder can be heat-inactivated for use as control.
2. Colonies are beta-hemolytic and may vary in morphology between strain types depending on the level of hyaluronic acid encapsulation.
3. All cell cultures should be undertaken in a culture hood under sterile conditions. It is highly recommended to have tissue culture facilities in an area physically separate from the bacterial work lab space.
4. Cell viability is assessed by diluting HL-60 cells in trypan blue followed by counting on a hemocytometer.
5. Cell concentrations must not exceed 5×10^5/ml at this stage.
6. It is highly recommended that HL-60 cells are tested regularly for mycoplasma contamination using a commercial polymerase chain reaction (PCR) detection kit (Universal Mycoplasma Detection Kit, ATCC or similar). Mycoplasma detection should be carried out before freezing down master aliquots and monthly for working stocks.
7. Following reconstitution from liquid nitrogen, HL-60 cells take some time to recover and for cell growth to accelerate. In the first 2 weeks following reconstitution, cells will typically

only need splitting every 5–7 days in order to maintain the cell density below 1×10^6 cells/ml. After this initial phase, splitting of HL-60 cells will be required every 3–4 days in order to maintain the cell density below 1×10^6 cells/ml. Once splitting is required every 3–4 days, HL-60 cells can be propagated in M2 medium.

8. It is recommended that HL-60 cells only be used for differentiation up to passage 35–40.
9. HL-60 cell viability must be ≥90% prior to differentiation.
10. The day HL-60 cells should be used following differentiation must be experimentally defined using flow cytometry and is typically day 3–6. Cells should be used when they show maximum expression of CD35, minimum expression of CD71, and fewest dead cells (*see* **Note 11** for details of flow cytometry analysis).
11. To determine whether the HL-60 cells are correctly differentiated, flow cytometry analysis is strongly recommended. Typically, the differentiated cultures should show 80% viability estimated using trypan blue. The cell surface markers that confirm the acquisition of the differentiated phenotype are maturation marker CD35 (Antibody clone DF1513, Bio-Rad) and proliferation markers CD71 and CD35 (Antibody clone E11, Bio-Rad) (Fig. 3). More than 55% of the cells in the cultures should express CD35 and less than 12% of the cells should maintain CD71 expression. The cultures should not have an abundance of dead cells, which can be determined using commercial viability staining kits (such as the LIVE/DEAD cell stain kits from Thermo Fisher Scientific). Typically, more than 85% of the cells should show negative staining with viability stain, indicating the cells are alive.

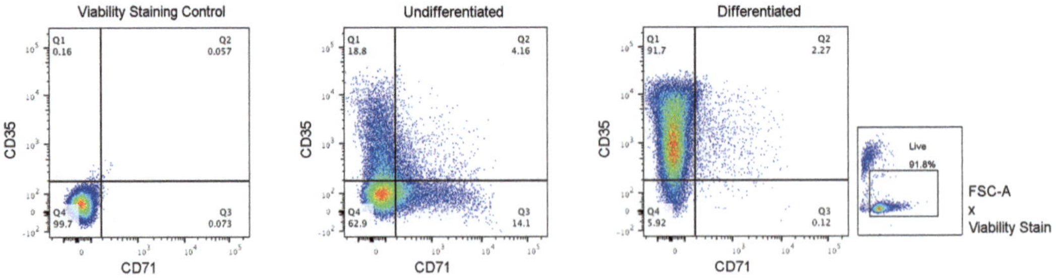

Fig. 3 Undifferentiated compared to differentiated HL-60 phenotype. The viability staining control shows a sample stained only with the viability stain and is used as a negative staining control to set gates for CD71 and CD35. In this example, the undifferentiated HL-60 cells have 18.8% CD35+ and 14.1% CD71+, while the differentiated cells have 91.7% CD35+ and 0.12% CD71+. In this example, the differentiated HL-60 cells show 91.8% live cell staining

12. Do not use HL-60 cells that exceed 1.5×10^6 cells/ml at initial count, as this may indicate a problem during differentiation.
13. HL-60 cell viability before and after HBSS washes should be greater than 80%, as assessed by trypan blue staining.
14. For some GAS strains, dilution of baby rabbit complement will need to be optimized by carrying out the optimum dilution assay with several different dilutions of complement. Some strains may require a higher concentration of complement, and it is recommended the highest concentration of complement, which gives a maximum of 35% nonspecific killing (NSK), is used for the OPK assay.
15. GAS bacteria are added at the concentration determined during the optimal dilution assay (Subheading 3.6).
16. Columns 1 and 2 are used as controls to calculate NSK induced by complement alone (average counts of column 2 containing active complement, compared to average counts of column 1 containing heat-inactivated complement). It is recommended this does not exceed 35% for the GAS OPK assay. In addition, the maximum bacterial growth with all assay components aside from serum is calculated from average counts of column 2. It is from this maximum that the 50% killing criteria are calculated and thus it is important to have numbers of colonies countable by an automated colony counter (*see* **Note 18 (b)**).
17. The incubation time of GAS with HL-60 cells is another parameter that can be adjusted depending on the GAS strain type. Incubation times of between 45 and 90 min are likely to enable phagocytosis. For longer incubation times, the dilution of complement added to the assay will need to be adjusted accordingly in order to prevent high NSK of >35%.
18. To facilitate data analysis, Opsotiter, an Excel®-based template, has been developed by researchers at the Bacterial Respiratory Pathogen Reference Laboratory at the University of Alabama at Birmingham (UABRF, Birmingham, AL, USA). A license to use this software can be requested from UABRF [12]. Data obtained from the colony counter can be directly inserted into the Opsotiter software that converts colony counts to an "opsonic index" (OI) as well as calculating NSK for the OPK assay and maximum killing for each serum sample tested. The OI is defined as the dilution of serum that kills 50% of bacteria and is determined using linear interpolation. For GAS OPK assays, the following acceptance criteria for determining an OI are recommended:

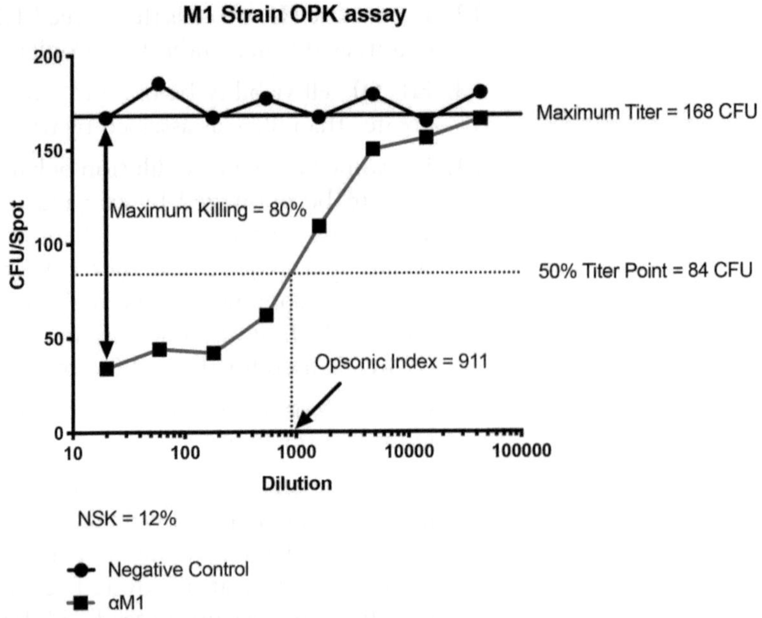

Fig. 4 Example output of M1 OPK assay. OPK assay with M1 GAS using vaccinated rabbit serum targeting the M1 protein (squares) and a negative control (circles). Indicated on graph are maximum average growth (solid horizontal line, calculated from column 2), 50% titer point (dashed horizontal line, half of maximum average growth), maximum killing (vertical double-headed arrow, the lowest CFU achieved by immune serum as the percentage of maximum average growth), and opsonic index (OI; arrow, dilution of serum that kills 50% of bacteria). Nonspecific killing (NSK; average counts of column 2, compared to average counts of column 1) is also calculated on Opsotiter spreadsheet and is indicated below the graph

(a) NSK of less than 35% (*see* **Note 14**).

(b) Maximum killing of at least 70% for test serum. This enables reliable linear interpolation of the 50th centile and determination of an accurate OI.

OIs can be compared between sera in assays to the same GAS strain type. Direct comparison of OIs between different GAS strain types is not advised. *See* Fig. 4 for an example result of a GAS OPK assay.

References

1. Ralph AP, Carapetis JR (2013) Group A streptococcal diseases and their global burden. Curr Top Microbiol Immunol 368:1–27
2. Sheel M, Moreland N, Fraser J et al (2015) Development of group a streptococcal vaccines: an unmet global health need. Expert Rev Vaccines 15:238–277
3. Lancefield RC (1957) Differentiation of group a streptococci with a common R antigen into three serological types, with special reference to the bactericidal test. J Exp Med 106:525–544
4. Reglinski M, Lynskey N, Sriskandan S (2016) Modification of the classical Lancefield assay of group a streptococcal killing to reduce inter-donor variation. J Microbiol Methods 124:69–71
5. Steer AC, Carapetis JR, Dale JB et al (2016) Status of research and development of vaccines for Streptococcus pyogenes. Vaccine 34:2953–2958
6. Osowicki J, Vekemans J, Kaslow DC et al (2018) WHO/IVI global stakeholder consultation on group a streptococcus vaccine development: report from a meeting held on 12-13 December 2016. Vaccine 36:3397–3405
7. Romero-Steiner S, Libutti D, Pais L et al (1997) Standardization of an opsonophagocytic assay for the measurement of functional antibody activity against Streptococcus pneumoniae using differentiated HL-60 cells. Clin Diagn Lab Immunol 4:415–422
8. Romero-Steiner S, Frasch CE, Carlone G et al (2006) Use of Opsonophagocytosis for serological evaluation of pneumococcal vaccines. Clin Vaccine Immunol 13:165–169
9. Fleck R, Romero-Steiner S, Nahm M (2005) Use of HL-60 cell line to measure opsonic capacity of pneumococcal antibodies. Clin Diagn Lab Immunol 12:19–27
10. Jones S, Moreland NJ, Zancolli M et al (2018) Development of an opsonophagocytic killing assay for group a streptococcus. Vaccine 36:3756–3763
11. Salehi S, Hohn CM, Penfound TA et al (2018) Development of an Opsonophagocytic killing assay using HL-60 cells for detection of functional antibodies against Streptococcus pyogenes. Msphere 3:e00617–e00618
12. Burton RL, Nahm MH (2006) Development and validation of a fourfold multiplexed Opsonization assay (MOPA4) for pneumococcal antibodies. Clin Vaccine Immunol 13:1004–1009

Part V

Group A Streptococcus Therapies

Chapter 27

Validation of Suitable Carrier Molecules and Target Genes for Antisense Therapy Using Peptide-Coupled Peptide Nucleic Acids (PNAs) in Streptococci

Gina Barkowsky, Bernd Kreikemeyer, and Nadja Patenge

Abstract

Antisense peptide nucleic acids (PNAs) targeting genes involved in metabolism or virulence are a possible means to treat infections or to investigate pathogenic bacteria. Potential targets include essential genes, virulence factor genes, or antibiotic resistance genes. For efficient cellular uptake, PNAs can be coupled to cell-penetrating peptides (CPPs). CPPs are peptides that serve as molecular transporters and are characterized by a comparably low cytotoxicity. So far, there is only limited information about CPPs that mediate PNA uptake by Gram-positive bacteria. Here, we describe two methods to identify suitable CPP–antisense PNA conjugates, novel carrier molecules, and efficient target genes for streptococcal species and to evaluate their antimicrobial efficiency.

Key words Peptide nucleic acids, Cell-penetrating peptides, PNA killing assay, MIC, *Streptococcus pyogenes*, *Streptococcus pneumoniae*, *Streptococcus salivarius*

1 Introduction

Peptide nucleic acids (PNAs) are currently gaining importance in diagnostics [1], food analytics [2], and nanotechnology [3]. PNAs are synthetic nucleic acid derivatives, in which nucleobases are bound to a pseudopeptide backbone with methyl carbonyl bonds. PNA molecules can serve as antisense agents inhibiting transcription or translation [3]. Today, a major goal of public health is to limit the application and distribution of antibiotics. In this context, the use of antisense PNAs targeting essential genes or antibiotic resistance genes is extremely interesting. Bacterial cell walls and membranes limit the translocation of PNAs into bacteria. To improve cellular uptake, PNAs are frequently coupled to cell-penetrating peptides (CPPs) [4]. These are naturally occurring or synthetic peptides composed of 5–30 amino acids that are classified by their properties into cationic, hydrophobic, and amphipathic CPPs [5]. However, not all CPPs support uptake into bacteria.

Thomas Proft and Jacelyn M. S. Loh (eds.), *Group A Streptococcus: Methods and Protocols*, Methods in Molecular Biology, vol. 2136, https://doi.org/10.1007/978-1-0716-0467-0_27, © Springer Science+Business Media, LLC, part of Springer Nature 2020

CPP-mediated translocation depends on the cell membrane and cell wall composition of the respective bacteria [6, 7]. Typical examples of CPPs employed in bacteria are the synthetic $(KFF)_3K$ and the HIV-1 Transactivator of Transcription (HIV-1 TAT) derived peptides. $(KFF)_3K$ facilitated the uptake of PNAs, among others, in *Escherichia coli* and *Staphylococcus aureus* [8, 9]. HIV-1 TAT was able to penetrate *Listeria monocytogenes*, *S. aureus*, and *Staphylococcus epidermidis* [10, 6]. We observed previously that HIV-1 TAT--coupled anti-*gyrA* PNAs were able to inhibit growth in *S. pyogenes* [11]. Recently, our group has identified two additional CPPs (K8, and $(RXR)_4XB$) for effective PNA uptake into *Streptococcus pyogenes* [12] and one CPP ($(RXR)_4XB$) for the uptake into *Streptococcus pneumoniae* and *Streptococcus salivarius* (unpublished data). Here, we introduce two methods to evaluate CPP–antisense PNA antimicrobial efficiency in Streptococci: (1) A PNA killing assay, which is employed to determine bacterial cell count reduction following CPP–PNA treatment. This experimental setup can also be used to investigate enzyme activity in the bacterial crude extract or to measure target gene transcript levels [13] following treatment. (2) A modified PNA minimal inhibitory concentration (MIC) assay. The MIC is used as standardized measure for the antimicrobial effect of drugs [14]. Since streptococci are fastidious organisms [15] and the solubility of CPP–PNA conjugates varies depending on the properties of the two components, a specific protocol for the PNA MIC assay was developed.

We recommend performing the PNA kill assay first, because less PNA is required. Thus, it is cheaper and allows an initial screening for efficient CPP–PNAs.

2 Materials

2.1 PNA Stock Solution

1. 1 mM CPP–PNA stock: Preheat heating block to 60 °C. Add *x* µl of sterile Molecular Biology Grade water and incubate for at least 10 min at room temperature (RT) for swelling. Vortex tube thoroughly and spin tube shortly to collect the material on the bottom. Incubate the tube for 10 min at 60 °C in the heating block, vortex thoroughly, and spin the tube shortly. Incubate the tube for 2 days at 4 °C to ensure complete dissolution of the PNA. Store stock solution at 4 °C (*see* **Notes 1** and **2**).
2. 10 mM PNA stock: Preheat heating block to 60 °C. Add *x* µl of sterile Molecular Biology Grade water and proceed as described in Subheading 2.1, **item 1**.

$$x\ \mu l = \frac{\text{nmol (CPP} - \text{PNA)}}{\text{mM(stock solution)}}$$

2.2 PNA Killing Assay

1. Selected Streptococcus strain (*see* **Note 3**).
2. Molecular Biology Grade water, sterile (referred to as water throughout the protocol).
3. PNA or CPP–PNA conjugate. Lyophilized powder. Prepare a 1 mM stock solution of the PNA or CPP–PNA of interest in water (*see* Subheading 2.1).
4. Phosphate-buffered saline (PBS): 137 mM NaCl, 2.7 mM KCl, 10 mM Na_2HPO_4 and 2 mM KH_2PO_4, pH 7,4. Sterilize by autoclaving. Store at room temperature (RT).
5. Brain Heart Infusion Broth (BHI): Prepare according to the manufacturer's instructions.
6. THY agar plates: Todd-Hewitt broth supplemented with 0.5% yeast extract and 15% agar.
7. Laboratory tube rotator placed in a 37 °C incubator.

2.3 PNA Minimal Inhibitory Concentration (MIC) Assay

1. Selected Streptococcus strain (*see* **Note 3**).
2. PNA or CPP–PNA conjugate. Lyophilized powder. Prepare a 10 mM stock solution of the PNA or CPP–PNA of interest in water (*see* Subheading 2.1).
3. Laked Horse Blood (LHB): 1 ml LHB is mixed with 1 ml sterile water. The solution is frozen and thawed five times (*see* **Note 4**). Following centrifugation at 1608 × *g* for 30 min, the supernatant is decanted and the pellet is stored at −20 °C.
4. Cation-adjusted Mueller–Hinton Broth (CAMHB).
5. LHB/CAMHB medium: Add 2 ml CAMHB to the LHB pellet and mix thoroughly.
6. Two percent acetic acid: Store at 4 °C.
7. Forty percent bovine serum albumin (BSA): Weigh 10 g BSA and transfer to a centrifuge tube (*see* **Note 5**). Add water to 25 ml and sterilize by filtration. Store at 4 °C.
8. Filter with a pore size of 0.2 μM or smaller.
9. Solution 1: 0.02% acetic acid and 0.4% BSA. Dilute 2% acetic acid and 40% BSA in LHB/CAMHB medium (1:100) directly before use. Add 3 μl of 2% acetic acid and 3 μl of 40% BSA solution to 294 μl LHB/CAMHB medium.
10. Solution 2: 0.01% acetic acid and 0.2% BSA. Dilute solution 1 with LHB/CAMHB medium (1:2).
11. 96-well clear polypropylene corner notch microplate.
12. THY agar plates. See above.
13. Phosphate-buffered saline (PBS) (*see* Subheading 2.2, **item 4**).

3 Methods

3.1 PNA Killing Assay

1. Overnight culture: Grow bacterial strains in Todd-Hewitt broth supplemented with 0.5% yeast extract (THY) or brain heart infusion broth (BHI) at 37 °C under a 5% CO_2/20% O_2 atmosphere.
2. Prepare from your PNA or CPP–PNA stock solution a 10× working solution in water.
3. Add 50 μl of 10× PNA solution, 10× CPP–PNA solution, or water (mock control) to a 2 ml microcentrifuge tube.
4. Dilute the bacterial overnight culture 1:2000 in PBS/BHI (*see* **Note 6**).
5. Transfer 450 μl of the bacterial suspension to each sample and mix thoroughly.
6. Remove 100 μl of your sample and prepare a tenfold dilution series. Spread appropriate dilutions evenly on the surface of the agar plates. Incubate overnight at 37 °C and 5% CO_2.
7. Incubate the remaining samples at 37 °C and 7 rpm (*see* **Note 7**).
8. Collect 100 μl samples at different time points throughout the experiment and at the end of your incubation. Prepare tenfold dilution series. Spread appropriate dilutions evenly on the surface of the agar plates. Incubate overnight at 37 °C and 5% CO_2.
9. After incubation, count plates from each series that contains 30–300 CFUs and calculate the CFU/ml or the LOG-reduction using the following calculation:

$$\frac{(\text{Average count})}{(\text{Dilution plated}) \times (\text{ml plated})} = \text{CFU/ml}$$

$$\log\left(\frac{\text{CFU}}{\text{ml}}\text{untreated}\right) - \log\left(\frac{\text{CFU}}{\text{ml}}\text{PNA treatment}\right) = \log - \text{reduction}$$

3.2 PNA MIC Assay

The experimental setup is shown schematically in Fig. 1.

1. Overnight culture: Culture bacterial strains in Todd-Hewitt broth supplemented with 0.5% yeast extract (THY) or brain heart infusion broth (BHI) at 37 °C under a 5% CO_2/20% O_2 atmosphere.
2. 10 μl of solution 1 is added to the first well of each row (Fig. 1, number 1) and 100 μl of solution 1 is added to one control well (Fig. 1, "-").

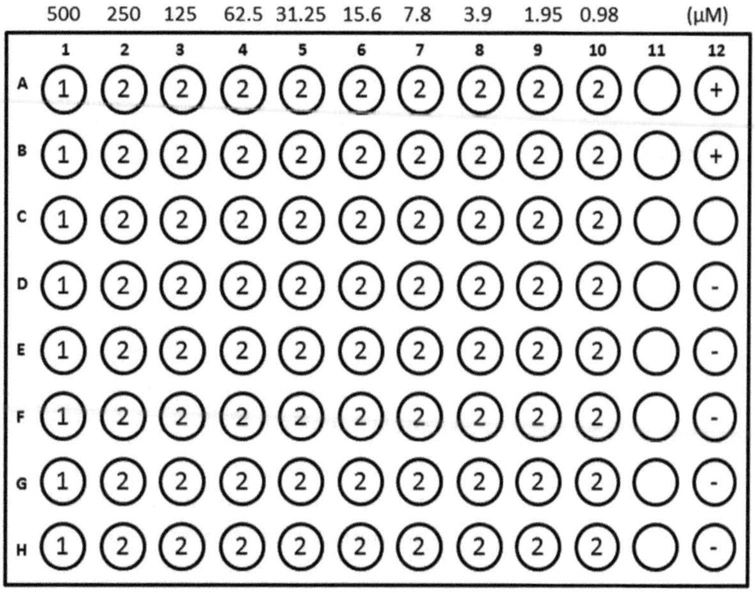

Fig. 1 Schematic illustration of the MIC assay experimental setup in a 96-well plate. Number 1: 10 μl of solution 1; Number 2: 10 μl of solution 2; (+): growth control, 100 μl of the bacterial suspension; (−): medium control, 100 μl of solution 1 or 100 μl of solution 2, respectively

3. 10 μl of solution 2 is added to the wells of rows 2–10 (Fig. 1, number 2) and 100 μl of solution 2 is added to one control well (Fig. 1, "-").
4. 10 μl PNA stock solution is added to the first well and mixed carefully. 10 μl of the mixture is transferred into the next well. This procedure is repeated for each following column. In row 10, 10 μl mixture is discarded.
5. Prepare the bacterial suspension: Dilute the bacteria in LHB/CAMHB medium to 10^5 CFU/ml (*see* **Note 8**).
6. 100 μl of the bacterial suspension is added to a growth control well (Fig. 1, "+"). 90 μl bacterial suspension is added into test wells containing PNA solution.
7. Use 100 μl of the bacterial suspension to prepare a tenfold dilution series. Spread appropriate dilutions evenly on the surface of THY agar plates. Incubate overnight at 37 °C and 5% CO_2. Calculate the CFU/ml of your inoculum.
8. Spread 100 μl LHB/CAMHB medium evenly on the surface of THY agar plates in order to rule out contaminations.
9. Incubate the 96-well plate for 16–20 h at 37 °C and 5% CO_2.
10. Determine the turbidity of the wells by eye. Compare the control wells and growth control wells with PNA-treated

wells. MICs are recorded as the lowest concentration where no turbidity is observed in the wells (*see* **Note 9**).

4 Notes

1. Spin the PNA tube shortly to collect the powder on the bottom.
2. If your PNA conjugate exhibits a low solubility in water, addition of organic solvents, for example DMSO, may be required. Stocks can be stored at −20 °C for long-term storage. Heating to 60 °C and vortexing before further use are recommended.
3. We used *Streptococcus pyogenes, Streptococcus pneumoniae*, and *Streptococcus salivarius* for our experiments. *S. pyogenes* and *S. salivarius* strains were cultured in Todd-Hewitt broth supplemented with 0.5% yeast extract (THY) at 37 °C under a 5% CO_2/20% O_2 atmosphere. *S. pneumoniae* strains were cultured for 8 h in brain heart infusion broth (BHI) at 37 °C under a 5% CO_2/20% O_2 atmosphere. You may need different culture conditions for your organism.
4. We are performing this process over 5 days.
5. Forty percent BSA takes long to dissolve completely. We recommend preparing the solution at least 1 day in advance.
6. You need to determine the optimal ratio of PBS and BHI for each bacterial strain. Bacteria should not grow or die over the course of your experiment. Mix PBS and BHI in different ratios, add bacteria to each mixture, and determine the CFUs/ml before and after incubation.
7. Rotation is crucial for the PNA kill assay. Incubation time may vary.
8. Before performing the MIC assay, you need to determine the optimal dilution of your bacteria in order to get an inoculum of 10^5 CFU/ml.
9. This step is very important. Both controls are required to be able to perceive a difference in turbidity between the wells.

Acknowledgments

The work of G.B. and B.K. was supported by a grant from the Ministerium für Bildung, Wissenschaft und Kultur, Mecklenburg-Vorpommern (ESF/14-BM-A55-0010/16). The work of NP was supported by the University Medicine Rostock (Forun 889008).

References

1. Dean DA (2000) Peptide nucleic acids: versatile tools for gene therapy strategies. Adv Drug Deliv Rev 44(2–3):81–95
2. Sforza S, Corradini R, Tedeschi T et al (2011) Food analysis and food authentication by peptide nucleic acid (PNA)-based technologies. Chem Soc Rev 40(1):221–232
3. Sharma C, Awasthi SK (2017) Versatility of peptide nucleic acids (PNAs): role in chemical biology, drug discovery, and origins of life. Chem Biol Drug Des 89(1):16–37
4. Trabulo S, Cardoso AL, Mano M et al (2010) Cell-penetrating peptides—mechanisms of cellular uptake and generation of delivery systems. Pharmaceuticals 3(4):961–993
5. Wang F, Wang Y, Zhang X et al (2014) Recent progress of cell-penetrating peptides as new carriers for intracellular cargo delivery. J Control Release 174:126–136
6. Abushahba MF, Mohammad H, Thangamani S et al (2016) Impact of different cell penetrating peptides on the efficacy of antisense therapeutics for targeting intracellular pathogens. Sci Rep 6:20832
7. Gomarasca M, Martins TF, Greune L et al (2017) Bacterium-derived cell-penetrating peptides deliver gentamicin to kill intracellular pathogens. Antimicrob Agents Chemother 61 (4):e02545–e02516
8. Eriksson M, Nielsen PE, Good L (2002) Cell permeabilization and uptake of antisense peptide-peptide nucleic acid (PNA) into *Escherichia coli*. J Biol Chem 277 (9):7144–7147
9. Nekhotiaeva N, Awasthi SK, Nielsen PE et al (2004) Inhibition of *Staphylococcus aureus* gene expression and growth using antisense peptide nucleic acids. Mol Ther 10(4):652–659
10. Abushahba MF, Mohammad H, Seleem MN (2016) Targeting multidrug-resistant staphylococci with an anti-rpoA peptide nucleic acid conjugated to the HIV-1 TAT cell penetrating peptide. Mol Ther Nucleic Acids 5(7):e339
11. Patenge N, Pappesch R, Krawack F et al (2013) Inhibition of growth and gene expression by PNA-peptide conjugates in *Streptococcus pyogenes*. Mol Ther Nucleic Acids e132:2
12. Barkowsky G, Lemster A-L, Pappesch R et al (2019) Influence of different cell-penetrating peptides on the antimicrobial efficiency of antisense PNAs in Streptococcus pyogenes. Mol Ther Nucleic Acids 18:444–454
13. Good L, Nielsen PE (1999) Peptide nucleic acid (PNA) antisense effects in Escherichia coli. Curr Issues Mol Biol 1(1–2):111–116
14. Wen X, Gehring R, Stallbaumer A et al (2016) Limitations of MIC as sole metric of pharmacodynamic response across the range of antimicrobial susceptibilities within a single bacterial species. Sci Rep 6:37907
15. Weinstein MP (2018) Methods for dilution antimicrobial susceptibility tests for bacteria that grow aerobically, vol 26. Clinical and Laboratory Standards Institute, Pittsburgh, pp 14–16

Chapter 28

Interfering with the Folding of Group A Streptococcal pili Proteins

Fernanda Contreras and Jaime Andrés Rivas-Pardo

Abstract

Gram-positive bacteria use their adhesive pili to attach to host cells during early stages of a bacterial infection. These extracellular hair-like appendages experience mechanical stresses of hundreds of picoNewtons; however, the presence of an internal isopeptide bond prevents the pilus protein from unfolding. Here, we describe a method to interfere with nascent pili proteins through a peptide that mimics one of the β-strands of the molecule. By using AFM-based force spectroscopy, we study the isopeptide bond formation and the effect of the peptide in the elasticity of the pilus protein. This method could be used to afford a new strategy for mechanically targeted antibiotics by simply blocking the folding of the bacterial pilus protein.

Key words Isopeptide bond, Single-molecule force spectroscopy, Protein folding

1 Introduction

During the early stages of the bacterial colonization, pathogens use specialized protein to adhere themselves to host cells and surfaces [1]. These protein structures, known as pili, withstand large pulling forces imposed by the mucus flow or other physiological mechanism implemented in the host [2, 3]. Whereas in Gram-negative bacteria, such as *E. coli*, the pili are made of hundreds of protein domains only stabilized by hydrogen bonds and other noncovalent interactions [4], and the pili from Gram-positive bacteria are assembled as a single polyprotein where each protein subunit is covalently linked through intermolecular bonds [5] (Fig. 1a). Moreover, Gram-positive bacteria have implemented a novel strategy to prevent the unfolding of the structural pili subunits during the pulling perturbation. Pili subunits form an intramolecular bond between a Lys residue from the first β-strand and an Asn residue located in the C-terminal portion of the pilus domain [6], establishing an isopeptide bond that prevents the mechanical unfolding of the molecule [7].

Thomas Proft and Jacelyn M. S. Loh (eds.), *Group A Streptococcus: Methods and Protocols*, Methods in Molecular Biology, vol. 2136, https://doi.org/10.1007/978-1-0716-0467-0_28,

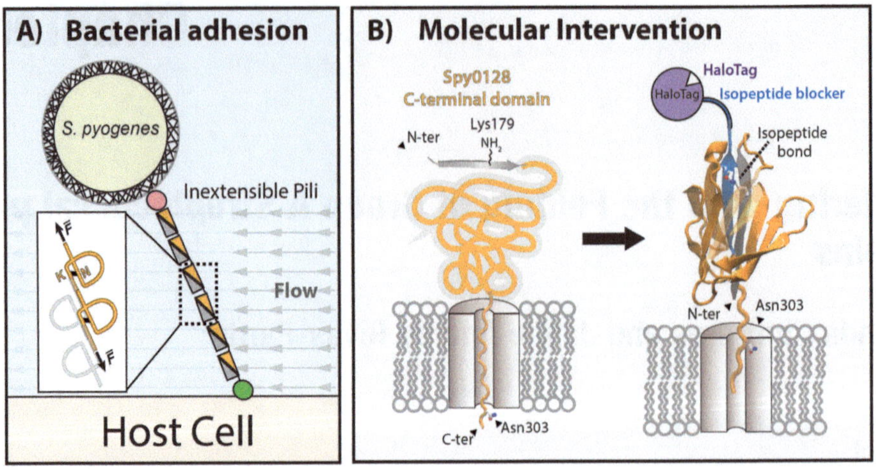

Fig. 1 Gram-positive bacterial adhesion. (**a**) The *Streptococcus pyogenes* coccoid use its micrometer-long pilus to attach to the host cell during the infection. The pilus is formed by three major proteins subunits: Spy0125 or adhesin (green subunit), which contains the binding domain to the host cell and is located at the end of the pilus; Spy0128 or shaft protein (gray/yellow subunit), which forms the rod of the pilus; and Spy130 or base (pink subunit), which anchors the entire micrometer-long filament to the cell wall [17]. The structural Spy0128 pilin subunit contains two domains, but only the C terminus is under force [6, 7]. The presence of the isopeptide bond between Lys179 and Asn303 prevents the mechanical unfolding. (**b**) The isopeptide blocker mimics one of the β-strands of the Spy0128 pilin, which contains the Asn residue required for the formation of the isopeptide bond. During the folding of the Spy0128 molecule, the β-strand-8 is the last structural element to emerge from the tunnel. Thus, the isopeptide blocker could be recognized for the rest of the pili Spy0128 in via of folding, forming an intermolecular isopeptide bond between the pilin domain and the isopeptide blocker [8]. The isopeptide blocker is expressed as a HaloTag fusion protein, which allows specific decoration during the protein purification (*see* Subheading 3)

Here, we describe a method to interfere with nascent pili molecules based on a peptide that impairs the formation of the intramolecular isopeptide bond [8]. The interfering peptide—*isopeptide blocker*—mimics one of the β-strands that forms the isopeptide bond. Moreover, the isopeptide blocker also contains a HaloTag for specific and straightforward protein labeling and a Tobacco Etch Virus (TEV) site to sever the protein tag [8] (Fig. 1b). Specifically, we use a system based on the structural pilus domain of *Streptococcus pyogenes* and an isopeptide blocker specially designed to interfere with the formation of the intramolecular bond between Lys179 and Ans303 [8]. The isopeptide blocker binds covalently to the pilus domain through the intermolecular isopeptide bond, not only turning the Spy0128 into an extensible protein but also interfering in the proper folding of the pilus molecule [8].

2 Materials

2.1 Protein Engineering and Isopeptide-Blocker Design

1. Plasmid pQE-80L (Amp^r) (Qiagen).
2. Plasmid pBAD18a (Kan^r) (lab customized, introduction of RBS sequence, *see* **Note 1**).
3. Plasmid pT7Blue (Amp^r) (Novagen).
4. Plasmid pFN18a TEV-HaloTag (Amp^r) (Promega).
5. Restriction enzymes *Eco*RI, *Dra*III, *Bam*HI, *Bgl*II, and *Kpn*I (with supplied buffers).
6. T4 DNA ligase (with supplied buffer).
7. High-fidelity DNA polymerase and buffer supplied by the vendor.
8. 10-mM dNTPs.
9. *E. coli* strains: XL10-Gold® (Stratagene), and BLR (DE3) pLysS (Novagen).
10. DNA primers at 100 μM. RBS-Fwd: 5′-AAT TGA TTA AAG AGG AGA AAT TAA CTA TGG AAT TCC CAT GGG GTA CCG-3′ and RBS-Rev: 5′-TCG ACG GTA CCC CAT GGG AAT TCC ATA GTT AAT TTC TCC TCT TTA ATC-3′. This pair introduces an RBS (underlined sequence) followed by a *Eco*RI and *Kpn*I site. IB-Fwd: 5′-GA TCC GAA TTC ACT GAT AAA GAT ATG ACC ATT ACT TTT ACA AAT AAA GAC GCT GAA CAC ATC GTG A-3′ and Rev: 5′-GA TCT CAC GAT GTG TTC AGC GTC TTT ATT TGT AAA AGT AAT GGT CAT ATC TTT ATC AGT GAA TTC G-3′. This pair of primers introduces the isopeptide blocker (underlined sequence) flanked by *Eco*RI in the 5′ and *Dra*III (screening purposes only) and *Bgl*II sites in the 3′-end.
11. DNA containing the pilus gene, either genomic DNA or plasmid DNA.
12. Gel DNA band purification kit.
13. Plasmid DNA purification kit.

2.2 Protein Expression

1. Luria-Bertani (LB) broth: 1% (w/v) bacto tryptone, 0.5% (w/v) yeast extract, and 1% (w/v) NaCl, diluted in distilled water and adjusted to pH 7.0.
2. LB agar plates: LB broth supplemented with 1% (w/v) agar.
3. Minimum medium: M9CA medium broth (VWR), supplemented with: 1.5% (w/v) glycerol, 40 μg/mL thymine, 2 mM $MgSO_4$, and 0.1 mM $CaCl_2$.

4. Antibiotics: 12.5 μg/mL tetracycline, 12.5 μg/mL chloramphenicol, 35 μg/mL kanamycin, and 100 μg/mL carbenicillin.
5. 0.2% L-arabinose.
6. 1-mM Isopropyl β-D-1-thiogalactopyranoside (IPTG).

2.3 Protein Purification

1. 10% w/v Triton X-100.
2. 100 mg/mL Lysozyme, diluted in 10 mM Tris–HCl pH 8.4.
3. 100-mM phenylmethylsulfonyl fluoride (PMSF), freshly diluted in 100% ethanol.
4. 1 mg/mL DNase I grade II, diluted in 20-mM Tris–HCl pH 7.5, 1-mM $MgCl_2$, 50% w/v glycerol.
5. 1 mg/mL RNase A, diluted in 10-mM Tris–HCl pH 7.5, 15-mM NaCl.
6. Elution/wash buffer: 50-mM Na-phosphate buffer pH 7.0, 300-mM NaCl.
7. Elution buffer: 50-mM Na-phosphate buffer pH 7.0, 300-mM NaCl, 150-mM Imidazole.
8. Hepes buffer: 10-mM Hepes pH 7.2, 150-mM NaCl, 1-mM EDTA.
9. 50-mM Leupeptin, diluted in elution/wash buffer.
10. Metal affinity resin, either cobalt(II) or nickel(II).
11. 0.2-M $NiSO_4$ or 0.2-M $CoCl_2$.
12. FPLC AKTA System including a Superdex-200 HR 10 × 300 GL column (GE).

2.4 Protein Labeling and SDS–PAGE

1. 1-μM Alexa488-Halo-ligand (Promega), diluted in Hepes buffer.
2. SDS–PAGE, resolving gel: 10% acrylamide/Bis (29:1), 390-mM Tris–HCl pH 8.8, 0.1% sodium dodecyl sulfate (SDS), 0.1% ammonium persulfate (APS), and 0.1% *N*,*N*,*N*,*N*′-tetramethylethylenediamine (TEMED).
3. SDS–PAGE, stacking gel: 4% acrylamide/Bis (29:1), 125-mM Tris–HCl pH 6.8, 0.1% SDS, 0.1% APS, and 0.1% TEMED.
4. Running buffer (TGS): 25-mM Tris–HCl pH 8.6, 192-mM glycine, and 0.1% SDS.
5. Loading buffer (5×): 60-mM Tris–HCl pH 6.8, 25% glycerol, 2% SDS, 5% β-mercaptoethanol, 1% bromophenol blue.
6. Staining solution: 40% methanol, 10% acetic acid, 0.1-g Coomassie Brilliant Blue R-250.
7. Distaining solution: 40% methanol, 10% acetic acid.

2.5 AFM-Based Force Spectroscopy

1. AFM tips: MCLT Bruker AFM probes, C cantilever (triangular).
2. Atomic Force Microscope: Luigs and Neumann AFM/AFS and custom-made AFM setup based on Bruker AFM head, PI piezoelectric P-363.3CD PicoCube, and lab-made electronic to control the microscope. Both microscopes operate using routines wrote in Igor 6.0 (Wavemetrics).
3. Data analysis is conducted using either Igor 6.0 (Wavemetrics) or other data analysis software.

3 Methods

3.1 Preparation of the Pili Protein and Isopeptide Blocker

To study the mechanical properties of a pilus protein, the strategy highly suggested is to cloned in a tandem manner [9], including or not force reporters [7, 10] (Fig. 2a). The tandem cloning strategy allows to identify that a single molecule has been tethered between the surface and the force probe, a repetitive unfolding signal reveals the presence of pili protein, avoiding spurious recordings that interfere with the proper data acquisition. Furthermore, inextensible molecules like Spy0128, where the presence of an intramolecular isopeptide bond prevents the unfolding, require to include force reporters in the polyprotein, which are used to confirm that a single molecule is tethered to the force probe [7]. The cloning of multiple pili domain yielding a repetitive pattern in the force recordings is known as the *fingerprint* of a single-molecule unfolding.

3.1.1 Pilus Polyprotein Design

Vector pT7Blue is used to clone the pilus gene. The pilus gene is amplified by PCR using a 5′-primer which matches precisely the N terminus of the gene but includes a hanging *Bam*HI recognition site (GGATCC); and a 3′-primer that matches the C terminus of the molecule and contains a *Bgl*II site (AGATCT), followed by two Cys codons (AGA-AGA), a stop codon (TAA), and a *Kpn*I restriction site (GGTACC) (*see* **Note 2**) [9, 11]. In the case that force reporters are needed, those should be amplified using the same strategy with 5′-primer-*Bam*HI and 3′-primer-*Bgl*II-Cys-Cys-Stop-*Kpn*I [11].

1. In a PCR tube, add 10 μL of 5× DNA polymerase buffer followed by 2.5 μL of each primer (10 μM), 1 μL of 10 mM dNTPs, and 0.5-μL DNA polymerase (0.02 U/μL). Add 100 ng of genomic DNA or 1-ng plasmid DNA containing the pilus gene. The final volume of the reaction is 50 μL.
2. The protocol for amplification includes: (1) *Denaturation step:* 98 °C for 30 s; (2) *Amplification cycles step*: 98 °C for 10 s, followed by (primer T_m—5) °C for 30 s, and 72 °C during 30 s per Kb; and (3) *Extension step*: 72 °C for 2 min.

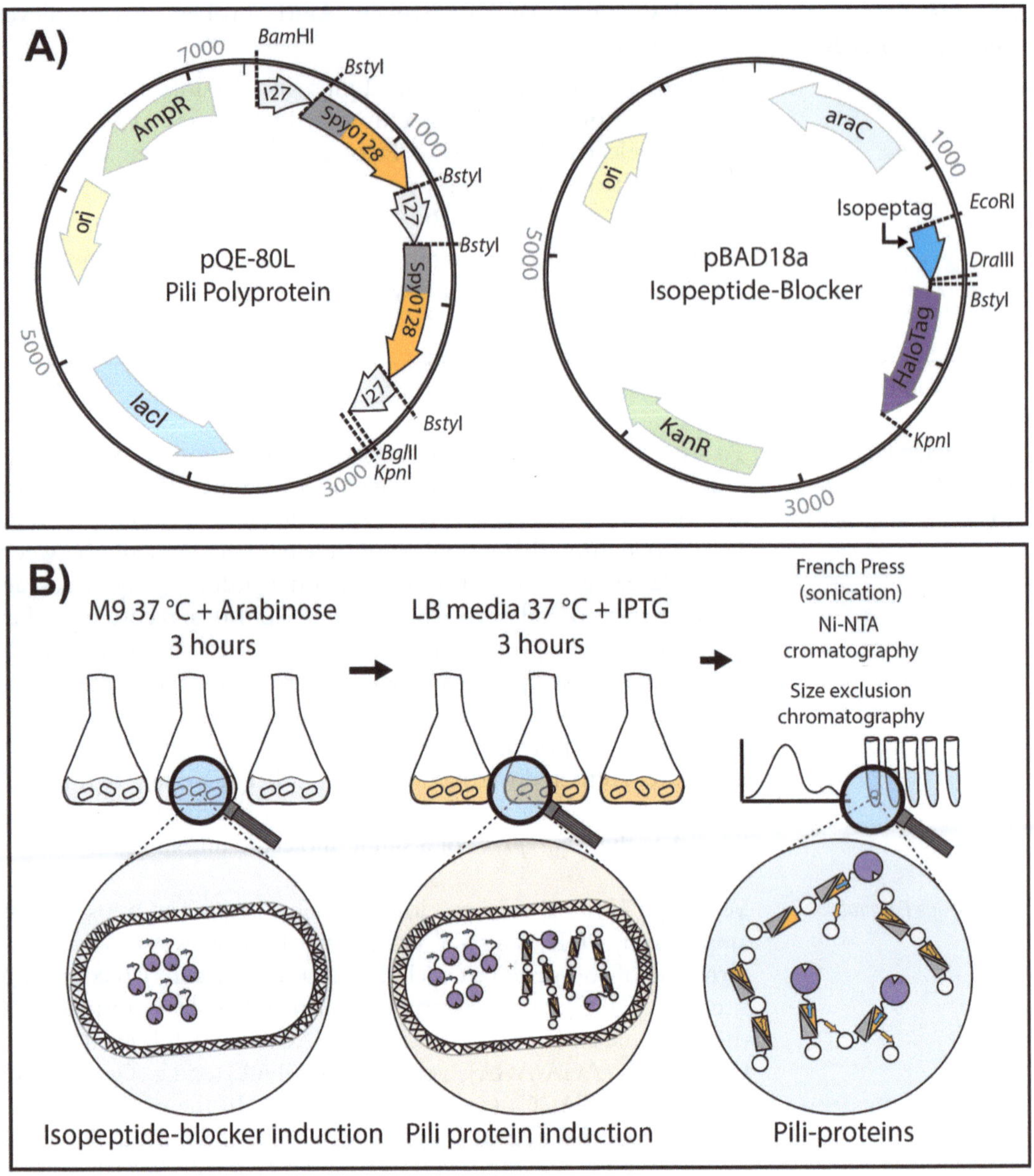

Fig. 2 Molecular Strategy for Blocking Isopeptide Bond Formation. (**a**) Vector design. The pili polyprotein DNA construct is clone into the pQE-80L following a tandem cloning strategy based on the use of the compatible restriction enzyme *Bam*HI and *Bgl*II [9]. Moreover, the construct includes two pili Spy0128 domains flanked by three titin I27 domains, which are used as force reporters in the smFS experiments. The constructs contain a HisTag in the N-terminal preceding the *Bam*HI site, and two terminal Cys residues before the *Kpn*I site. The pBAD18a vector confers kanamycin resistance, and the cloned gene is under the control of arabinose. The isopeptide-blocker DNA construct (blue arrow) is followed by a sequence that code for a TEV site sequence and the HaloTag gene. (**b**) Protein expression and blocking nascent pilin protein. First, the bacteria are cultured in minimum media to allow the expression of the isopeptide blocker (purple circle + blue arrow) by inducing with arabinose. After 3 hours, the bacteria culture is washed and transferred to LB media with IPTG for the induction of the pili Spy0128 protein (gray circle + gray: yellow triangle). The purification includes three steps: cell disruption, metal (HisTag) affinity chromatography, and size exclusion. Protein fractions from the metal affinity and size exclusion chromatography are analyzed through SDS–PAGE

3. Digest an aliquot containing 1 μg of PCR product using 10 U of *Bam*HI and 10 U of *Kpn*I. An appropriated volume for this reaction is 20 μL.
4. Incubate the digestion reaction at 37 °C for at least 4 h.
5. Purify the digested PCR product through a gel band purification kit according to manufacturer's instructions.
6. Digest approximately 200 ng of DNA vector with 10 U of *Bam*HI and 10 U of *Kpn*I in 10 μL. Incubate at 37 °C for at least 4 h.
7. Purify the digested vector DNA using the PCR gel band kit (as in **step 5**).
8. Ligate 50 ng of previously digested pT7Blue vector with a ~10× molar excess of the insert (*see* **Note 3**). The ligation mix includes 2 μL of 10× T4 DNA ligase buffer, the insert and vector, and 1 μL of T4 DNA ligase. The total volume of the reaction is 20 μL.
9. Incubate at 16 °C overnight.
10. Transform 100 μL of XL10-Gold® *E. coli* (Amp^-) cloning competent cells with an aliquot of 1 μL of ligation reaction.
11. Plate on agar plates containing 100 μg/mL of carbenicillin.
12. Purify plasmid DNA using a plasmid purification kit according to manufacturer's instructions.
13. Digest ~100 ng of purified plasmid DNA with *Bam*HI and *Bgl*II in the recommended buffer to cut out the pilus gene.
14. Sequence the pilus DNA construct (*see* **Note 4**).
15. To accomplish the sequential cloning of multiple pilus genes, digest ~400 ng of the vector including the pilus insert with 10 U of *Bgl*II.
16. Purify the digested pilus gene using a PCR gel band kit.
17. Treat ~200 ng of the *Bgl*II digested vector-insert with 10 U of *Kpn*I, and also purified through gel band purification kit (*see* **Notes 5–7**). The appropriate volume for these reactions varies between 20 and 50 μL, and both digestion reactions need to be incubated at 37 °C for at least 4 h.
18. Ligate the double-digested *Bgl*II-*Kpn*I vector-monomer with the purified pilus gene monomer from **step 6**, calculating a new cloning factor (*see* **Note 3**). The ligation is accomplished using the same conditions than **steps 8** and **9**.
19. Sequence the new construct (pilus *dimer*) (*see* **Note 4**).
20. Repeat this tandem cloning process as many pilus domains are needed in the DNA construct (*see* **Note 8**).

21. The pT7Blue vector containing the pilus construct is digested with *Bam*HI and *Kpn*I, releasing the polyprotein pilus gene segment following the same conditions as **step 6**.
22. Digest the expression vector pQE-80L using the same conditions as **step 6**.
23. Clone the polyprotein gene segment into the pQE-80L following the same strategy than **steps 8–12**. The DNA pili protein construct is stored at −20 °C until further experimentation or analysis is required.

3.1.2 Isopeptide-Blocker Design

To interfere with the folding and formation of the isopeptide bond on a pili protein domain, it is necessary to control the concentration of the isopeptide blocker within the cell [8]. For that, the peptide should be cloned in a vector that is inducible using a different strategy than pQE-80L-pili polyprotein construct, and thus, we have chosen pBAD18a vector which confers kanamycin resistance, and the promotor is under the control of arabinose [8] (Fig. 2a).

1. The two single primers containing the isopeptide-blocker sequence are heat at 70 °C in a thermocycler and let them cool slowly at RT for 1 h.
2. Digested pT7 vector containing the TEV-HaloTag sequence, following the conditions from Subheading 3.1.1, **step 6**.
3. The DNA duplex fragment is later ligated to the digested pT7-TEV-HaloTag vector, using the same conditions than Subheading 3.1.1, **steps 8** and **9**.
4. Digest the construct with *Eco*RI and *Kpn*I using conditions in Subheading 3.1.1, **step 6**.
5. The free insert (*isopeptide blocker*) is ligated into the pBAD18a Kan^R vector, following Subheading 3.1.1, **steps 8–12**. The DNA of the construct isopeptide-blocker-pBAD18a is stored at −20 °C until further analysis.

3.1.3 Isopeptide-Blocker + Pili-Protein System

Both vectors should be used to transform expression *E. coli* BLR (DE3) pLysS cells [12].

1. Transform 50–100 μL of BLR (DE3) pLysS *E. coli* (Amp^-/Kan^-) expression competent cells with an aliquot of 1 μL of pQE-80L vector containing the pilus polyprotein construct.
2. Plate on agar plates containing carbenicillin, tetracycline, and chloramphenicol.
3. Test the transformed *E. coli* BLR (DE3) pLysS pQE-80L for pilus protein expression (*see* **Note 9**), and select the best colony to continue with the preparation of chemically competent cells (see protocol for protein expression below).

4. Prepare chemically competent cells using the previously selected colony. Store the cells supplemented with glycerol 50%, in aliquots of 50–100 μL in the freezer at −80 ° C.
5. Transform an aliquot of 50–100 μL of competent cells BLR (DE3) pLysS pQE-80L *E. coli*, previously selected for expression, with 1 μL of pBAD18a vector containing the isopeptide-blocker gene segment.
6. Plate the transforming BLR (DE3) pLysS pQE-80L pBAD18a *E. coli*, on agar plates containing carbenicillin, kanamycin, tetracycline, and chloramphenicol.
7. Test the new transforming cells by their ability to express the isopeptide blocker (*see* **Note 10**).
8. The cells containing the pQE-80L and pBAD18a vectors are ready for the pilus protein expression and isopeptide interfering. These cells are stored at −80 °C supplemented with glycerol 30% until further analysis.

3.2 Culturing and Protein Purification

The *E. coli* BLR (DE3) pLysS cells containing both plasmids, pQE-80L and pBAD18a, are sequentially induced for protein expression starting with arabinose followed by IPTG [8]. Following this strategy, the genes will be sequentially turned on within the bacterium, ensuring that at the moment that the pili polyprotein is expressed, a high concentration of the isopeptide blocker is already in the cell [8] (Fig. 2b). Later, the cells are disrupted by sonication, and the pili protein decorated with the isopeptide blocker is purified through two chromatographic steps (Fig. 2b). The expression and purification could be done following the next steps.

3.2.1 Expression of the Isopeptide-Blocker-Pili System

1. The *E. coli* BLR (DE3) pLysS cells including the vectors pQE-80L pBAD18a are initially cultured using the standard methods provided by the manufacturing using the appropriate amounts of antibiotics (QIAgen-The QIAexpressionist Manual).
2. Inoculate the *E. coli* cells into 400 mL of minimum media M9CA supplemented with the appropriate amounts of antibiotics. The cells are incubated at 37 °C in an orbital shaker.
3. Add the appropriate amount of arabinose to the medium (*see* **Note 10**), after that the culture reaches an optical density (OD_{600}) of 0.4–0.6. The protein induction is conducted at 37 °C in an orbital shaker.
4. Transfer the cells from M9CA medium to LB broth supplemented with the antibiotics through centrifugation at 3000 × *g* in a refrigerated centrifuge (equilibrated at 25 °C). The cells are incubated at 37 °C in an orbital shaker for 15 min to equilibrate the cells to the new medium.

5. Induce the pilus protein by adding 1-mM IPTG to the LB medium and incubating in the orbital shaker at 37 °C (*see* **Note 9**).
6. Centrifuge the cells at 4000 × *g* for 10 min in a refrigerated centrifuge equilibrated at 4 °C.
7. Wash the cells with <50 mL of elution/wash (E/W) buffer, and either use to continue with the purification of the pilus protein or store at −80 °C with 30% glycerol.

3.2.2 Purification of the Pilus Protein

1. Suspend the pellet of bacteria from 400 mL of culture in 16 mL of E/W buffer prechilled on ice (*see* **Note 11**).
2. Add 160 μL of 100 mg/mL lysozyme, and 3.2 μL of 50 mM leupeptin and 160 μL of 100 mM PMSF.
3. Incubate the cell suspension on ice for 30 min.
4. Transfer to a rocking platform for 10 min at 4 °C.
5. Add 1.6 mL 10% w/v Triton X-100, 80 μL 1 mg/mL DNase I, and 80 μL 1 mg/mL RNase A.
6. Incubate for 10 min in the rocking platform at 4 °C.
7. Disrupt the cell integrity by ultrasound (*see* **Note 12**).
8. The cell debris is removed at 39,000 × *g* for 40 min at 4 °C.
9. Add >20 mL E/W buffer to 2 mL of 50% Ni^{2+}-NTA or Co^{2+}-NTA resin (*see* **Note 13**).
10. Transfer the supernatant (or lysate) from **step 8** to a packed and equilibrated metal affinity resin (*see* **Notes 14** and **15**).
11. Wash the resin with a minimum of 20 column volumes of E/W buffer.
12. Collect aliquots of 50–100 μL to monitor the absorbance at 280 nm, which should be not more than 0.02.
13. Add the elution buffer to the resin and collect fractions of ~300 μL on microcentrifuge tubes.
14. Quantify the fractions obtained by absorbance at 280 nm.
15. Analyze the protein fractions through SDS-PAGE (*see* Subheading 3.2.3).
16. Pool the selected fractions and load them into a centrifugal concentrator up to a volume of ~600 μL.
17. Inject the pooled concentrated protein on the Superdex-200 HR 10 × 300 GL connected to a FPLC AKTA system.
18. Collect fractions of 250 μL on microcentrifuge tubes.
19. Quantify the protein fractions by absorbance at 280 nm.
20. Analyze the protein fractions through SDS–PAGE (*see* Subheading 3.2.3).

3.2.3 SDS–Polyacrylamide Gel Electrophoresis

1. Label aliquots of 15 μL with 1 mM of dye Alexa488-Halo-ligand.
2. Incubate the protein+dye at RT during >1 h using amber microcentrifuge tubes or tubes covered with aluminum foil.
3. Prepare and mount the SDS–polyacrylamide gel.
4. Denature the labeled fractions by adding 5× loading for 5 min at 100 °C.
5. Centrifuge the tubes to maximum speed for 10 min.
6. Load a volume of 15 μL of protein into each well gel.
7. Run the electrophoresis at 80 volts for approximately 3 h in running buffer, protecting the entire tank with aluminum foil.
8. Remove the gel from the cassette and wash with distaining solution (*see* **Note 16**).
9. Transfer the gel to a gel tray and image using epi blue light in a documentation gel station (Fig. 3a–d).
10. Stain the gel with staining solution (Coomassie blue) for 1 or 2 h.
11. Wash several times the gel with distaining solution.

3.3 Single-Molecule Force Spectroscopy (smFS)

Force spectroscopy is a suitable strategy to investigate the mechanical properties of proteins, including the pili protein from GAS microorganism. Atomic force microscope (AFM) is one of the most popular techniques within the force spectroscopy methods, due to its high precision in position and ability to apply high forces on protein substrates [13]. Multiple force protocols can be used to characterize the mechanical properties of proteins; nevertheless, the pulling at constant velocity turn to be very useful to determine the unfolding force and the contour length after unfolding.

3.3.1 AFM Measurements

1. Deposit a volume of ~10–20 μL of the purified pili protein on top the gold-coated glass disc coverslip previously mounted on the piezo.
2. Incubated for 20 min, allowing the sample to get dry at the edges of the sample drop.
3. Mount the MCLT cantilever in the fluid chamber and fill the chamber with Hepes buffer.
4. Transfer the fluid chamber-cantilever into the AFM-head and align the laser in the tip of the C-cantilever (*see* **Note 17**).
5. Equilibrate the system for 10–20 min.
6. Transfer the AFM-head, including the fluid chamber, to the microscope and seal the fluid chamber by moving the piezo stage.
7. Equilibrate the system for at least 15 min.

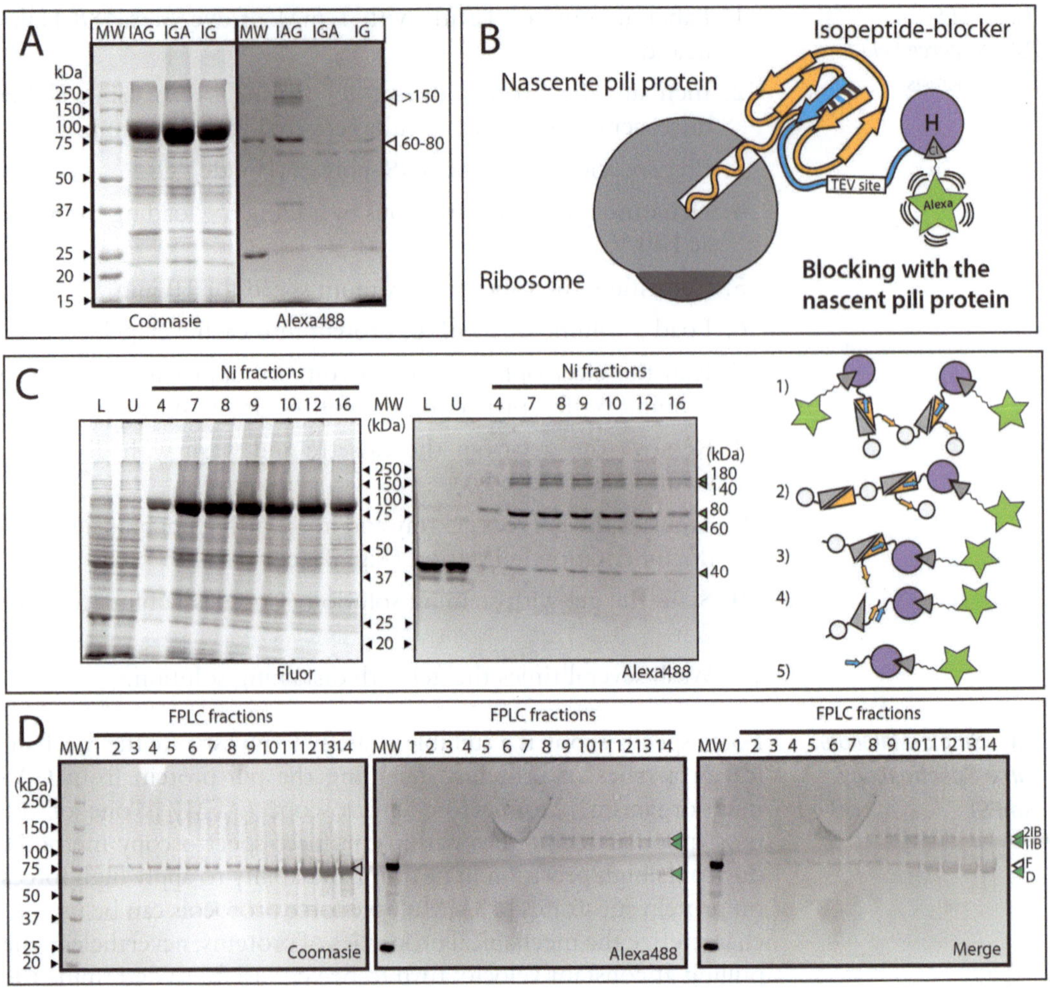

Fig. 3 Protein analysis, labeling, and detection. (**a**) The sequential induction of isopeptide blocker followed by the pili polyprotein yields a fluorescently labeled protein at high molecular weights when the protein preparation is incubated with Alexa488-Halo-ligand and analyzed through SDS–PAGE (IAG lane, induction with arabinose followed by IPTG). Whereas the induction with IPTG alone (IG lane) or IPTG followed by arabinose (IGA lane, induction with IPTG followed by arabinose), only faint bands at 80 kDa and 60 kDa are detected. (**b**) The nascent Spy0128 pili protein offers the right scaffolding for the isopeptide blocker, which could establish a proper H-bond network together with the intermolecular isopeptide bond. Pili polyproteins bound to the isopeptide blocker include a HaloTag which recognizes the Alexa488-HaloTag-ligand allowing the specific fluorescent label. (**c**) Protein fractions from the metal affinity chromatography. The fractions including the lysate (L), unbound (U), and fractions from 4–16 are incubated with Alexa488-HaloTag-ligand and analyzed by SDS–PAGE. The L and U content high concentration of the halo-tagged isopeptide blocker (band above 37 kDa), which does not bind to the resin, and it is washed during the chromatography. Protein fractions in addition to the main pili polyprotein at ~90 kDa (imaged with the fluorescent protein dye), in the gel imaged for Alexa488 several bands above 140 kDa and below 80 kDa, are present, including a small fraction of halo-tagged isopeptide blocker (40 kDa). The interpretation of each of these bands is offered at the right of the gel [8]. (**d**) Protein fractions from the size exclusion chromatography. The fractions are also labeled with Alexa488-HaloTag-ligand, showing a similar pattern than the Ni-fractions but without the contaminant proteins. Protein fractions with high fluorescent label at high molecular weight can be used for the AFM experiments. Empty arrowheads indicate the location of the pili polyprotein, while the green arrowheads indicate the migration of the multiple fluorescently labeled Spy0128 pili domains

8. Calibrate the cantilever through power spectrum and tip deflection. Corroborate that the spring constant obtained is within the range. For MCLT C-cantilevers, the spring constant is ~15 pN nm^{-1}.
9. Select the force-extension pulling protocol. Start with the data acquisition setting the *contact force*—or peaking force—at 2 nN and the *pulling amplitude*, which should be larger than the theoretical full extension of the pili protein (~400 nm should be enough for polyproteins of 8-mer) (Fig. 4).

3.3.2 Data Analysis

From the total of the pulling traces recorded in the force experiments, only a fraction will contain useful information. A large proportion of the recordings will be incomplete, and it will show no unfolding of the force reporters or will show unfolding peaks hard to interpret no matching with the expected extension and unfolding forces. Thus, the data analysis is based on strong fundaments which will allow extracting high-quality information without bias the conclusion of the experiments.

1. In the case of pili proteins flanked by force reporters, only data showing I27 domain unfolding will be considered for the analysis. Thus, data containing two or more force reporters are analyzed (*see* **Note 18**).
2. Correct the cantilever's deflection in each trace following the protocol implemented by Alegre-Cebollada et al. [7], which established that the origin in the extension (0 nm) corresponds at +100 pN of pushing force.
3. Fit the unfolding traces to a polymer elasticity model such as the worm-like chain [14], determining the initial extension (L_i) (or extension of the molecule before the unfolding of the reporters or pili domain), contour length (L_C) (or extension of the unfolded molecule), the final extension (L_F) (or full extension of the polyprotein), and the unfolding force (F_U) [8, 11] (Fig. 4).
4. Build histograms with the data acquired (L_i, L_C, L_F, F_U) summarizing quantitatively all the information collected in the FX experiments (*see* **Note 19**) (Fig. 4).

4 Notes

1. The pBAD18a (Kan^r) vector could lack the ribosome-binding site (RBS) upstream of the multiple cloning site. In the past, we have introduced an RBS in the pQE-80L through short DNA segments, including appropriate restriction enzyme sites [8].
2. It is necessary to confirm that the pili protein gene lacks any restriction enzyme sites included in both primers: *Bam*HI (present in the forward primer), and *Bgl*II and *Kpn*I (present

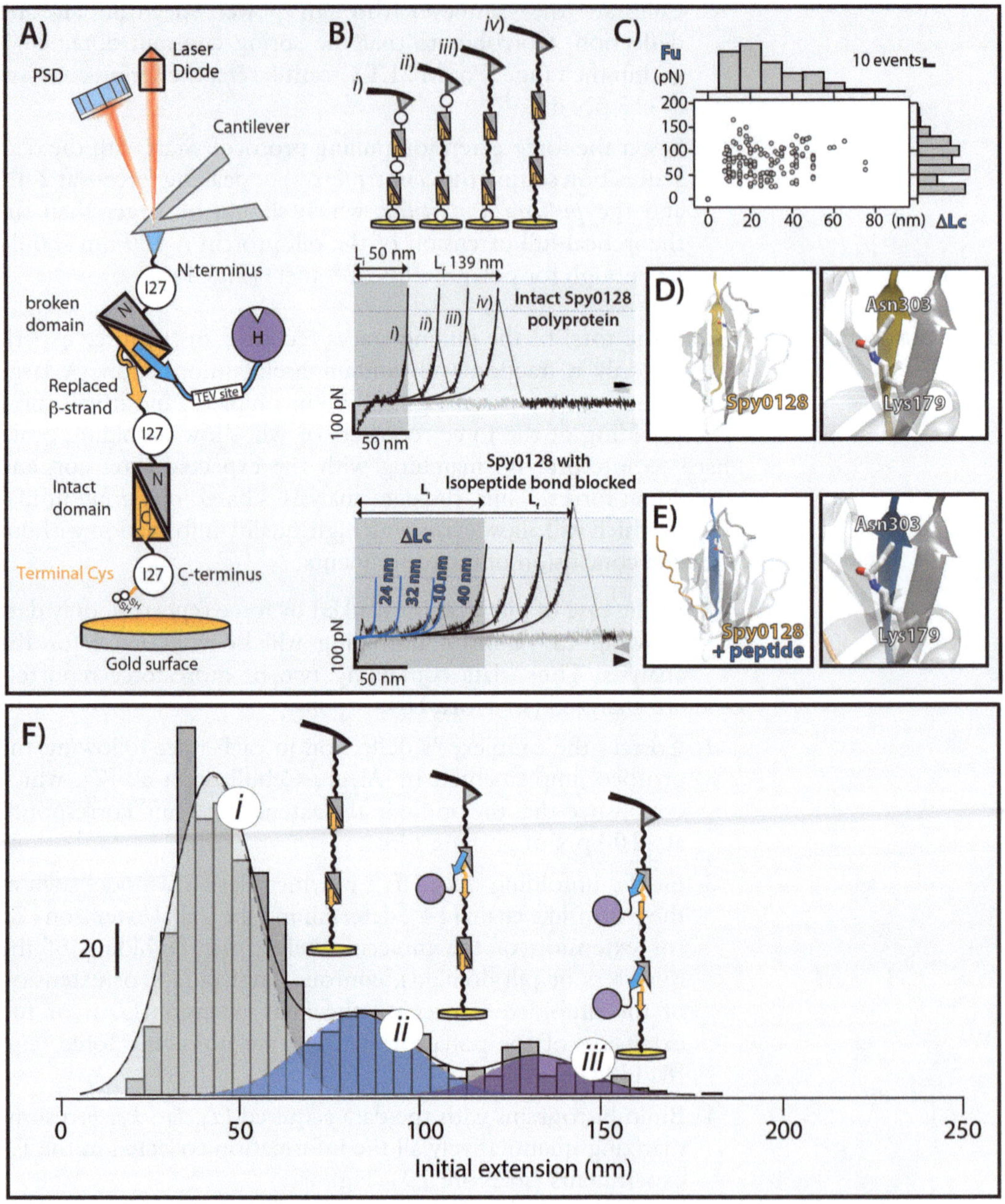

Fig. 4 AFM-based force spectroscopy. (**a**) Scheme of the AFM experiments used to conduct single-molecule Spy0128 mechanics. The pili polyprotein is immobilized on top a gold-coated cover slide through thiol chemistry. The cantilever pull from the polyprotein, changing the angle of the laser beam reflected in the detector (PSD). (**b**) Top, cartoon showing a single-molecule pili polyprotein being pull with AFM. The pili polyprotein is inextensible, and only the force reporters unfold when the molecule is stretched. Middle, recording showing an intact pili polyprotein. In this trace, only the unfolding of I27 force reporters are registered. Bottom, the initial extension can be used as a proxy for identifying if the Spy0128 are decorated with the isopeptide blocker. In the last recording, the initial extension, Li, is larger than the Spy0128 without isopeptide blocker. Moreover, a series of unfolding peaks are found preceding the unfolding of the I27

in the reverse primer). In the case that the gene contains any of these sites, they need to be removed by a synonymous mutation through PCR. Also, both primers should include one or two extra nucleotides at the 5′ end to facilitate the action of the restriction enzyme in the followed step.

3. Because the vector and insert have different molecular weights (nucleotide length), a proportional factor can be calculated:

$$\text{Cloning Factor} = \frac{\text{Vector nucleotide base length}}{\text{Insert nucleotide base length}}$$

$$\text{Amount of Insert} = \frac{500 \text{ ng}}{\text{factor}}$$

4. The sequencing should confirm the presence of the cloning sites, the absence of mutation, and the right orientation of the inserted gene.
5. The double digestion with *Bgl*II and *Kpn*I removes the Cys residues and stops codons from the construct, at the time a compatible site is offered to ligate with the *Bam*HI of the next insert to clone.
6. *Bgl*II (A′GATCT) and *Bam*HI (G′GATCC) restriction sites are compatible cohesive. Thus, when the DNA vector containing an insert is digested with, *Bgl*II could be ligated with the pilus protein insert previously digested with *Bam*HI. Both sites, *Bgl*II and *Bam*HI, yield a new restriction site: *Bst*YI (GGATCT).
7. Because *Bgl*II is not active in the same buffer as *Kpn*I, it is necessary to conduct the double digestion in two rounds of reactions. Nevertheless, it is possible to perform the *Bgl*II digestion in buffer 3.1, and add the *Kpn*I enzyme and the time the solution is diluted up to half. By doing this, the NaCl and the buffer will be diluted, diminishing the inhibitory effect over the *Kpn*I enzyme.

Fig. 4 (continued) domains [8]. (**c**) Contour length and force double histogram for the experiment in the presence of the isopeptide blocker. Although a large proportion of the recording is not affected by the isopeptide blocker (see below, panel **f**), the traces that show a larger Li also include mechanical intermediates. Nevertheless, these intermediates are quite variable in length and force, in contrast with the observation of the Spy0128 E258A which is extensible but shows an unfolding force above 290 pN and contour length of 50 nm. (**d** and **e**) Molecular models showing the isopeptide bond between Lys179 and Asn303 and between Lys179 and Asn from the isopeptide blocker (blue arrow) (model based on Spy0128 PDB 3B2M). (**f**) The initial extension histogram shows three populations based on the fit using worm-like chain model. The three Gaussian populations are centered at 41 (gray population, (i) intact Spy0128, 80 (blue population, (ii) one Spy0128 intervened with isopeptide blocker, and 130 nm (purple population, (iii) both Spy0128 intervened with isopeptide blocker

8. The tandem cloning of pilus domain genes could be made by only pilus protein domains or alternated with the cloning of force reporters (I27 titin gene for example) [7, 10, 11]. Nevertheless, the DNA construct should not exceed a coding protein with molecular weight above 80–100 kDa, due to expression systems based on *E. coli* are not able to express successfully proteins above this numbers; however, using large culture volumes, it has been possible to produce polyproteins up to 150 and more kDa [15, 16].
9. It is known that not all the clones after a DNA transformation have the same ability to express a protein; therefore, following the first transformation with pQE-80L, several clones were used to test for protein expression. A simple IPTG induction changing the incubation temperature and time is used to select the best clone for protein expression. SDS–PAGE analysis is enough to determine which clone express larger amounts of the pili polyprotein.
10. Following with the idea from **Note 9** that not all the clones express the same amount of the inducible protein, *E. coli* BLR (DE3) pLysS containing pQE-80L and pBAD18a are tested for isopeptide-blocker expression. Due to the protein expression with arabinose could be highly dependable of the concentration of the sugar, it is necessary to assay different clones, induction times, and a multiple concentration (0.02–2% w/v). The protein expression is analyzed by SDS–PAGE.
11. Several protein purification steps should be done using prechilled buffers, avoiding that the pellet and lysate reach RT. By doing this, the activity of bacterial proteases will be reduced during the protein preparation.
12. The protocol for sonication needs to be tuned for each sonicator system. For a Branson cell disrupter (200 W), 40% of power and pulses of 1 s followed by 1 s of pause. A total between 30 and 40 pulses of sonication is enough to reach a high recovery of the protein from a suspension of cells (*E. coli* suspension from 400 mL of culture and suspended in 16 mL of E/W buffer).
13. The equilibration of the resin can be completed either on a plastic column, passing the >20 mL through the packed resin, or a conical tube where the resin is suspended and separated by gravity or centrifugation at 1000 × *g* for 3 min.
14. To ensure a high His-tagged protein retention on the resin, the lysate can be passed through the column several times, or the resin can be incubated for longer periods of time (>1 h). Nevertheless, extending the time could affect the quality of the purification due to the action of proteases, and thus, this step needs to be completed at 4 °C.

15. The binding of His-tagged proteins depends on the temperature. Thus, in the case of low yield due to the binding, the incubation with the resin could be done at RT. It could be necessary to increase the protease inhibitors.
16. During the Alexa488-Halo-ligand imaging using epi blue light, it is possible to visualize the multiple bands from the prestained molecular weight marker by washing the gel with distaining solution.
17. Adding cold buffer to the fluid chamber will affect the stability of the experiment, increasing the drift of the force measurements remarkably. To avoid it, is recommendable to have Hepes buffer on the bench equilibrate at RT, avoiding any delay during the sample mounting, calibration, and data collection. The Hepes buffer could be supplemented with 0.001% sodium azide to prevent the growth of microorganisms.
18. In the case of pili proteins lacking of force reporters, the classification of data needs to be based on the presence of repetitive unfolding patterns, meaning the appearance of sawtooth pattern with equivalent unfolding force and contour length [10].
19. For the interfering with the isopeptide bond experiments, the L_i and L_F histogram could be very useful to evaluate the success of the experiments, due to these histograms will show if the extension before of the force reporters and extension of the entire polyprotein is larger by adding the isopeptide blocker, respectively [8]. Moreover, the L_C of mechanical events not related to the reporters and F_U are useful tools to evaluate the mechanical stability of the pili domain [7, 8, 11]. An isopeptide blocker bound to the pili domain will prevent the formation of the intramolecular bond, turning the Spy0128 in an extensible molecule. Thus, the L_C will show the extension of the mechanical intermediates present in the blocked pili domain, together with the force at which intermediates unfold.

Acknowledgments

This work was supported by FONDECYT 11180705 (J.A.R-P.). We thank all the members of the Rivas-Pardo's lab and the Inmunogenomics and Applied Genomics Laboratories for their helpful discussions.

References

1. Persat A, Nadell CD, Kim MK et al (2015) The mechanical world of bacteria. Cell 161:988–997. https://doi.org/10.1016/j.cell.2015.05.005
2. Maier B, Potter L, So M et al (2002) Single pilus motor forces exceed 100 pN. Proc Natl Acad Sci U S A 99:16012–16017. https://doi.org/10.1073/pnas.242523299
3. Telford JL, Barocchi MA, Margarit I et al (2006) Pili in Gram-positive pathogens. Nat Rev Microbiol 4:509–519. https://doi.org/10.1038/nrmicro1443
4. Hospenthal MK, Costa TRD, Waksman G (2017) A comprehensive guide to pilus biogenesis in Gram-negative bacteria. Nat Rev Microbiol 15:365–379. https://doi.org/10.1038/nrmicro.2017.40
5. Proft T, Baker EN (2009) Pili in Gram-negative and Gram-positive bacteria—structure, assembly and their role in disease. Cell Mol Life Sci 66:613–635. https://doi.org/10.1007/s00018-008-8477-4
6. Kang HJ, Coulibaly F, Clow F et al (2007) Stabilizing Isopeptide bonds revealed in Gram-positive bacterial Pilus structure. Science 318:1625–1628. https://doi.org/10.1126/science.1145806
7. Alegre-Cebollada J, Badilla CL, Fernández JM (2010) Isopeptide bonds block the mechanical extension of pili in pathogenic Streptococcus pyogenes. J Biol Chem 285:11235–11242. https://doi.org/10.1074/jbc.M110.102962
8. Rivas-Pardo JA, Badilla CL, Tapia-Rojo R et al (2018) Molecular strategy for blocking isopeptide bond formation in nascent pilin proteins. Proc Natl Acad Sci 115:9222–9227. https://doi.org/10.1073/pnas.1807689115
9. Carrion-Vazquez M, Oberhauser AF, Fowler SB et al (1999) Mechanical and chemical unfolding of a single protein: a comparison. Proc Natl Acad Sci U S A 96:3694–3699. https://doi.org/10.1073/pnas.96.7.3694
10. Echelman DJ, Alegre-Cebollada J, Badilla CL et al (2016) CnaA domains in bacterial pili are efficient dissipaters of large mechanical shocks. Proc Natl Acad Sci U S A 113:2490–2495. https://doi.org/10.1073/pnas.1522946113
11. Rivas-Pardo JA, Alegre-Cebollada J, Ramírez-Sarmiento CA et al (2015) Identifying sequential substrate binding at the single-molecule level by enzyme mechanical stabilization. ACS Nano 9:3996–4005. https://doi.org/10.1021/nn507480v
12. Moffatt BA, Studier FW (1987) T7 lysozyme inhibits transcription by T7 RNA polymerase. Cell 49:221–227
13. Valle-Orero J, Rivas-Pardo JA, Popa I (2017) Multidomain proteins under force. Nanotechnology 28:174003. https://doi.org/10.1088/1361-6528/aa655e
14. Bustamante C, Marko JF, Siggia ED, Smith S (1994) Entropic elasticity of lambda-phage DNA. Science 265:1599–1600
15. Haining AWM, von Essen M, Attwood SJ et al (2016) All subdomains of the Talin rod are mechanically vulnerable and may contribute to cellular Mechanosensing. ACS Nano 10:6648–6658. https://doi.org/10.1021/acsnano.6b01658
16. Aubin-Tam M-E, Olivares AO, Sauer RT et al (2011) Single-molecule protein unfolding and translocation by an ATP-fueled Proteolytic machine. Cell 145:257–267. https://doi.org/10.1016/j.cell.2011.03.036
17. Mandlik A, Swierczynski A, Das A, Ton-That H (2008) Pili in Gram-positive bacteria: assembly, involvement in colonization and biofilm development. Trends Microbiol 16:33–40. https://doi.org/10.1016/j.tim.2007.10.010

Part VI

Application of Group A Streptococcus-Derived Methods

Chapter 29

Use of Streptolysin O (SLO) to Study the Function of Lipid Rafts

Heather E. Miller

Abstract

Group A Streptococcus (GAS) produces the pore-forming toxin, streptolysin O (SLO). SLO sequesters cholesterol and induces a plasma membrane repair process that removes the pores via a lipid raft-mediated endocytosis. The impact SLO has on membranes makes it an effective toxin for investigating the function of lipid rafts in cellular processes. Lipid rafts are essential for B-cell activation. Indeed, antigen-stimulated B-cell receptors (BCRs) require localization with lipid rafts for efficient signaling and internalization. SLO treatment impairs BCR activation by competing for lipid rafts. Here, disrupting lipid rafts using SLO and assessing the effects on BCR activation by fluorescence microscopy and flow cytometry are described.

Key words Lipid rafts, Membrane rafts, Streptolysin O (SLO), Cholesterol-dependent cytolysin, B-cell activation, Membrane repair

1 Introduction

Lipid rafts are membrane domains of organized lipids and proteins that generate platforms for supporting cellular processes, such as signaling, endocytosis, and membrane fusion. Lipid rafts are heterogeneous, consisting of cholesterol, sphingolipids, gangliosides, and various membrane-associated proteins [1, 2]. Activation of immune cells depends on lipid rafts. Indeed, B cells require lipid rafts for an efficient antigen-induced signaling response. B-cell receptors (BCRs) are transmembrane proteins that bind antigen on the cell surface. Upon binding to antigen, BCRs enter lipid rafts and activate raft-resident signaling molecules. These molecules promote intracellular signaling to activate pathways involved in B-cell effector functions, such as cytokine and antibody production [3–5]. BCR signaling is significantly enhanced by lipid raft-mediated BCR clustering, which forms the immunological synapse, also called the BCR cap [6, 7]. Additionally, lipid rafts are required for endocytosis of antigen bound BCRs, whereby the antigens are

Thomas Proft and Jacelyn M. S. Loh (eds.), *Group A Streptococcus: Methods and Protocols*, Methods in Molecular Biology, vol. 2136, https://doi.org/10.1007/978-1-0716-0467-0_29,

then processed and presented on the B-cell surface to stimulate T cells [8].

Streptolysin O (SLO) is a pore-forming toxin produced by Group A Streptococcus (GAS). SLO is a cholesterol-dependent cytolysin that creates 30-nm diameter pores within target membranes [9]. The virulence of GAS is enhanced by SLO, which protects the bacteria from phagocytic killing by inducing apoptosis in neutrophils and macrophages [10, 11]. B-cell plasma membranes damaged by SLO undergo a repair process that removes the pores via lipid raft-mediated endocytosis. The repair pathway begins with a calcium influx via SLO pores. This triggers exocytosis of lysosomes and release of acid sphingomyelinase (ASM) to the cell surface. The ASM breakdown of sphingomyelin increases endocytosis of lipid rafts along with the injured membrane. If this repair process occurs during BCR stimulation, BCRs lose lipid raft support and BCR clustering, signaling, and endocytosis is impaired (Fig. 1) [12].

Here, the use of SLO to study the function of lipid rafts in BCR stimulation is described. The process involves SLO treatment of stimulated mouse splenic B cells, followed by assessing BCR activation using microscopy and flow cytometry. The techniques involve fluorescence staining to observe BCR/lipid raft formation, B-cell signaling, and BCR endocytosis. For BCR and lipid raft staining, BCRs are labeled/stimulated with a fluorophore conjugated cross-linking antibody and lipid rafts with a fluorophore conjugated cholera toxin subunit B (CTB). For detecting BCR signaling, an anti-phosphotyrosine antibody is used to label total tyrosine phosphorylation in the cell. For assaying BCR endocytosis, a biotinylated cross-linking antibody is used to induce BCR endocytosis, followed by detecting remaining surface BCRs with a fluorophore-conjugated streptavidin. Additionally, how to monitor the amount of SLO pores per cell by adding propidium iodide during the SLO treatment is described. These methods may also be applied to explore the function of lipid rafts in the activation of T lymphocytes, macrophages, and other immune cells.

2 Materials

B cells from cell lines or purified from spleen, lymph nodes, or blood may be used. The methods described below pertain to splenic naïve mouse B cells. However, this will be applicable to human cells as well. Below are suggested fluorophore conjugates for staining cells, but these may be substituted with preferred fluorophores.

2.1 BCR Stimulation and SLO Treatment

1. Serum-free Dulbecco's Modified Eagle Medium (DMEM) with 20-mM HEPES (pH 7.0) (*see* **Note 1**).

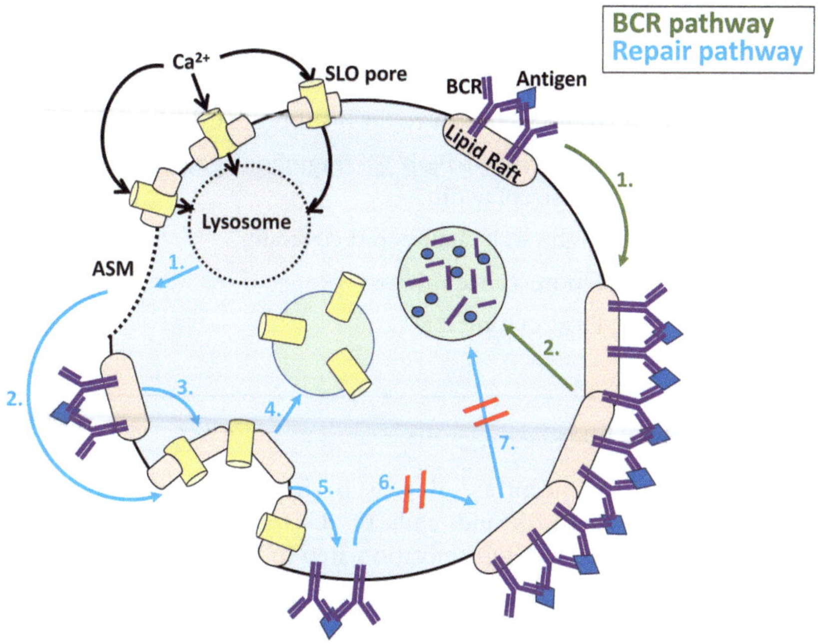

Fig. 1 Predicted model of SLO-mediated effects on BCR activation. The general BCR activation pathway proceeds as depicted by green arrows and numbers. (1) BCRs stimulated by antigen move into lipid raft-rich domains. These signaling BCRs within lipid rafts cluster and form a polarized BCR cap. (2) Lipid rafts and activated BCRs regulate endocytosis of antigen/BCR complexes, whereby antigens are processed for presentation. The SLO-induced membrane repair pathway and inhibition of BCR activation proceeds as shown by blue arrows and numbers. (1) Influx of calcium via SLO pores results in lysosomes fusing with the plasma membrane and release of acid sphingomyelinase (ASM). (2) ASM activity on membrane lipids initiates endocytosis. (3) Lipid rafts required for endocytosis of damaged membrane are taken from BCRs. (4) As endocytosis commences, the pores are removed from the plasma membrane. (5) The level of damaged membrane correlates with the amount of BCRs lacking lipid rafts. (6) Antigen-stimulated BCRs lacking lipid rafts are impaired in clustering and forming BCR caps. (7) Impaired BCR clustering and signaling prevents endocytosis required for antigen processing and presentation

2. A fluorophore conjugated cross-linking antibody for labeling and stimulating BCRs. For splenic naïve mouse B cells, use AF546-F(ab′)$_2$ fragment goat anti-mouse IgM (*see* **Note 2**).
3. Streptolysin O (SLO) (*see* **Note 3**).
4. Propidium iodide (PI), if monitoring the level of SLO pores per cell.

2.2 Cell Fixation and Fluorescent Staining

1. Phosphate-buffered saline solution (PBS).
2. 8% paraformaldehyde (PFA) in PBS for fixing cells.
3. Permeabilization and blocking solution: 0.05% saponin and 1% BSA in PBS.

4. Mouse IgG_{2b} anti-phosphotyrosine monoclonal antibody (clone 4G10) and fluorescently conjugated secondary antibody, AF488-goat anti-mouse IgG_{2b}.
5. AF488-cholera toxin B subunit (CTB) (*see* **Note 4**).
6. Biotinylated-F(ab′)$_2$ fragment goat anti-mouse IgM and PE-streptavidin.
7. Eight-well chambered coverslip.
8. Fluorescence microscope.
9. Flow cytometer.

3 Methods

3.1 SLO Treatment and Labeling BCRs and Lipid Rafts for Microscopy

1. Concentrate B cells to 5×10^6 cells/ml in DMEM with HEPES and chill to 4 °C. Keep cells at 4 °C until the BCR/SLO activation step of warming to 37 °C (*see* **Note 5**).
2. Take 500 μl of cells in a microfuge tube, add AF546-F(ab′)$_2$ fragment goat anti-mouse IgM (5 μg/ml), and incubate for 30 min.
3. Label lipid rafts by adding AF488-CTB (3 μg/ml) and incubate for 10 min.
4. Wash cells to remove unbound antibody and CTB by centrifuging cells at 4 °C, aspirating supernatant, and resuspending the pellet in 500-μl cold DMEM with HEPES. Wash cells three times.
5. Add SLO (200 ng/ml) to cells and incubate for 5 min to allow SLO to bind to membranes (*see* **Note 6**).
6. Warm cells to 37 °C for 5 min, and this will activate BCRs and SLO.
7. Fix cells with 4% PFA by adding 500 μl of 8% PFA to cells. Fix for 20 min at RT.
8. Wash cells three times with PBS to remove PFA.
9. Add 300 μl of cells to a well of an eight-well chambered coverslip.
10. Image cells using a fluorescence microscope. Expected observations are shown in Fig. 2.

3.2 Tyrosine Phosphorylation Staining for Microscopy

1. Follow **steps 1–8** from Subheading 3.1 but remove **step 3**, do not label with CTB.
2. Add permeabilization/blocking solution to cells for 20 min. Continue to use this solution for antibody incubations and washes.

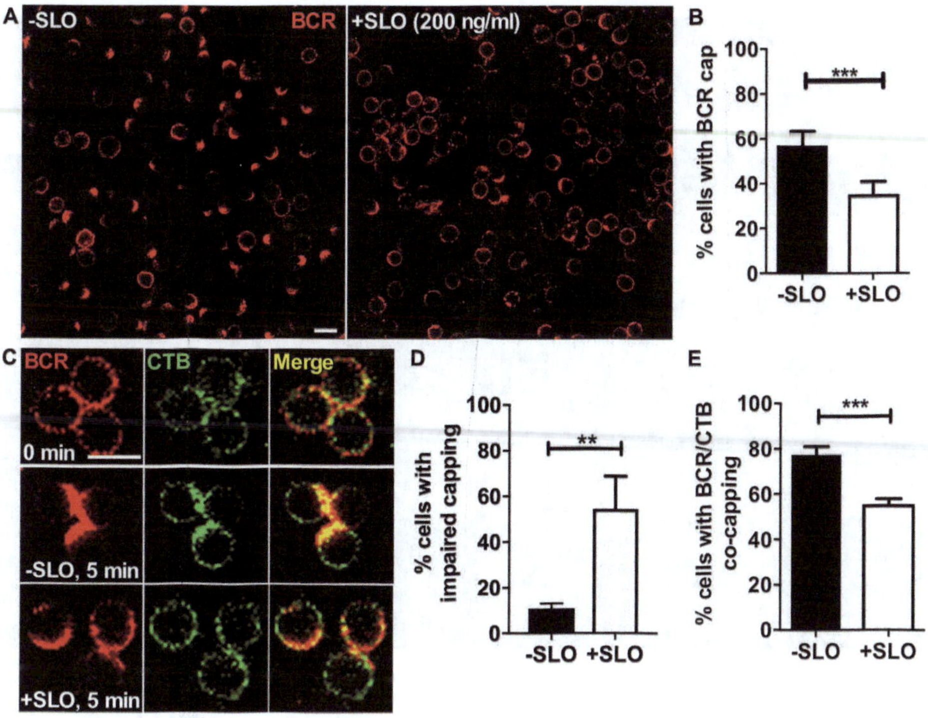

Fig. 2 SLO treatment during BCR stimulation disrupts lipid raft formation required for BCR-polarized clustering. (**a** and **b**) SLO treatment of stimulated splenic mouse B cells reduces BCR capping. BCRs were labeled with AF546-F(ab′)$_2$ anti-mouse IgM for 30 min at 4 °C, washed and incubated with SLO (0 or 200 ng/ml) for 5 min, and then warmed to 37 °C for 5 min, followed by fixation. Images were taken using a confocal fluorescence microscope, and the percentage of B cells with BCR capping was quantified. (**c**–**e**) SLO treatment impairs BCR capping and reduces colocalization of BCR caps with lipid rafts. Impaired BCR caps were observed as elongated BCR caps rather than condensed caps. Colocalization of BCRs with lipid rafts was determined by CTB co-capping with BCRs. BCRs were labeled with AF546-F(ab′)$_2$ anti-mouse IgM for 30 min at 4 °C, and lipid rafts were labeled with AF488-CTB for 10 min. Cells were washed and incubated with SLO (0 or 200 ng/ml) for 5 min at 4 °C and then warmed to 37 °C for 5 min, followed by fixation. Images were taken using a confocal fluorescence microscope, and the percentage of BCR capped B cells with impaired capping or BCR/CTB co-capping was quantified. Graphs represent the means ± SD of ≥80 cells from three independent experiments. Statistical significance determined by Student's *t*-test ($^{**}P < 0.01$, $^{***}P < 0.001$). Scale bar, 10 μm. ((**b**, **d**, and **e**) reproduced from ref. 12 with permission from Rockefeller University Press Journals)

3. Incubate cells with mouse IgG_{2b} anti-phosphotyrosine monoclonal antibody (1 μg/ml) for 30 min and then wash three times.
4. Incubate cells with AF488-goat anti-mouse IgG_{2b} (5 μg/ml) for 30 min and then wash three times.
5. Wash cells three times with PBS.
6. Add 300 μl of cells to a well of an eight-well chambered coverslip. Image cells using a fluorescence microscope. Expected observations are shown in Fig. 3a.

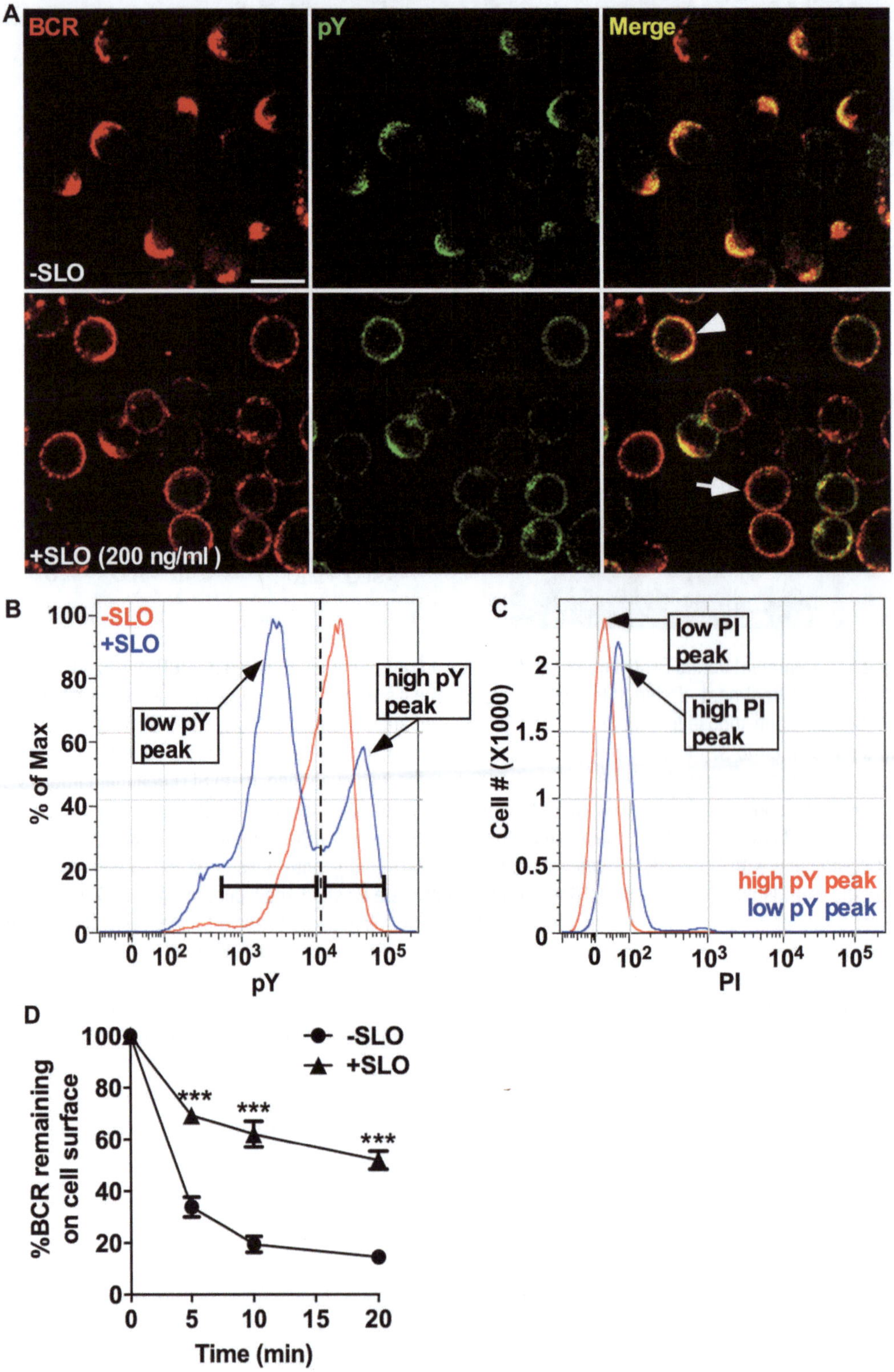

Fig. 3 SLO treatment of stimulated B cells induces tyrosine phosphorylation intensities that correlate with the level of SLO-mediated pores. (**a**) SLO treatment during BCR stimulation produces different tyrosine

3.3 Tyrosine Phosphorylation Staining for Flow Cytometry

1. Follow **steps 1–5** from Subheading 3.2, but before the 5-min activation step of SLO at 37 °C, add PI (250 μg/ml) to monitor the level of SLO pores formed (*see* **Note 7**).
2. Analyze cells on a flow cytometer. Expected results are shown in Fig. 3b and c.

3.4 BCR Endocytosis Assay Using Flow Cytometry

1. Concentrate B cells to 5×10^6 cells/ml in DMEM and chill to 4 °C. Keep cells at 4 °C until the activation step of warming to 37 °C.
2. Take 500 μl of cells in a microfuge tube, add biotinylated F(ab′)$_2$ anti-mouse IgM (5 μg/ml), and incubate for 30 min.
3. Wash cells three times to remove unbound antibody.
4. Add SLO (200 ng/ml) to cells and incubate for 5 min.
5. Warm cells to 37 °C for desired times. For BCR endocytosis, 5, 10, and 20 min are adequate.
6. Fix cells with 4% PFA by adding 500 μl of 8% PFA to cells. Fix for 20 min at RT.
7. Wash cells three times with PBS to remove PFA.
8. Add PE-streptavidin (3 μg/ml) in PBS to detect remaining surface BCRs.
9. Wash to remove free streptavidin and analyze samples on a flow cytometer. Expected results are shown in Fig. 3d.

Fig. 3 (continued) phosphorylation (pY) intensities. In stimulated B cells, the pY co-caps with BCRs. With SLO treatment, pY staining is uniform around the cell and is either low intensity (arrow) or high intensity (arrowhead). BCRs were labeled at 4 °C with AF546-F(ab′)$_2$ anti-mouse IgM, washed and incubated with SLO (0 or 200 ng/ml) for 5 min, and then warmed to 37 °C for 5 min, followed by fixation. Cells were permeabilized and stained for pY using mouse anti-phosphotyrosine monoclonal antibody (4G10), followed by AF488-goat anti-mouse IgG$_{2b}$ antibody. (**b** and **c**) SLO treatment during BCR stimulation creates two B-cell populations of pY intensity that correlates with the level of SLO pores, whereby more pores per cell decrease BCR signaling. B cells were treated as in (**a**) but with the addition of PI during the 5-min incubation at 37 °C to monitor the level of SLO-mediated pores. Histograms of pY intensity show activated B cells have one peak, whereas SLO treated have two peaks. Gating on each peak (indicated by brackets) and creating respective PI intensity histograms show cells from the high pY peak have low PI intensity (low amount of pores per cell) and cells from the low pY peak have high PI intensity (high amount of pores per cell). (**d**) SLO treatment decreases BCR endocytosis. B cells were labeled at 4 °C with biotinylated-F(ab′)$_2$ anti-mouse IgM, washed and incubated with SLO (0 or 200 ng/ml) for 5 min, and then warmed to 37 °C and fixed at indicated times. Cells were then labeled with PE-streptavidin and measured for the fluorescence intensity of the remaining surface BCRs by flow cytometry. Graphs represent the means ± SD from three independent experiments. Statistical significance determined by Student's *t*-test ($***P < 0.001$). Scale bar, 10 μm. ((**d**) reproduced from ref. 12 with permission from Rockefeller University Press Journals)

4 Notes

1. For consistent SLO damage to cells, use serum-free medium at pH 7.0, because serum and pH may interfere with SLO binding to membranes.
2. $F(ab')_2$ fragment is used to stimulate BCRs, as the F_c portion of a whole antibody binds to F_c receptors on B cells, which may inhibit B-cell activation. Naïve mouse B cells have BCRs of the IgM isotype and, thus, require anti-mouse IgM for stimulation; however, anti-mouse IgG also works.
3. For the experiments described, a mutant SLO with a cysteine deletion (Rodney Tweten, University of Oklahoma) was used. If not using the mutant SLO, a thiol activation step may be required, as SLO is oxidized in solution. Sigma–Aldrich recommends activating SLO with 20 mM cysteine or 10 mM dithiothreitol before use.
4. CTB binds ganglioside GM1 and is commonly used to label lipid rafts.
5. Cells must remain at 4 °C during the labeling procedures. This will prevent BCR and SLO activation before desired.
6. The SLO concentration used will have to be optimized for samples being assayed. Test with SLO concentrations 100, 200, 300, and 400 ng/ml. Following SLO treatment for 5 min at 37 °C, put cells on ice and add PI to stain dying/dead cells. Run on a flow cytometer and determine the percentage of PI+ cells. Look for a SLO concentration that gives around 25–30% PI+ cells, and this ensures that the majority of cells were attacked by SLO. For detailed instructions on determining SLO concentrations and cell membrane repair, *see* Ref. 13. Alternatively, if a flow cytometer is not available, following PI treatment, wash cells to remove PI, and fix with 4% PFA. Image using a fluorescence microscope and count PI+ cells to determine percentages. If the lowest concentration of SLO appears to kill too many cells, increase the concentration of cells used.
7. PI is a cell impermeable dye and stains cells with damaged, unrepaired plasma membranes, which are usually dying/dead cells. However, the presence of PI during SLO-mediated injury and repair can monitor the level of pores that were formed on a cell.

References

1. Lingwood D, Simons K (2010) Lipid rafts as a membrane-organizing principle. Science 327:46–50
2. Sezgin E, Levental I, Mayor S et al (2017) The mystery of membrane organization: composition, regulation and roles of lipid rafts. Nat Rev Mol Cell Biol 18:361–374
3. Cheng PC, Dykstra ML, Mitchell RN et al (1999) A role for lipid rafts in B cell antigen receptor signaling and antigen targeting. J Exp Med 190:1549–1560
4. Gupta N, DeFranco AL (2007) Lipid rafts and B cell signaling. Semin Cell Dev Biol 18:616–626
5. Gupta N, DeFranco AL (2003) Visualizing lipid raft dynamics and early signaling events during antigen receptor-mediated B-lymphocyte activation. Mol Biol Cell 14:432–444
6. Sohn HW, Pierce SK, Tzeng SJ (2008) Live cell imaging reveals that the inhibitory FcgammaR-IIB destabilizes B cell receptor membrane-lipid interactions and blocks immune synapse formation. J Immunol 180:793–799
7. Phee H, Rodgers W, Coggeshall KM (2001) Visualization of negative signaling in B cells by quantitative confocal microscopy. Mol Cell Biol 21:8615–8625
8. Stoddart A, Dykstra ML, Brown BK et al (2002) Lipid rafts unite signaling cascades with clathrin to regulate BCR internalization. Immunity 17:451–462
9. Bhakdi S, Tranum-Jensen J, Sziegoleit A (1985) Mechanism of membrane damage by streptolysin-O. Infect Immun 47:52–60
10. Timmer AM, Timmer JC, Pence MA et al (2009) Streptolysin O promotes group A Streptococcus immune evasion by accelerated macrophage apoptosis. J Biol Chem 284:862–871
11. Uchiyama S, Dohrmann S, Timmer AM et al (2015) Streptolysin O rapidly impairs neutrophil oxidative burst and antibacterial responses to group A Streptococcus. Front Immunol 6:581
12. Miller H, Castro-Gomes T, Corrotte M et al (2015) Lipid raft-dependent plasma membrane repair interferes with the activation of B lymphocytes. J Cell Biol 211:1193–1205
13. Miller H, Song W (2018) Use of Streptolysin O-induced membrane damage as a method of studying the function of lipid rafts during B cell activation. Methods Mol Biol 1707:235–241

Chapter 30

Engineering of Group A Streptococcus Isopeptide Bonds into Immunoglobulin-Like Protein Domains

Paul G. Young and Edward N. Baker

Abstract

Intramolecular isopeptide bonds, formed autocatalytically between Lys and Asn/Asp side chains, are widely present in the immunoglobulin-like domains of Gram-positive bacterial adhesins, including Group A Streptococcus, and confer considerable mechanical and chemical stability. These properties make them attractive for applications in biotechnology. Here, we detail the practical considerations that are involved in engineering isopeptide bonds into Ig-like proteins, including the choice of a site where bond-forming residues could be introduced and the appropriate methodology for mutagenesis. We specify how to determine whether an isopeptide bond has formed, what strategies can be adopted to overcome problems, and how to monitor the stability of the engineered protein.

Key words Isopeptide bond, Protein engineering, Protein stability, Immunoglobulin, Ig-like domains, Covalent crosslink, Gram-positive bacteria

1 Introduction

Since the first discovery of intramolecular isopeptide bond crosslinks in the pili of the M1 strain of *Streptococcus pyogenes* [1], many more examples have been found, in the pili of other strains of *S. pyogenes* [2, 3] and in the pili and cell surface proteins of numerous other Gram-positive bacteria [4–7]. So far they have been found exclusively in Ig-like β-sheet domains with folds of the CnaA or CnaB types [8]. These isopeptide bonds are covalent bonds that are formed autocatalytically between lysine and asparagine/aspartic acid side chains by a single-turnover reaction, catalyzed by a neighboring carboxyl group from either an aspartic acid residue (CnaA domains) or glutamic acid residue (CnaB domains). The minimal requirement for bond formation is an isopeptide triad (Lys, Asn/Asp, and the catalytic Asp/Glu), located in a hydrophobic core environment [8].

The importance of the hydrophobic environment for intramolecular isopeptide bond formation has been well documented in the

Thomas Proft and Jacelyn M. S. Loh (eds.), *Group A Streptococcus: Methods and Protocols*, Methods in Molecular Biology, vol. 2136, https://doi.org/10.1007/978-1-0716-0467-0_30,

literature [1, 5, 8, 9]. The hydrophobic residues surrounding the isopeptide bond are critical, both to position the residues of the isopeptide triad in a precise orientation and to favor their uncharged forms, for formation of an isopeptide bond. Examination of crystal structures with isopeptide bonds shows that the bond-forming residues require certain geometric restrictions in the angles and distances between the residues. These requirements are different between the CnaA and CnaB folds, as the position of the isopeptide bond-forming residues is different in each fold [6–8]. In the CnaA fold, isopeptide bond forms between first and second last β-strands in the Ig-fold, whereas in the CnaB fold, they form between the first and last Ig β-strands (Fig. 1).

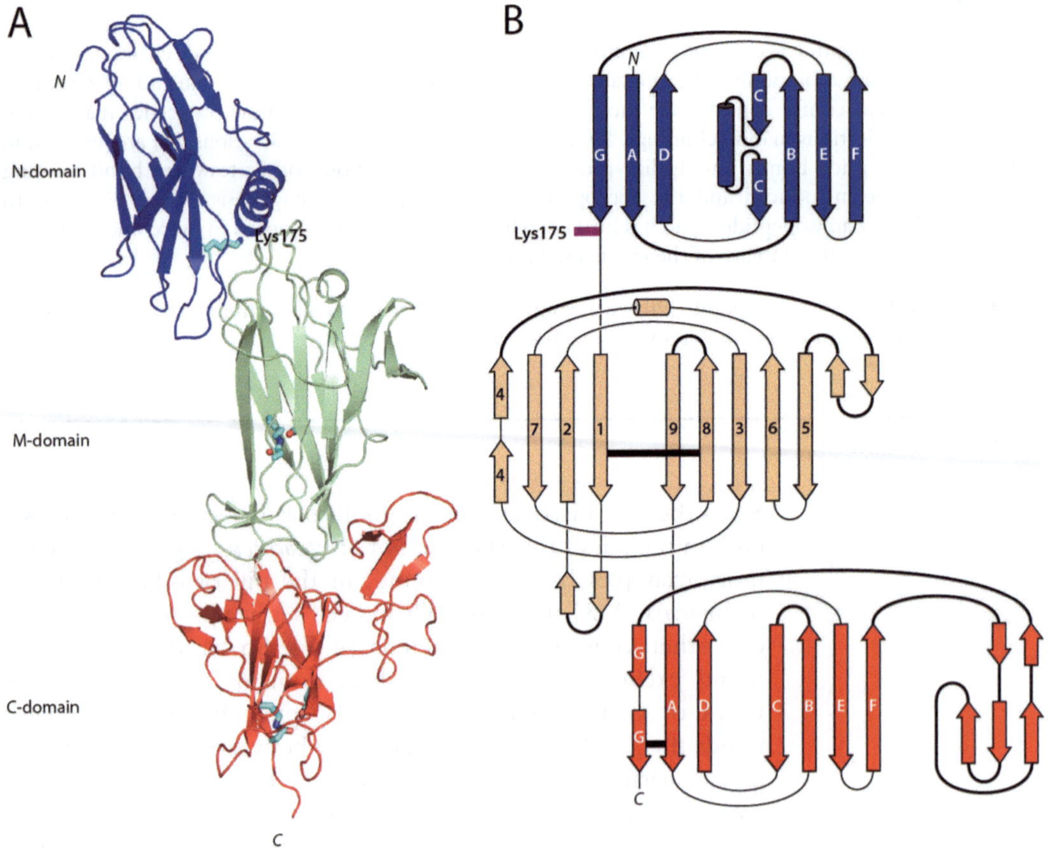

Fig. 1 Structure of the T6 backbone pilin [2] showing both CnaA and CnaB Ig-like domains with isopeptide bonds. (**a**) Ribbon diagram depicting the three domains of T6 colored in red (N terminus), green (middle), and red (C terminus). Residues involved in isopeptide bond formation and the N-domain pilin lysine (Lys175) are shown in stick form and colored cyan. (**b**) Topology diagram is color coded the same as the ribbon diagram. The CnaB domain β-strands are labeled A–G, while the CnaA domain β-strands are labeled 1–9. Each domain is decorated with extended loops or variable regions (unlabeled), which would be trimmed from a model before submitting for structural alignment. Isopeptide bond position is depicted as horizontal black lines (the N-domain lacks an intramolecular isopeptide bond)

The isopeptide bonds in naturally occurring cell-surface proteins of Gram-positive bacteria have been shown to provide extensive stability against external mechanical or chemical stress [10–12]. This has suggested that isopeptide bonds could be engineered into other protein domains for added stability [13], similarly to the introduction of nonnative disulfide bonds [14].

Below, we describe the methodology you would use to determine whether it is feasible to engineer an isopeptide bond into a preselected Ig-like fold. We recommend a template-based approach, whereby an isopeptide bond is introduced into an Ig domain using the closest structurally homologous Ig-fold that contains an intramolecular isopeptide bond as a model. We use our successful engineering of an isopeptide bond into the protein FctB [13] as an exemplar. We refer to the original (native) protein as FctB and the engineered form, with isopeptide bond, as FctBiso. We outline the molecular techniques required for cloning and expression. We discuss the difficulties involved in engineering charged residues within the hydrophobic core of a protein and provide methodology to determine the extent to which the engineered crosslink enhances protein stability.

2 Materials

Prepare all solutions using ultrapure water and store at room temperature unless otherwise indicated.

2.1 Modeling Software and Web Tools

1. Coot (Crystallographic Object-Oriented Toolkit) (https://www2.mrc-lmb.cam.ac.uk/Personal/pemsley/coot/).
2. PDBeFold (http://www.ebi.ac.uk).
3. MolProbity (http://molprobity.biochem.duke.edu).
4. Research Collaboratory for Structural Bioinformatics Protein Data Bank (https://www.rcsb.org/).

2.2 Cloning

1. Miniprep plasmid DNA purification kit.
2. SfoI restriction enzyme and buffer. Store at −20 °C.
3. InFusion cloning kit including Stellar chemically competent cells, Takara Bio USA, Inc. (CloneTech).
4. A high-fidelity polymerase such as Pfu Ultra II fusion HS (Stratagene).
5. 0.6% w/v agarose gels: Mix 0.6 g of agarose with 100 mL 1× TAE buffer.
6. Tris–EDTA buffer (TE; 50×): Dissolve 242 g Tris base in H_2O, add 57.1 mL glacial acetic acid, and 100 mL of 500 mM EDTA (pH 8.0) solution. Make up to 1 L with H_2O.

7. Luria-Bertani medium (LB broth): Add 10 g Bacto-tryptone, 5 g yeast extract, 10 g NaCl to 800 mL H_2O. Adjust pH to 7.5 with NaOH and bring volume up to 1 L with H_2O. Autoclave to sterilize.
8. LB agar: Include 20 g of agar to LB media before autoclaving.
9. Ampicillin (1000× stock): Add 100 mg ampicillin to 10 mL H_2O. Filter sterilize and store at −20 °C.

2.3 Expression and Purification

1. Luria-Bertani (LB) medium (*see* Subheading 2.2, **item** 7).
2. 1 M Isopropyl β-D-1-thiogalactopyranoside (IPTG). Filter sterilize and store at −20 °C.
3. Protease inhibitor tablets, for example, cOmplete™, Mini Protease Inhibitor Cocktail Tablets.
4. IMAC Lysis buffer: 100 mM Tris–HCl, pH 8.0, 300 mM NaCl, 10 mM imidazole (pH 8.0), 2% v/v glycerol, protease inhibitors. Store at 4 °C.
5. IMAC wash buffer: 100 mM Tris–HCl, pH 8.0, 300 mM NaCl, 20 mM imidazole (pH 8.0).
6. IMAC elution buffer: 100 mM Tris–HCl, pH 8.0, 300 mM NaCl, 500 mM imidazole (pH 8.0).
7. FPLC instrument, for example, ÄKTA prime system (GE healthcare). Typical flow rates of 1 mL/min.
8. Cell disruptor (Constant Cell Disruption Systems).
9. Tobacco Etch Virus protease (rTEV). Store at −80 °C.
10. rTEV digest buffer: 25 mM Tris–HCl, pH 8.0, 100 mM NaCl, 1 mM β-mercaptoethanol.
11. Size-exclusion buffer (SEC): 10 mM Tris–HCl, pH 8.0, 100 mM NaCl.
12. HiTrap IMAC column (5 mL), GE Healthcare Life Sciences.
13. Size-exclusion chromatography Superdex S200 10/300 column, GE Healthcare Life Sciences.
14. Spin concentrator columns (10-kDa MWCO).
15. Liquid nitrogen.

2.4 SDS–Page

1. Precast 4–12% Bis-Tris gradient polyacrylamide gels (ThermoFisher).
2. MES SDS running buffer (20×) (ThermoFisher).
3. SDS loading buffer (5×): 0.3 M Tris–HCl (pH 6.8), 10% SDS, 25% β-mercaptoethanol, 0.1% bromophenol blue, 45% glycerol. Store the remaining aliquots at −20 °C.
4. Molecular weight protein ladder standard.
5. SDS–PAGE mini gel tank and power supply.

2.5 Mass Spectrometry

1. Electrospray Ionization-Mass Spectrometry buffer: 50% v/v acetonitrile and 0.1% v/v formic acid in water.
2. LC-MS/MS buffer: 50% v/v acetonitrile and 50 mM ammonium bicarbonate.
3. Trypsin (recombinant proteomics grade).

2.6 DSF

1. SYPRO Orange dye.
2. 96-well plates specific for real-time PCR instrument.
3. Optically clear sealing film for 96-well plates.
4. iCycler iQ Real-Time Detection System.

3 Methods

3.1 Domain Searching

1. We recommend that you prepare your protein of interest by editing its atomic coordinates (represented in a .pdb file) such that the structure now contains only the core Ig-like fold, removing any additional domains and long loop regions (Fig. 1).
2. Open the .pdb file in Coot (Crystallographic Object-Oriented Toolkit). Coot [15] is free software used to display and manipulate atomic models and is widely used by the structural biology community. Remove all domains except the Ig domain you are targeting for isopeptide bond engineering (*see* **Note 1**). We removed the polyproline tail from FctB.
3. Once you have your model, you can search the Worldwide Protein Data Bank (wwPDB) database for structural homologs to your Ig domain. For this search, we recommend using the program PDBeFold (http://www.ebi.ac.uk), which searches the wwPDB database for structures similar to your template protein.
4. Open PDBeFold and select *Coordinate file* for the query. Choose your edited coordinate file and select upload. Leave the target as the *Whole PDB archive*.
5. Select the polypeptide chain you wish to search the PDB with (i.e., the Ig domain of your target protein). If you have edited your template, then it will typically be chain A.
6. Set the *Lowest acceptable match (%)* for the query and target to be 50% and 10%, respectively (*see* **Note 2**).
7. We recommend leaving the rest of the form as the default parameters for the first run. If no suitable matches are found, then first deselect the *match connectivity* box, and then *match individual chains* to expand the range of potential matching targets selected.

8. Submit your query. Your results will appear in the same search window.
9. The results will appear as a list of structural alignments listed from highest to lowest Q-score. High Q-score structures have the lowest root-mean-square difference (RMSD) alignment of Cα-atoms over the largest proportion of the protein (*see* **Note 3**).
10. Inspect the list of structurally aligned proteins. Sort them by RMSD as you should aim to focus on the regions with the closest alignment. Some alignments may have a low RMSD but cover only a small region of the protein. These alignments that only match a small proportion of the protein will be less useful.
11. Look at the title of the aligned target structures. A promising candidate will very likely have pilin, fimbriae, or adhesin in the title. These proteins often contain intramolecular isopeptide bonds.
12. Click on the number in the left column to view the alignment. This opens another window where you can view the sequence alignment and also the superimposed structures. This gives you an idea of the extent to which the proteins are matched.
13. For FctB, we selected the major pilin protein from *Streptococcus pyogenes* as the template, Spy0128 (PDB code: 3b2m, chain A). The structural alignment between FctB and Spy0128 has a Q-score of 0.19 and a RMSD of 1.75 Å over 94 residues (*see* **Note 4**).
14. An alternative approach to searching the entire wwPDB database is to limit the alignment to a group of structures known to contain intramolecular isopeptide bonds. These can be selected by searching the wwPDB with an advanced text search for "isopeptide bond" and "adhesion" (*see* **Note 5**).
15. Make a list of all the PDB files that contain intramolecular isopeptide bonds in a text file. Use text edit or Gedit (*see* **Note 6**).
16. Open PDBeFold and select *coordinate file* for the query. Choose your edited coordinate file and select upload. For the target source, select a *List of PDB codes.* Choose your saved list of PDB codes and repeat the **steps 6–9**.
17. Select your top three aligned structures.

3.2 Isopeptide Bond Modeling

Once potential protein templates have been selected, the structural alignments can be viewed and it can be determined whether the polypeptide configuration around the template isopeptide bond is suitable to accommodate an isopeptide bond.

1. Download the selected template structures and open in Coot, or directly import into Coot using *Get PDB using Accession code* in the File tab.
2. Align each of the templates with your selected Ig query (your target protein) using *SSM Superimpose* in the Calculate tab. You may need to trim your template structure back to the single domain that was aligned in the PDBeFold session (*see* **Note** 7).
3. Find the catalytic isopeptide triad in the template. Use the *Rotate Translate zone/chain/molecule* in the Coot toolbar to position the Cα atoms of the isopeptide triad residues over the residues that have matching positions in your target Ig-fold. In FctB, the residues Asn 13, Gln 67, and Pro 117 were found to match Spy0128 residues Lys 36, Glu 117, and Asn 168, respectively [13]. Use the *Rotate Translate "molecule"* to visually align the two domains around the three residues (*see* **Note 8**).
4. Measure the distances between the Cα atoms of the residues that overlay the isopeptide triad. Open *Distances and angles* in the *Measures* tab. Select *Distance* and click on each of the Cα atoms to be measured (Fig. 2). The distances should be ideally be within approximately 1.0 Å of the template distances.

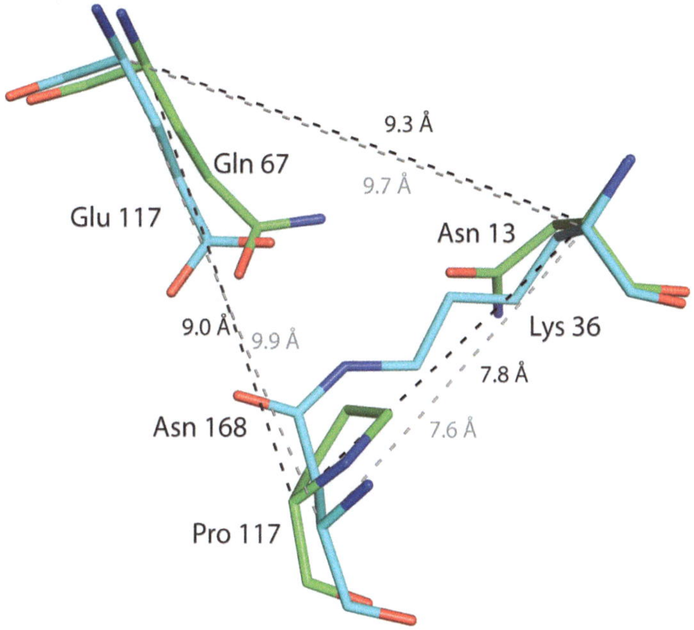

Fig. 2 Structural superposition of the isopeptide triad residues from Spy0128 (cyan, PDB ID code: 3B2M) with residues that match their positions in FctB (green, PDB ID code: 3KLQ). Residues are shown in stick form with the distances between the Cα atoms for FctB in black and Spy0128 in gray

5. Another key feature of isopeptide bonds is the linear order of the residues in the peptide chain. In the amino acid sequence, Lys is first, followed by Glu (or Asp for CnaA domains) and then the Asn (or Asp) residue. This suggests that the Asn/Asp residue, which resides on the last or second-to-last β-strand in the Ig domain, is the final one to be brought into proximity of the first two residues as the protein folds. This is likely an important aspect of isopeptide bond formation (*see* **Note 9**).
6. Mutate the residues in your query Ig domain to the corresponding isopeptide-forming residues. In Coot, select the *Simple mutate* button in the tool bar (radioactive symbol) and click on your residue to bring up a menu of residues. For example, for Asn 168 in FctB, we selected Lys, which makes the N168K change.
7. Select the *Rotamer* selection button in the tool bar and select a rotamer that closely resembles that of the Lys residue in the template structure. Repeat for the other two residues.
8. The residues may need to be finely positioned to match the residues of the template isopeptide bond as closely as possible. Select the *Regularize zone* button in the tool bar and double click on your residue. This can then be dragged into a position that resembles the template residue (*see* **Note 10**).
9. Position the $-NH_2$ group of the Asn side chain carboxyamide next to the ε-amino group of Lys side chain. Then, delete the Asn amino group by selecting *Delete item* (trashcan) and then *delete atom* from the submenu. Click on the $-NH_2$ group to delete it. Pull down the *Extensions/Modeling* menu tab and select the *Link atoms* tab. Link the amino group of Lys (NZ) to the Cγ carbon atom of Asn (CG) by clicking on each atom (*see* **Note 11**).
10. The Carboxyl group of the Glu should be in a position to hydrogen bond to the atoms of the isopeptide unit. If the distance between the Glu and the isopeptide bond is too short, try modeling an Asp in place of the Glu.
11. Select the *Environment Distances* in the Measures tab. The bonding networks between the residue you click on and neighboring residues will be shown (Fig. 3). Take note of any residue clashes, or any region where there is a void within the core of the protein. These will destabilize the protein.
12. We recommend that you use MolProbity (http://molprobity.biochem.duke.edu) to evaluate the reliability of your altered structural model. Click on *Evaluate X-ray structure* and run through the steps of *Choose a structure*, *Add hydrogens,* and *Analyze sterics & geometry.*

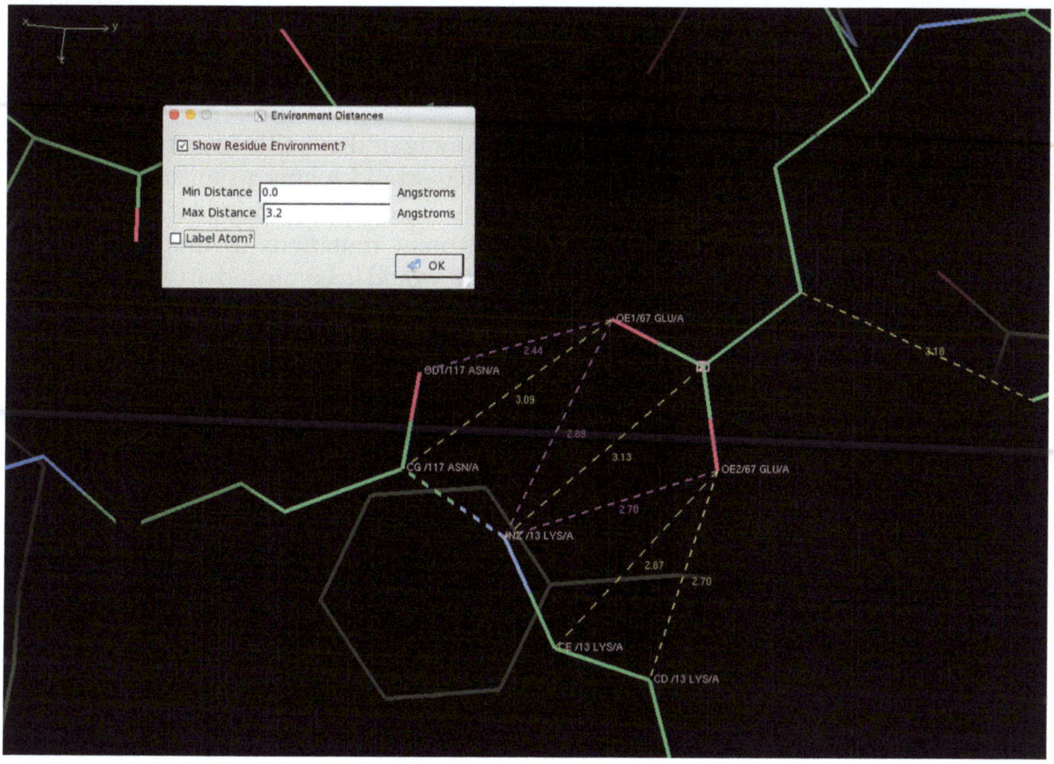

Fig. 3 Screen shot from Coot showing the environmental distances for Glu 67 in the FctBiso model. The "link" between Lys 13 and Asn 117 (NZ-CG atoms, respectively) is represented as a thicker dashed line. Hydrogen bonds and their distances are represented as blue dashed lines, while nonpolar contacts are shown as green dashes. Note, in addition to the hydrogen bond to Lys NZ, the Glu OE2 atom is in close proximity to Lys CD and CE carbon atoms. These distances, which are under 3 Å, should be considered as clashes

13. The results can be viewed in a multicriterion chart. Among geometric output can be found a list of residues that clash, together with Ramachandran and rotamer outliers. Any residue that is highlighted in red must be adjusted to satisfy regular protein geometry. MolProbity is a fast way to evaluate the changes you have made, to ensure that they are within the normal limits of protein conformation (*see* **Note 12**).
14. Upload the structure with hydrogens added from MolProbity into Coot. This helps to visualize the residue clashes. The majority of these clashes are likely to be between the isopeptide triad and existing nonpolar residues in the protein core. Mutate the residues that clash with the triad to smaller nonpolar residues, for example, changing isoleucine to valine. Always try to select a common (favorable) rotamer for any mutated residues.
15. If large aromatic residues were removed during the isopeptide triad mutagenesis, then surrounding residues may need to be mutated to larger amino acids to fill the void. The mutations will be specific for each protein.

16. Repeat the analysis of any mutations you make in Coot with MolProbity until you are satisfied with the isopeptide engineered model.

3.3 Cloning

Herein, we describe the cloning of the FctBiso construct described in Ref. 13. Other constructs would follow a similar experimental flow, albeit with conditions unique to each individual construct. While we originally used point mutations to mutate individual residues one at a time, we would recommend to have the gene synthesized and cloned by a commercial company such as GenScript.com (https://www.genscript.com) or a gBlock (gene block) synthesized by Integrated DNA Technologies (https://sg.idtdna.com). Gene synthesis or gBlocks are more practical and economical when considering large numbers of point mutations.

1. Design the InFusion™ cloning inserts as illustrated in Fig. 4. We use the pProExHta plasmid vector (*see* **Note 13**). An initial gBlock containing the minimum point mutations to introduce the isopeptide triad into the desired position in the

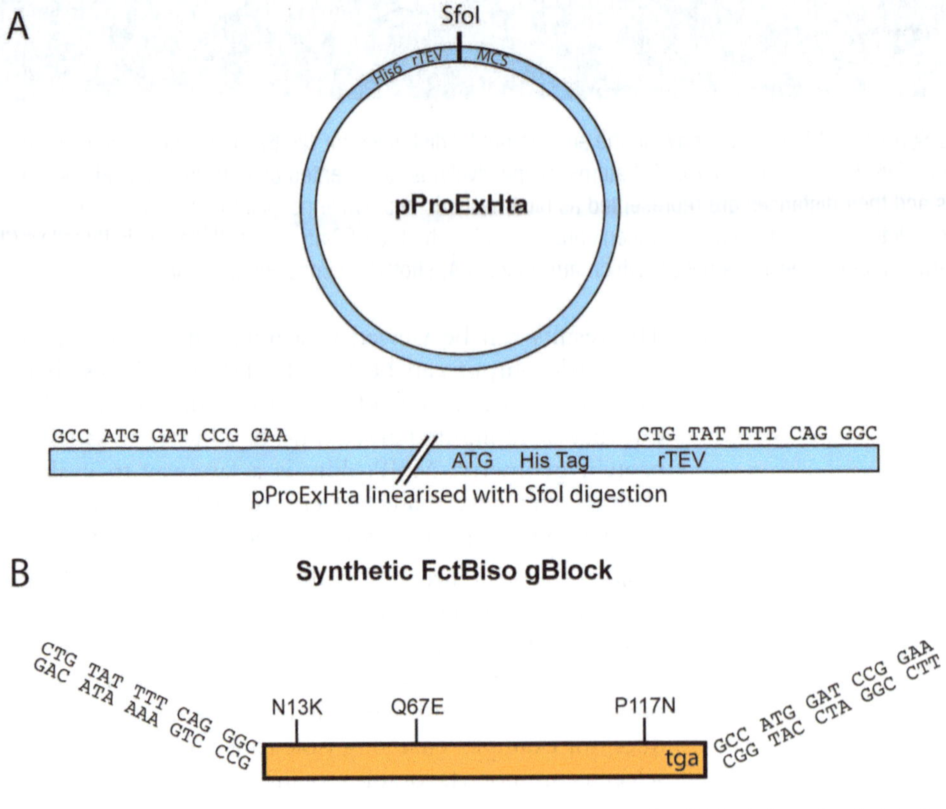

Fig. 4 Schematic diagram of InFusion cloning. (**a**) The expression vector pProExHta is linearized with digestion by the restriction enzyme SfoI, which cleaves at the beginning of the multiple cloning site (MCS). (**b**) Design of the FctBiso gBlock insertion sequence incorporating point mutations for N13K, Q67E, and P117N. Ensure a stop codon is included at the end of the gene. The construct has 15 bp of sequence at the 5′- and 3′-end that is homologous to the linearized vector

hydrophobic core of the protein is synthesized containing an additional 15-bp complementary to the vector each side of the SfoI restriction site in pProExHta as consistent with InFusion™ cloning protocols (*see* **Note 14**).

2. Purify pProExHta plasmid from *E. coli* (DH5α) using a miniprep plasmid DNA purification kit according to manufacturer's instruction.
3. Linearize 1 μg of pProExHta by digestion with SfoI (*see* **Note 15**). For each microgram of vector, add 10 units of SfoI. Use Cutsmart buffer, NEB.
4. Subject 1 μg of the SfoI digested plasmid to DNA agarose gel electrophoresis, visualize the cut plasmid using UV light, and excise using a scalpel or gel excision tool. Purify the DNA from the extracted gel using a gel clean up kit (*see* **Note 16**).
5. Mix 100 ng of linearized pProExHta and 200 ng of synthesized gBlock construct in an InFusion™ reaction mix as described in the manufacturer's protocols. Incubate at 52 °C for 15 min and then cool on ice. Include a control without InFusion™ mix to determine the background from parental vector.
6. Transform 2 μL of plasmid into the InFusion™ supplied Stellar *E. coli* competent cells (50 μL) using supplier's heat shock and positive clone selection protocols. All growth media contains the required selection antibiotic for the pProExHta vector, 100 μg/mL ampicillin.
7. Pick single colonies and culture overnight in 5 mL of LB medium with 100 μg/mL ampicillin. Extract and purify plasmid DNA using a miniprep plasmid DNA purification kit according to manufacturer's instruction.
8. Confirm that your clone contains the correct sequence and that the point mutations are present using a DNA sequencing service (*see* **Note 17**). Proceed to the expression and analysis of the protein.

3.4 Expression

1. Transform sequence-verified plasmids into an *E. coli* expression strain such as BL21 (DE3) (*see* **Note 18**).
2. Seed 1 L of LB with 10 mL of an overnight culture of the transformed expression strain and incubate at 37 °C, shaking at 180 rpm until the optical density of the culture reaches $OD_{600} = 0.5–0.7$ (*see* **Note 19**).
3. Induce protein expression with the addition of 1 mM IPTG, and transfer the flasks to 18 °C and incubate for 16 h with shaking at 180 rpm.

4. Harvest cells by centrifugation (4000 × *g* for 15 min), decant supernatant, and discard. Add lysis buffer to the cell pellet, 20 mL for each 1 L culture and resuspend bacteria. Protease inhibitors are included to prevent protein degradation. Store the pelleted cells at −20 °C until needed.
5. Thaw and lyse the bacterial suspension using a cell disruptor at 18 kpsi and remove insoluble debris by sedimentation at 30,000 × *g*.
6. Equilibrate an IMAC column with 20 mL of lysis buffer and load the soluble lysate onto the column. Attach the loaded column to an FPLC instrument and wash with 100 mL of wash buffer. Elute bound protein with a gradient of 20 to 500 mM imidazole (Elution buffer) over 60 mL, collecting 1 mL fractions.
7. Analyze eluted protein fractions by SDS–PAGE according to the manufacturer's protocols to assess yield and purity of the purified protein (*see* **Note 20**).
8. Pool fractions containing recombinant protein and cleave the His_6 affinity tag from the recombinant protein with the addition of a 1:50 (molar) ratio of rTEV-His_6 to recombinant protein, concurrent with dialysis against 1 L of rTEV digestion buffer at 4 °C for 16 h.
9. Separate cleaved recombinant protein, rTEV-His_6 protease, and uncleaved protein by immobilized metal affinity chromatography (IMAC) step as described in **step 6**, collecting the cleaved recombinant protein in the column flow through.
10. Concentrate the cleaved recombinant protein by centrifugation to 0.5 mL using a 10 kDa molecular-mass cut-off protein spin concentrator.
11. Finally, purify the recombinant protein using size-exclusion chromatography on a Superdex S200 10/300 column and in SEC buffer (*see* **Note 21**). Concentrate the protein as described in **step 10** to a final concentration of ~10 mg/mL, aliquot 50 μL volumes into cryogenic vials, and flash cool by emersion in liquid nitrogen with subsequent storage at −80 °C (*see* **Note 22**).

3.5 Characterization of Isopeptide Bond Formation by Mass Spectrometry

The presence of an isopeptide bond can be confirmed by electrospray ionization-mass spectrometry to obtain an accurate intact molecular weight [16]. The formation of a Lys-Asn isopeptide bond is accompanied by the loss of NH_3 and a decrease of mass of 17 Da when compared to the mass of the wild-type protein. Chromatography tandem mass spectrometry (LC-MS/MS) can then be used to substantiate the presence of the isopeptide cross-linked between peptides from a trypsin digest of the protein (described, e.g., in [16]).

1. To obtain an accurate intact mass of the purified protein, dilute the purified protein in 50% acetonitrile and 0.1% formic acid to a concentration of ~50 nmol/mL. Submit this sample (~100 μL) to your mass spectrometry facility for processing. The raw data can be deconvoluted using the Bayesian Protein Reconstruct tool in BioAnalyst software (Applied Biosystem).
2. For LC-MS/MS sample preparation, it is best to start with the highest sample purity possible. Ensure complete homogeneity by running a sample of the purified protein on an SDS–PAGE gel.
3. Cleave the appropriate band from the gel with a clean scalpel.
4. Wash the diced gel band(s) with 50% acetonitrile and 50 mM ammonium bicarbonate to remove staining compounds.
5. Add trypsin (10 ng/μL) dissolved in 25 mM ammonium bicarbonate to the gel pieces (typically 200 ng of protease) and incubate at 37 °C for 16 h.
6. Collect the supernatant containing digested peptide fragments and analyze by LC-MS/MS at your mass spectrometry facility.
7. The MS/MS data are used to search for matches against the predicted trypsin-digested peptide spectrum of the wild-type protein sequence using Mascot software (Matrix Science).
8. As the engineered isopeptide bond links two nonadjacent peptides, select the peptides with no matches from the initial Mascot search.
9. Manually analyze the spectra of these peptides to find a fragment that corresponds to a covalent isopeptide linkage between two peptide fragments (*see* **Note 23**).

3.6 Thermal Stability by Differential Scanning Fluorimetry (DSF)

DSF is a method for determining the thermal stability of proteins [17]. It uses a fluorophore (SyproOrange) that becomes highly fluorescent when exposed to a nonpolar environment, such as the hydrophobic regions of unfolded proteins. As a protein unfolds the interaction of SyproOrange with the exposed hydrophobic protein, moieties will lead to a corresponding increase in fluorescence emission. Ensure all experiments are done in triplicate and always compare the mutated protein against the wild-type protein in each experiment.

1. Fresh working stock of SyproOrange is prepared from a 1:25 dilution of the original stock solution (5000× concentrate in DMSO).
2. Protein solution (30 μM) is mixed with 5 μL of working stock SyproOrange and made up to 25 μL final volume with protein buffer. Mix samples well by inversion before transferring to a 96-well plate specific for real-time PCR instrument.

3. Seal the plate with optically clear sealing film to prevent evaporation and cross-contamination.
4. Place the plate directly in the iCycler iQ real-time detection system to protect from light and reduce photobleaching.
5. Set instrument excitation filter from 530 to 560 nm and the emission filter from 575 to 595 nm (*see* **Note 24**).
6. Heat the plate linearly from 25 to 95 °C in 1 °C/min increments. Monitor thermal denaturation following the increase of SYPRO Orange fluorescence intensity associated with protein unfolding process.
7. After the run, the fluorescence imaging data can be analyzed on the DSF Analysis (version 2.6) spreadsheet. Fluorescence intensity is plotted as a function of temperature, resulting in a sigmoidal curve that can be described by a two state transition.
8. If the plot is a sigmoidal curve, determine the inflection point of the curve, which is the melting temperature (T_m, °C) of the protein. The melting temperature can be calculated using GraphPad Prism™ (GraphPad software).

3.7 X-Ray Crystallography

A structure-guided approach can be essential for an iterative process such as protein engineering. X-ray crystallography can give you the definitive visual proof of the presence of an engineered isopeptide bond and can guide further mutagenesis if initial point mutations do not lead to bond formation. A protein structure with an unformed bond can provide an explanation as to why the bond is not forming and what interactions the mutated residues are making. This experimental information can inform on what further changes need to be made to ensure isopeptide bond formation.

Crystallography requires specialist knowledge and equipment and is beyond the scope of this chapter. We recommend that you seek a collaboration with an experienced X-ray crystallographer to help you with both the protein design process and the experimental crystallography.

4 Notes

1. This can be done either in Coot using the "delete range" function. Alternatively, if you note what regions you wish to delete, this can be done in text edit. Select and delete the residues you do not want in your final model and ensure that you save as a .pdb file and not a .pdb.txt file.
2. For the query protein (the protein you wish to engineer), this selects the minimal matching region to cover more than 50% of the domain or protein. For possible template proteins (i.e., the proteins in the wwPDB database), we select a much lower value

as some of them may have multiple domains, with only a single domain that matches well with the query protein. If you select a higher value, you can often miss these highly similar domains because they are only a small part of a much larger structure.

3. RMSD stands for the root mean square difference, calculated between Cα-atoms of matched residues of the query and target structures. The RMSD is presented in angstroms (Å), with a larger RMSD meaning the query structure is generally less similar to aligned target. It is often advantageous to sort by RMSD.
4. The Q-score is low because Spy0128 is a two-domain protein in which only one domain, the N-terminal domain, matches well with the single domain of FctB. This is why RMSD is a better selection parameter for ranking your query protein against the target. Additional pilin proteins have been deposited that are of higher similarity to FctB since we published the engineered structure (FctBiso). Another pilin protein with the PDB code 6BBT would now be the preferred template protein.
5. Searching with only the term "isopeptide bond" retrieves multiple ubiquitin-related structures that are enzyme catalyzed intermolecular isopeptide bonds. These are not relevant, so we narrow the search by adding an additional search term of "adhesion" since most intramolecular isopeptide bonds are found in pili or fimbriae that are associated with bacterial cell adhesion.
6. Each four-letter accession code needs to be followed by a coma and a space, for example, 2WOY, 2WQS, 2WZA, 2XI9, 2XIC, 2XID, 3B2M, 3GLD, 3GLE, 3HR6, 3HTL, 3PF2, 3PG2, 3PHS, 3QDH, 3RKP, 3RPK, 4BUG, 4HSQ, 4HSS, 4MLI, 4MLS, 4P0D, 5MKC, 5TTD, 5U5O, 5U6F, 6BBT, 6BBW, and 6N0A. These 30 structures were found with the two-term search of "isopeptide bond" and "adhesion," or an additional search with "isopeptide bond" and "*Streptococcus*" as a search term. These are by no means all the protein structures with intramolecular isopeptide bonds, and the number of structures is certain to increase over time.
7. Superposition of Ig domains is difficult and not always accurate using superposition PDBeFold (SSM). Ig domains are often aligned in the reverse topology with the N terminus of the query aligned to the C terminus of the template. The topology of the Ig domains needs to be matched. We have found that the ASH structural superposition algorithm, which can be found at https://sysimm.org/ash/, is particular good at aligning Ig-folds.

8. This can take a few iterations to get as close as possible. Alternatively, you can use the SSQ function in the "Calculate" tab. Here you align a short sequence between the two domains, for example, Residues 12–14 of FctB with residues 35–37 of Spy0128. This will accurately position the three residues from each domain, which may be sufficient to overlay two similar domains. If no satisfactory alignment is found for your protein, you may be able to visually select an appropriate site, measuring distances between strands at different positions in the protein. Search for the positioning of the Lys-Asn residues at ~7.8 Å first and then determine if there is an appropriate strand to place a Glu or Asp adjacent to this site. Lastly, ensure that the residues are buried in the hydrophobic core of the protein.
9. It is likely that the Lys and Glu residues are loosely accommodated in the core of the folding protein. These residues may be directed into their final position by hydrophobic residues that "zip" onto the core as the last strand carrying the Asn folds into the domain. The binding of the last strand would "seal" the charged residues in the hydrophobic core bringing Asn into proximity of Lys and Glu to catalyze the formation of the isopeptide bond.
10. When adjusting the residue position, you can select individual atoms to drag by holding down the control key as you click and drag the atom with the mouse. Ensure any adjustments are within the normal residue geometry in terms of bond angles, distances, etc., which can be monitored in the regularization popup window. Click "accept" once you have finished.
11. The bond distance of the link should be ~1.3 Å and planar between Asn carboxyl and the Lys ε-carbon.
12. There are excellent resources to help you interpret the results and how to fix them. http://kinemage.biochem.duke.edu/teaching/workshop/CSHL2012/.
13. The vector includes a hexahistidine tag (His6) sequence for affinity purification by immobilized metal affinity chromatography (IMAC) followed by an rTEV protease cleavage site for removal of the affinity tag if required. The gene is inserted in the multiple cloning cassette at the SfoI restriction site to limit the number of vector-derived residues in the rTEV-cleaved protein to glycine and alanine.
14. While we used the inverse PCR mutagenesis protocol for all our mutations of FctBiso, we recommend using gene synthesis to produce the desired mutations. This speeds the cloning and allows for coding optimization and negates the need to obtain the DNA from the host organism. A stop codon is included at the 3′-end of the gene to prevent transcription into the vector DNA.

15. This cuts the vector at the beginning of the multiple cloning site immediately after the rTEV protease recognition site leaving blunt ends. By using this site, the recombinant protein can be cleaved from the His_6 affinity tag, leaving only a glycine and alanine vector-derived residues at the N terminus of your recombinant protein.
16. It is recommended that the vector is purified by agarose gel purification to limit the amount of uncut parental vector in the InFusion™ cloning reactions. Ethidium bromide is included in the agarose to aid visualization of the DNA under UV light.
17. The sequencing primers we use for pProExHta are as follows: pProEX-HTa Fwdseq+50 GCT GTT GAC AAT TAA TCA TCC G, pProEXHT Revseq CAG ACC GCT TCT GCG TTC TG.
18. For simplicity and ease of fermentation, we recommend expression in an *E. coli* host strain. If the Ig-isopeptide construct also contains disulfide bonds, aggregation can sometimes occur during expression and the construct may need to be expressed in *E. coli* with a trxB mutation or exported to the oxidizing environment of the periplasmic space using a leader sequence. We describe the expression and purification of FctBiso [13] as an exemplar.
19. Individual expression conditions may need to be optimized for each construct. We find that expression at lower temperatures leads to higher yields. If your protein is expressed at low levels, you can increase the OD to 1.0–1.2 before induction to increase cell mass.
20. Determine the purity of the protein by assessing the number of bands on the SDS–PAGE gel. A single band of the expected size, as judged by comparing to a molecular weight protein ladder standard, is a good measure of purity. However, a unique aspect of proteins containing intramolecular isopeptide bonds is that they appear to run at a smaller than expected molecular size in SDS–PAGE. This is because the stability afforded by the covalent isopeptide bond prevents the protein from fully unfolding; hence, they run further into the gel because they have a smaller radius of gyration. In cases where the isopeptide bond is slow to form, you will have two populations, with and without the isopeptide bond. These two populations will separate on SDS–PAGE to give two bands. We recommend running a sample of purified wild-type protein on the same gel for a comparison. Protein degradation will also produce smaller protein bands; hence, we recommend mass spectrophotometry to confirm the presence of an engineered isopeptide bond.

21. Determine the protein concentration (mg/mL) using a UV spectrophotometer at 280 nm before loading on the SEC column. Do not load more than 25 mg of protein on a 10/300 column at one time. If your protein concentration is higher than 50 mg/mL or very viscous, then dilute your protein in buffer and load a smaller volume.
22. Storing purified protein in small aliquots avoids repeated freeze thawing cycles that can lead to protein degradation. Most techniques for protein characterization, with the exception of protein crystallography, require relatively small amounts of purified protein.
23. We recommend using the expertise of your local mass spectrometry service for the experimentation and analysis of the data.
24. SYPRO Orange excitation and emission wavelengths are, respectively, 490 and 575 nm.

Acknowledgments

The authors acknowledge Hanna Kwon and Christopher Squire for their work in engineering the FctBiso construct.

References

1. Kang HJ, Coulibaly F, Clow F et al (2007) Stabilizing isopeptide bonds revealed in Gram-positive bacterial pilus structure. Science 318:1625–1628
2. Young PG, Moreland NJ, Loh JM et al (2014) Structural conservation, variability and immunogenicity of the T6 backbone pilin of serotype M6 *Streptococcus pyogenes*. Infect Immun 82:2949–2957
3. Young PG, Raynes JM, Loh JM et al (2019) Group A Streptococcus T-antigens have a highly conserved structure concealed under a heterogeneous surface that has implications for vaccine design. Infect Immun 87(6): e00205–e00219
4. Kang HJ, Baker EN (2012) Structure and assembly of Gram-positive bacterial pili: unique covalent polymers. Curr Opin Struct Biol 22:200–207
5. Hagan RM, Bjornsson R, McMahon SA et al (2010) NMR and theoretical analysis of a spontaneously formed Lys-asp isopeptide bond. Angew Chem Int Ed Engl 49:8421–8425
6. Vengadesan K, Narayana SVL (2011) Structural biology of Gram-positive bacterial adhesins. Protein Sci 20:759–772
7. Baker EN, Squire CJ, Young PG (2015) Self-generated covalent cross-links in the cell-surface adhesins of gram-positive bacteria. Biochem Soc Trans 43:787–794
8. Kang HJ, Baker EN (2010) Intramolecular isopeptide bonds: protein crosslinks built for stress? Trends Biochem Sci 36:229–237
9. Hu X, Hu H, Melvin JA et al (2011) Autocatalytic intramolecular isopeptide bond formation in gram-positive pili: a QM/MM simulation. J Am Chem Soc 133:478–485
10. Kang HJ, Baker EN (2009) Intramolecular isopeptide bonds give thermodynamic and proteolytic stability to the major pilin protein of *Streptococcus pyogenes*. J Biol Chem 284:20729–20737
11. Alegre-Cebollada J, Badilla CL, Fernandez JM (2010) Isopeptide bonds block the mechanical extension of pili in pathogenic *Streptococcus pyogenes*. J Biol Chem 285:11235–11242
12. Echelman DJ, Alegre-Cebollada J, Badilla CL et al (2016) Cna-A domains in bacterial pili are efficient dissipaters of large mechanical shocks. Proc Natl Acad Sci U S A 113:2490–2495
13. Kwon H, Young PG, Squire CJ et al (2017) Engineering a Lys-Asn isopeptide bond into an

immunoglobulin-like protein domain enhances its stability. Sci Rep 7:42753
14. Matsumura M, Becktel W, Levitt M et al (1989) Stabilization of T4 phage lysozyme by engineered disulfide bonds. Proc Natl Acad Sci U S A 86:6562–6566
15. Emsley P, Cowtan K (2004) Coot: model-building tools for molecular graphics. Acta Crystallogr D Biol Crystallogr 60:2126–2132
16. Kang HJ, Middleditch M, Proft T et al (2009) Isopeptide bonds in bacterial pili and their characterization by X-ray crystallography and mass spectrometry. Biopolymers 91:1126–1134
17. Niesen FH, Berglud H, Vedadi M (2007) The use of differential scanning fluorimetry to detect ligand interactions that promote protein stability. Nat Protoc 2:2212–2221

Index

Thomas Proft and Jacelyn M. S. Loh (eds.), *Group A Streptococcus: Methods and Protocols*, Methods in Molecular Biology, vol. 2136, https://doi.org/10.1007/978-1-0716-0467-0,

F

G

H

I

T

U

V

W

X

MIX
Papier aus verantwortungsvollen Quellen
Paper from responsible sources
FSC® C105338

If you have any concerns about our products,
you can contact us on
ProductSafety@springernature.com

In case Publisher is established outside the EU,
the EU authorized representative is:
Springer Nature Customer Service Center GmbH
Europaplatz 3, 69115 Heidelberg, Germany

Printed by Libri Plureos GmbH
in Hamburg, Germany